Surface Enhanced Raman Spectroscopy: Biosensing and Diagnostic Technique for Healthcare Applications

Edited by

Swati Jain
Amity Institute of Nanotechnology,
Amity University,
Noida, UP,
India

&

Sruti Chattopadhyay
Center for Biomedical Engineering,
Indian Institute of Technology Delhi
New Delhi,
India

Surface Enhanced Raman Spectroscopy: Biosensing and Diagnostic Technique for Healthcare Applications

Editors: Swati Jain and Sruti Chattopadhyay

ISBN (Online): 978-981-5039-11-5

ISBN (Print): 978-981-5039-12-2

ISBN (Paperback): 978-981-5039-13-9

©2021, Bentham Books imprint.

Published by Bentham Science Publishers Pte. Ltd. Singapore. All Rights Reserved.

need for a court order if at any point you breach any terms of this License Agreement. In no event will any delay or failure by Bentham Science Publishers in enforcing your compliance with this License Agreement constitute a waiver of any of its rights.

3. You acknowledge that you have read this License Agreement, and agree to be bound by its terms and conditions. To the extent that any other terms and conditions presented on any website of Bentham Science Publishers conflict with, or are inconsistent with, the terms and conditions set out in this License Agreement, you acknowledge that the terms and conditions set out in this License Agreement shall prevail.

Bentham Science Publishers Pte. Ltd.
80 Robinson Road #02-00
Singapore 068898
Singapore
Email: subscriptions@benthamscience.net

**BENTHAM
SCIENCE**

CONTENTS

FOREWORD

"The fundamental importance of the subject of molecular diffraction came first to be recognized through the theoretical work of the late Lord Rayleigh on the blue light of the sky, which he showed to be the result of the scattering of sunlight by the gases of the atmosphere." CV Raman.

"…the highly interesting result that the colour of sunlight scattered in a highly purified sample of glycerine was a brilliant green instead of the usual blue."

Sir CV Raman

The quest to find the unknown drives scientists working in all possible conditions to unravel the mystery created through the fusion of natural phenomena. Sir Chandrasekhara Venkata Raman's path breaking discovery and its theoretical prediction conceptualized into a spectroscopic technique impertinently used in the chemical world. Raman spectroscopy is like panning for gold. A wealth of information is there if you can just sift through the rock, dirt, and sand obscuring it. In the 1970's, another astounding revelation was made about light scattering. Experiments conducted by Fleischmann from University of Southampton led to a chance discovery of SERS phenomenon where the unexpected rise in Raman signals was reported when the sample of interest was added onto silver metallic particles. This phenomenon was later explained and detailed by Prof. Richard Van Duyne whois credited with the discovery of SERS technique – explanation of magnificent rise of Raman signals, postulation of enhancement factor - electromagnetic theory for SERS and its eventual applications in analytical and bioanalytical chemistry. Over the years, the perspective in SERS has dramatically changed owing to highly advanced synthetic methods developed for the preparation of novel nanomaterials, which become the bed for the creation of 'hot-spot' suitable for signal enhancement. Thus, SERS has gained equitable momentum in the biological world as chemical analysis, where the possibility of single molecule detection has been proposed.

This e-book is intended as a nodal point for budding researchers, senior academicians and scientists who will gain a different perspective in the field of SERS based biosensing. It gave me immense pleasure when Dr. Swati and Dr. Sruti approached me for writing the foreword of this book as this book intends to bridge the fissure between physical and biological aspects of SERS and thus is easily readable for people from different scientific backgrounds. This book feature 'updated and recent', details on colloids and nanostructures, their fabrication, surface engineering and immobilization methods, all in context to SERS based biosensing.

The framework of the book is in the order of hierarchy with different sections devoted for each subject in the SERS technique. Each section comprising of 1-2 chapters, is designed to move from basic concepts of technique towards its real biological applications. The preface sets the stage, which duals as introductory chapter moves to chapter 1 about Raman vibrational spectroscopy technique authored by Dr. Mittal, who has long experience with the technique and its chemical analysis. This chapter also introduces SERS to readers. The second chapter is devoted to the theoretical understanding of SERS and enhancement factors. The creation of hot-spots and two-photon excited SERS are also explained in this chapter. Section B begins with 3rd chapter highlighting the role of novel nanoarchitecture materials, plasmonic and non-plasmonic nanomaterials and progression in nanofabrication procedures for the development of nanomaterials written in detail by Dr. Ranu Nayak. Proceeding chapters included in section C include biomedical and biosensing applications of SERS. In the first chapter of the section, consecrated efforts are directed to illustrate the significance of SERS for amyloid research by Dr. Nakul Maity. Amyloids have become ubiquitous in understanding disease pathogenesis and hence are quite useful for understanding disease and therapeutic interventions. The proceeding chapters orient the readers toward SERS biosensing of pathogens and other clinically relevant biomolecules for real applicability of the technique in healthcare settings as detailed in Chapters 5 and 6. This section concludes with the design of smartphone biosensors with SERS technique for point of care testing devices. Last section D of the e-book elucidates problems encountered by researchers in academia and companies working towards realising SERS technique for making commercially feasible tools. It also details improvising strategies such as tip enhanced, interference enhanced (TIRS); shell isolated nanoparticle-enhanced (SINRS) and spatially offset Raman spectroscopy (SORS) by authors hailing from both academic institutes and companies.

I hope this book stimulates thought and presents engaging rationale of SERS and its multiple diverse applications primarily in the biological system. The e-Book has the potential to channelize a deeper understanding of basic science as well as its applications in analysis and sensing, in nano sciences, surface chemistry, as a biological probe as well as in daughter disciplines such as plasmonics and near-field optics.

Dr. Surender M. Kharbanda

Department of Adult Oncology,
Dana Farber Cancer Institute
Harvard Medical School
Boston, MA 02115, USA

Scientific Founder, President &
Chief Scientific Officer
Genus Oncology LLC
Boston, MA 02118, USA

PREFACE

Million-fold enhancement of characteristic Raman signal of molecules presented as a monolayer on the surface of rough nanostructured metals refuelled the interest in Raman spectroscopy which initially was reserved for pure sample analysis. The amplification in signal now referred to as surface enhanced Raman spectroscopy (SERS) has been explored for numerous applications in physical, analytical, chemical, material, surface/topographical and biomedical sciences. The aim of this book is to comprehensively understand the concept of biological applications using SERS technique for sensing and imaging various analytes in *in-vitro* as well as *in-vivo* conditions.

Individual bonds in molecules give rise to unique vibrations by inelastic scattering resulting in molecularly specific spectra, namely Raman spectra. These inherently weak signals were later on developed by researchers into highly intense peaks using metallic nanostructures and this SERS phenomenon gained popularity, particularly in healthcare and medical applications. SERS offers high sensitivity, fingerprint analysis, optimization towards near infra-red signal, minimization of photo-bleaching and photo-degradation.

Nowadays, dramatic emphasis is devoted towards rapid and sensitive detection methodologies as well as gaining insight into molecular dynamics in *in-vivo* conditions through imaging. SERS based nanomaterials and devices, including novel plasmonic and non-plasmonic nanostructures and the development of stable Raman Reporter Molecules (RRMs) have propelled amended signalling attributes in SERS biosensing and diagnostic procedures. The instrument design has also changed focus towards SERS hand-held devices, smart phone integration and point-care-devices applicable in remote and intermittent sensing of target analytes.

Clearly, the time for a book is appropriate that summarizes basic notions and trends about thinking of SERS as a device for bioanalytical and biosensing tool defining what we know and understanding the deficiency in the technique in a way to harness this understanding into opportunities for the betterment of SERS and its biological applications. This is our ambition for assembling this e-book. International researchers in their respective sub-fields of SERS have contributed to this book. Their diverse background and training ranging from physics to inorganic chemistry to biomedical engineering, in my opinion, directly reflects the justification towards the multidisciplinary nature of SERS and its biological applications. The e-book is intended as a reference book for researchers and academicians working in SERS. It will also provide comprehensive concepts to newcomers starting to work in this field irrespective of their background in a simple manner.

Updated and recent analysis of materials and processes are detailed, all in context with SERS spectroscopy and its applicability in biomedical and healthcare fields. The book is planned in a hierarchical scale with discussions on theoretical beginnings of Raman and SERS spectroscopy moving towards chemical structures in SERS. Hence, the selection of topics covered in the preceding 8 chapters is highly subjective. The e-book is categorically differentiated into specific sections, each containing chapters catering to various aspects of SERS technique and biosensing applications. The sections move from basic physics of Raman spectroscopy and SERS towards plasmonic colloids and rough metal nanostructures, highlighting their synthesis as well as advancement in nano-assemblies. This includes active nanomaterials and nanodevices, including plasmonic and non-plasmonic nanostructures as well as Raman Reporter Molecules (RRMs). The largest section is, however, reserved for biosensing and diagnostic applications of SERS in biology and medicine. We have also put efforts towards understanding the concept of SERS for ultimately gaining perspective in developing an improvised biosensing system in a clinical setting. The book covers all, from basic knowledge to new exciting research and development in the field of SERS and its application for biosensing, diagnostics and imaging techniques.

Lastly, this book also summarizes lacunae of SERS technique, highlighting the need for optimization of signal acquisition parameters to prepare commercially viable and field deployable instruments. Remedial measures adopted for developing biosensing methodologies are also discussed with improved versions of SERS coming to the fore.

We cordially thank all our authors for their hard work and commitment to this book that they have invested in, writing highly relevant as well as excellent chapters. This international project would not have been possible without their efforts and dedication. We thank Ms. Humaira Hashmi at Bentham Publications, who suggested initiating this project to edit this SERS based e-Book. Finally, Dr. Swati Jain and Dr. Sruti are extremely thankful and grateful to their families for the continual support and motivation. We have tremendous faith that this e-book has the potential to stimulate thought processes leading to in-depth understanding of SERS so as to fully exploit this technique in innumerable biological applications.

Swati Jain

Amity Institute of Nanotechnology
Amity University
Noida, UP
India

Department of Science & Technology
Technology Bhawan
New Mehrauli Road
New Delhi, India

Sruti Chattopadhyay

Center for Biomedical Engineering
Indian Institute of Technology Delhi
New Delhi, India

List of Contributors

Ahmad Sheikh Bashir	Department of Bioresources, School of Biological Sciences, University of Kashmir, Srinagar 190006, India
Amar Ghosh	Department of Chemistry, Behala College, University of Calcutta, Kolkata, India
Animesh Mondal	Department of Bioresources, School of Biological Sciences, University of Kashmir, Srinagar 190006, India
Harpal Singh	Center for Biomedical Engineering (CBME), Indian Institute of Technology Delhi (IIT D), Hauz Khas, New Delhi, India
Harsimran Singh Bindra	School of Biotechnology, S.K. University of Agricultural Sciences and Technology of Jammu, Jammu and Kashmir, India
Jagjiwan Mittal	Amity Institute of Nanotechnology, Amity University, Noida, Uttar Pradesh, India
Kaushik Bera	Structural Biology and Bioinformatics Division, Indian Institute of Chemical Biology, Council of Scientific and Industrial Research, Kolkata 700032, India
Krishnendu Khamaru	Structural Biology & Bioinformatics Division , CSIR-Indian Institute of Chemical Biology, Council of Scientific and Industrial Research, 4, Raja S.C. Mullick Road, Kolkata 700032, India
Kriti Arya	Computational Instrumentation, CSIR-CSIO, Chandigarh, India
Manika Khanuja	Centre for Nanoscience and Nanotechnology, Jamia Millia Islamia, New Delhi- 110025, India
Nakul C Maiti	Department of Receptor Biology and Tumor Metastasis, Chittaranjan National Cancer Institute, 37, S.P. Mukherjee Road, Kolkata-700026, India
Pragya Agar Palod	Department of Physics, Shri Vaishnav Vidyapeeth Vishwavidyalaya, Indore, India
Rajasekhar Chokkareddy	Department of Chemistry, Aditya College of Engineering and Technology, Surampalem-533437, Andhra Pradesh, India
Ranu Nayak	Amity Institute of Nanotechnology, Amity University, Noida, UP, India
Richa Jackeray	Independent Contributor, Active Professional in Healthcare Industry, India
Robin Kumar	Amity Institute of Nanotechnology, Amity University, Noida, Uttar Pradesh, India
Sandip Dolui	Structural Biology and Bioinformatics Division, Indian Institute of Chemical Biology, Council of Scientific and Industrial Research, Kolkata 700032, India
Sinha Dona	Department of Receptor Biology and Tumor Metastasis, Chittaranjan National Cancer Institute, 37, S.P. Mukherjee Road, Kolkata-700026, India
Sruti Chattopadhyay	Center for Biomedical Engineering, Indian Institute of Technology Delhi (IITD), New Delhi, India
Surendra Thakur	SkillsCoLab, Durban University of Technology, Durban, India

Suvardhan Kanchi   Department of Chemistry, Sambhram Institute of Technology, M.S. Palya, Jalahalli East, Bengaluru 560097, India

Swati Jain   Amity Institute of Nanotechnology, Amity University, Noida, UP, India
Current affiliation: Department of Science & Technology, Technology Bhavan, New Mehrauli Road, New Delhi, India

Ujjal Kumar Sur   Department of Chemistry, Behala College, University of Calcutta, Kolkata-60, India

Venkatasubba Naidu Nuthalapati   Department of Chemistry Sri, Venkateswara University, Tirupati, Andhra Pradesh, India

Zainul Abid CKV   Independent Contributor, External Expert in Spectroscopy, India

State of the Art: Raman Vibrational Spectroscopy and Surface Enhanced Raman Spectroscopy

Jagjiwan Mittal[1,*] and **Robin Kumar**[1]

[1] *Amity Institute of Nanotechnology, Amity University, Sector125, Noida, Uttar Pradesh 201313, India*

Abstract: Raman spectroscopy depends on inelastic scattering of photons, known as Raman scattering. It uses monochromatic light using a laser and determines vibrational modes of molecules. This technique is commonly used for the identification of molecules by providing its structural fingerprint. Due to very low inelastic scattering, however, signals obtained by Raman spectroscopy are inherently weak and the problem is more with visible light. These weak Raman signals can be used by amplifying them by the method known as surface enhanced Raman spectroscopy (SERS). SERS is a powerful vibrational spectroscopy technique that allows for highly sensitive structural detection of low concentration analytes. The current chapter summarizes the basics of Raman spectroscopy and SERS, instrumentation, mechanisms differences and applications.

Keywords: Raman Scattering, SERS, Surface Enhanced Resonance Raman Spectroscopy SERRS, Vibrational Spectroscopy.

1.1. INTRODUCTION RAMAN VIBRATIONAL SPECTROSCOPY

1.1.1. History

Elastic light scattering has been observed since the 19th century by famous physicists Lord Rayleigh. Adolf Smekai [1]and for the first time, inelastic scattering of light was theoretically predicted, in 1923.Indian scientists C.V. Raman and K.S. Krishnan observed this effect in organic liquids by sunlight in 1928 [2, 3]. The effect was named as Raman effect. Due to this discovery, C.V. Raman won the Nobel prize in Physics in 1930.

The same inelastic scattering phenomenon was observed by Grigory Landsberg and Leonid Mandelstam in inorganic crystals [4]. Franco Rasetti observed the

---

* **Corresponding author Jagjiwan Mittal:** Amity Institute of Nanotechnology, Amity University, Sector125, Noida, Uttar Pradesh 201313, India; Tel: +919899010491; E-mail: jmittal@amity.edu

Raman effect in gases using ultraviolet light from a mercury vapour lamp. George Plazek developed the systematic pioneering theory of the Raman effect [5].

Despite the discovery of the Raman effect in 1930, its commercial application started after 1960 when the first laser was developed by T. Maiman [6]. Before laser as a source, Raman Spectroscopy suffered from the low intensity of the inelastic scattering (Raman scattering) and the much larger intensity of the Rayleigh scattering.

A lot of effort was required to get Raman spectra due to the low sensitivity. The sample was kept in a long tube and exposed along its length with a beam of monochromatic light by using filters gas discharge Lamp was used as a source. The use of Laser simplified Raman spectroscopy method and increased the sensitivity of the technique. Laser proved to be an ideal excitation source for getting enough Raman scattering due to its brilliance, monochromaticity and coherence to use it as spectroscopy. This makes Raman spectroscopy as a common analytical technique

1.1.2. Basic Theory

When radiations, either monochromatic or in the narrow frequency band pass through a transparent substance, almost all of the scattered energy will consist of the radiations of incident frequency. This scattering is known as Rayleigh scattering [7]. However, certain discrete frequencies above and below incident frequency are also scattered. This is known as Raman scattering.

In terms of quantum theory, if the collisions of photons having energy $h\nu$ (h is plank constant) and molecules are perfectly elastic, then there is no change in the energy of deflected photons. However, during inelastic collisions, energy is exchanged between photons and molecules. As a result, molecules may gain or lose energy, ΔE according to the difference in energy between allowed states. This energy is due to change in vibrational and/or rotational energy of a molecule. If molecules lose energy, photon will be scattered with $h\nu-\Delta E$. Otherwise, the energy of the photon will be $h\nu+\Delta E$. Radiations scattered with frequency lower than the incident radiations are referred to as Stokes radiations, whereas higher frequency radiations are known as anti-Stokes radiations. Stokes radiations are generally more intense than anti-stokes radiations.

In terms of classical theory, when a molecule is put into a static electric field, it suffers some distortion. This causes induced electric dipole moment in the molecule and results in the polarization of the molecule. The size of induced dipole μ depends on the applied filed E. Therefore, relation between μ and E is

$$\mu = \alpha E \tag{1}$$

here α is polarizability of the molecule.

During exposure of a sample to radiations of frequency ν, the electric field experienced by the molecule is:

$$E = E_0 \sin 2\pi\nu t \tag{2}$$

Here, E_0 is the applied electric field

When we put the value of E in equation (2) in equation (1), induced dipoles become:

$$\mu = \alpha\, E_0 \sin 2\pi\nu t \tag{3}$$

Since the frequency is the same during emission, this equation is true for Rayleigh scattering. In case of additional motion like vibrational or rotational in the molecule its polarizability changes. Polarization is due to vibration ν_{vib} is:

$$\mu = \alpha_0\, E_0 \sin 2\pi\nu t + \tfrac{1}{2}\,\beta E_0[\cos 2\pi\,(\nu - \nu_{vib})t - \cos 2\pi\,(\nu + \nu_{vib})t] \tag{4}$$

Here, α_0 is equilibrium polarizability and β is the rate of change of polarizability with the vibration. The oscillating dipole has frequency components $(\nu + \nu_{vib})$ or $(\nu - \nu_{vib})$.

If the vibration does not change the polarizability of the molecule, then $\beta = 0$ and dipole oscillate at same frequency of the incident radiation. Therefore, for any vibration in a molecule to be Raman active, it must cause some change in a component of the molecular polarizability.

1.1.3. Raman Active Vibrations

If a molecule in its structure has no symmetry then all of its vibrational modes are Raman active. However, when any symmetry exists in the structure of the molecule then the Raman activity of each vibration depends on the change in polarizability.

An asymmetric molecule water H_2O has three modes of vibrations, namely symmetric stretching, asymmetric stretching and bending. It is observed that polarizability in the molecule changes during the application of electric field in all three modes. Therefore, all the vibrations in the water molecule are Raman active.

On the other hand, CO_2 molecule has a two-perpendicular axis of symmetry. Therefore, each vibration in the molecule has to be considered for its Raman activity. CO_2 has four modes of vibrations, namely one symmetric stretching, one asymmetric stretching and two bending. It is observed that the polarizability changes during symmetric stretching and therefore, it is Raman active. However, other vibrations are Raman inactive because there is no change in polarizability.

1.1.4. Rule of Mutual Exclusion

After observing various compounds, a general rule [8, 9] is formulated for Raman and Infrared activities of the molecules.

"In a molecule that possesses a centre of symmetry, Infrared active vibrations are Raman inactive whereas Raman active vibrations are infrared inactive. However, if there is no centre of symmetry in the molecule, some or all vibrations may be both Raman and infrared active".

It is also observed in the Raman spectrum that symmetric vibrations show intense Raman lines whereas asymmetric vibrations are usually weak or unobservable.

1.1.5. Raman Spectrometer Instrumentation

There are four components of a Raman spectrometer:

1. Radiation source which can excite the molecules. Laser is used for this purpose.
2. Sample illumination system and light collection optics: laser energy is transmitted to and collected from the sample by fibre optics cables.
3. Filters or spectrometer is used for the selection of wavelength. A notch or edge filter is used to eliminate Rayleigh and anti-Stokes scattering and the remaining Stokes scattered light is passed on to a dispersion element, typically a holographic grating.
4. CCD, PMT, Photodiode array detector.

In the Raman spectrometer, a sample is normally exposed with a laser beam in the ultraviolet (UV), visible (Vis) or near infrared (NIR) range. Since Raman spectroscopy depends on its ability to measure a shift in wavelength, it is a must that a monochromatic excitation source should be used. Raman peaks are directly affected by the sharpness and stability of the excitation source. In earlier studies a mercury lamp was used for getting of spectra which took hours or even days to acquire due to weak light sources. However at present, lasers are used. A laser is the best excitation source but not all lasers are suitable for Raman spectroscopy. It is essential that the laser frequency is extremely stable and does not mode hop. (Fig. **1**) shows a sketch of the Raman spectrometer.

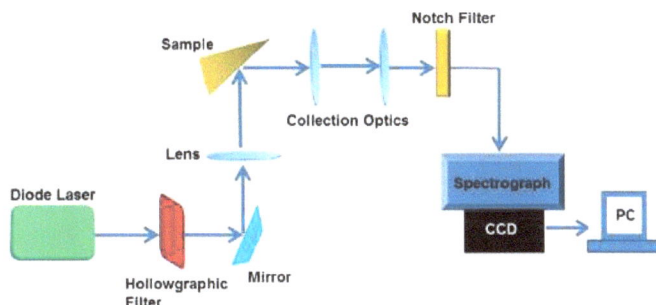

Fig. (1). Schematic illustration of general Raman spectrometer.

Solid state lasers with wavelengths of 532, 785, 830 and 1064 nm are used as the excitation source. The shorter wavelength lasers have higher Raman scattering cross-sections which will provide a greater signal, but fluorescence increases at shorter wavelength.

Scattered light from the sample is collected with a lens and using interference filter or spectrophotometer, Raman spectrum of the sample is obtained. As mentioned earlier, Raman scattering is very weak in comparison to intense Rayleigh scattering Very small amount of the incident light produces inelastic Raman signal. Spontaneous Raman scattering is very weak and special measures should be taken to distinguish it from the predominant Rayleigh scattering. Instruments such as notch filters, tuneable filters, laser stop apertures, double and triple spectrometric systems are used to reduce Rayleigh scattering and obtain high-quality Raman spectra. Multi-channel detectors like Photodiode Arrays (PDA) or, Charge-Coupled Devices (CCD) are used.

It is very necessary that high-quality, optically well-matched components should be used for getting good quality Raman spectrometer. Various ways are used for improving sample preparation, sample illumination or scattered light detection such as stimulated Raman using irradiation with a very strong laser pulse and coherent Anti-Stokes Raman, CARS using the two lasers.

1.1.6. Raman Spectrum

Raman spectrum is drawn between Raman shift and intensity. Raman shifts are typically reported in wavenumber, which have units of inverse length, as this value is directly related to energy. Most commonly, the unit chosen for expressing wavenumber in Raman spectra is inverse centimetres (cm^{-1}). An example of the Raman spectrum of multilayer graphene is shown in Fig. (**2**).

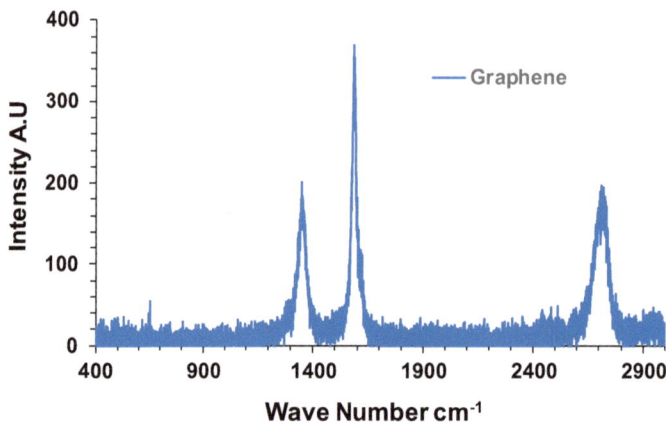

Fig. (2). Raman spectrum of graphene.

(Fig. **2**) displays three distinct peaks in the Raman spectrum of multilayer graphene as D peak, G peak, and second order G` peak [10]. D peak originate from zone-boundary phonons. This peak is not observed in first order Raman spectra of defect-free graphite. Such phonons give rise to a peak at 1350 cm^{-1} in defected graphite. G peak appears near ~ 1590 cm^{-1} and is due to the bond stretching vibration of all pairs of sp^2 atoms in rings and chains both. G` band (popularly known 2D band) appears at 2770 cm^{-1} is due to second order of zone-boundary phonons.

1.1.7. Applications of Raman Spectroscopy

Raman spectroscopy is a non-destructive technique and is used for qualitative or quantitative analysis in many varied fields. This technique can provide key information easily and quickly. Raman can be used to rapidly characterise the chemical composition and structure of solid, liquid, gas, gel, slurry or powder sample.

1.1.7.1. Chemical Analysis

Raman spectroscopy can be used to identify molecules and chemical bonding and intramolecular bonds. It is known that vibrational frequencies are specific to a chemical bonds and symmetry. Raman spectrum provides a fingerprinting the wavenumber range 500–1500 cm^{-1}to identify molecules. For example, Raman in combination with IR spectra were used to determine the vibrational frequencies of SiO, Si_2O_2, and Si_3O_3 [11]. Technique is also used to study the addition of a substrate to an enzyme.

1.1.7.2. Solid State Physics

Raman spectroscopy is used to characterize materials, population of a phonon mode. Later information is provided by the ratio of the Stokes and anti-Stokes intensity of Raman signal. This technique is also useful for observing plasmons, magnons superconducting gap excitations. Raman-shifted backscatter is used to determine the temperature along optical fibres.

1.1.7.3. Nanotechnology

Graphene, carbon nanotube, filled carbon nanotube, nanowire, nanoparticles etc., are extensively researched for various applications [12 - 16]. Raman can be used to understand their structures nanowires. For example, the radial breathing mode of carbon nanotubes is commonly used to evaluate their diameter. This technique also identifies the filling inside the carbon nanotubes [17]. It also helps in determining the number of layers in graphene.

1.1.7.4. Bio-pharmaceutical Industry

Raman spectroscopy has extensive applications in biology and medicine. It is used for identifying active pharmaceutical ingredients (APIs) and their polymorphic forms, confirmations for the existence of low-frequency phonons [18] in proteins, DNA [19 - 24], real-time, *in-situ* biochemical characterization of wounds, measurement of progress in wound healing progress [25] and to identify the counterfeit drugs. Raman spectroscopy is extensively used for studying bio-minerals [26]. Gas analysers using Raman spectroscopy are applied in real-time monitoring of anaesthetic and respiratory gas mixtures during surgery.

1.1.7.5. Study of Historical Painting and Documents

Raman spectroscopy is a non-destructive technique [27] which can be used for the study of historical paints and documents. The study includes information about the original state of the painting, pigments degraded with age, individual pigments in paintings and their degradation products, the chemical composition of historical documents and determining the best method of their preservation.

1.1.7.6. Explosives

Raman spectroscopy can be used to detect explosives safely using laser beams [28, 29].

1.1.7.7. Sensing Based on Raman Spectroscopy

Detection of low concentration gases especially polluted gases [30 - 34], is a requirement for the health of the human and natural world. Raman spectroscopy has great potential as a process for the identification and quantification of the composition of gaseous samples. Raman spectroscopy can also be used for the analysis. Studies have shown highly-sensitive quantitative Raman detection of various gases (nitrogen, oxygen, carbon dioxide, toluene, acetone and 1,1,1-trichloroethane) using a photonic crystal fibre probe [35].

1.2. SURFACE-ENHANCED RAMAN SCATTERING (SERS)

As mentioned above, due to very low inelastic scattering, signals obtained by Raman spectroscopy are inherently weak. This problem is exuberated when using visible light is used for excitation. This results in the unavailability of a small number of scattered photons for detection.

These weak Raman signals can be used by amplifying them. The method known as surface enhanced Raman scattering (SERS) uses nanoscale roughened metal surfaces classically made of gold (Au) or silver (Ag). Excitation of these roughened metal nanostructures using laser resonantly drives the surface charges generating a highly localized (plasmonic) light field. When a molecule is absorbed or lies close to the enhanced field at the surface, Raman signals are greatly enhanced by several orders of magnitude than normal Raman scattering. These results indicate the possibility of detecting low concentrations (10^{-11}) without the need for fluorescent labelling.

Therefore, SERS is a surface-sensitive technique that enhances Raman scattering by molecules adsorbed on rough metal surfaces or by nanostructures such as plasmonic-magnetic silica nanotubes [36]. The enhancement factor can be as much as 10^{10} to 10^{11} [37, 38], which means the technique may detect single molecule.

Signals can be further amplified when the roughened metal surface is used with laser light that is matched to the absorption maxima of the molecule. This method is known as surface-enhanced resonance Raman scattering (SERRS). (Fig. **3**) illustrates the basic of Surface-Enhanced Raman Scattering.

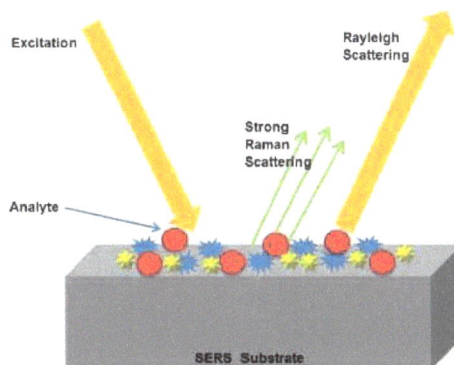

Fig. (3). Schematic illustration of Surface-Enhanced Raman Scattering.

1.2.1. Short History

First observations of the Raman spectra of pyridine adsorbed on electrochemically roughened Ag were done in 1974 [39] by Fleischmann, Hendra and McQuillan. However, the authors did not distinguish that this spectrum was enhanced and unusual. Later, two groups independently noted that the enhanced signal was not related to concentration of scattering species. Jeanmaire and Duyne [40] proposed an electromagnetic effect, whereas Albrecht and Creighton [41] proposed a charge-transfer effect for unusual enhancement. Ritchie predicted the existence of the surface plasmon [42].

1.2.2. SERS Mechanism

There are two mechanisms, electromagnetic and chemical mechanisms were proposed for the enhancement in the Raman signals. The main contributor to most SERS processes is the electromagnetic enhancement mechanism [43].

The electromagnetic enhancement comes from the amplification of the light by the excitation of localized surface plasmon resonances (LSPRs). Light is concentrated favourably in the gaps or crevices of plasmonic materials *e.g.*, Nano silver, nano gold, and nano copper. Enhancement depends on the structure of the supporting plasmonic material can reach factors of $\sim 10^{10} - 10^{11}$ [44].

Chemical enhancement involves the charge transfer mechanisms. This excitation wavelength is resonant with the metal-molecule charge transfer electronic states. Magnitudes of enhancement through charge transfer transitions are highly molecule specific [45, 46]. Theoretically, chemical enhancement can be $<10^3$ but experimentally, it is found as ~5-10 [47, 48].

Total enhancement in signals is the product of electromagnetic and chemical enhancement. For highly optimized surfaces, total enhancement may be ~ 10^{10} – 10^{11} [49].

1.2.3. SERS Technique

SERS is a non-destructive technique and is useful for determining the chemical and structural information of molecules. Substrates can be nanorods to three-dimensional colloidal solutions with tunable plasmon resonances and average enhancement factors. Since the maximum SERS enhancing region decreases extremely rapidly with the increase in the distance [10], the highest enhancements are observed in the nearest (few nanometres) to the substrate surface.

Excitation sources would produce efficient excitation of the plasmon resonance. For this laser is tuned to the peak of the plasmon resonance, for a substrate with a single peak in its LSPR spectrum. Maximum enhancement is observed with the shift of laser wavelength to the blue of the plasmon resonance [50]. Maximum signal is obtained when the plasmon frequency is slightly red-shifted from the laser wavelength.

For detection process, a filter is used to absorb or reflect any Rayleigh scattering while allowing for transmission of the Raman signal. Spectrograph and detector are used for getting the Raman spectra.

1.2.4. SERS Substrates

SERS spectrum also depends on the interaction between adsorbed molecules and the surface of plasmonic nanostructures. For a material to be described as plasmonic, it must have a negative real component and a small positive imaginary component of the dielectric constant. SERS should be measured on various molecules.

SERS substrate is chosen on the basis of the type and form of samples. Substrates used for analysis are either colloidal metal solution, or metal layer deposited on top. Colloidal substrates are 20-100 nm diameter metal nanoparticles suspended in solution. Samples are either deposited on a substrate, or mixed with the colloidal solution for analysis.

For assuring that the system is not undergoing a resonance Raman effect, SERS spectra are collected for non-resonant molecules. Gold (Au) or silver (Ag) and Copper (Cu) are ideal metals for use in SERS. All these three metals have localized surface plasmons resonance (LSPRs) that belong, where most Raman measurements wavelength range in visible and are near infrared regions. Au and Ag are mostly used as substrates because of higher air stability. Cu is more reactive, so is less preferred.

Research has been done in the development of SERS substrates [51], using the nanoparticles, including Ag and Au nanoparticles with various shapes. Other than Ag and Au, metals such as Li, Na, K, Rb, Cs, Al, Ga, In, Pt, Rh, and various metal alloys [52] have been studied as plasmonic substrates for SERS.

Advanced SERS techniques involve the modification of the metal surface either by chemical reaction or by adding functionalization of metal.

Research is going for [53 - 58] using semiconductors, quantum dots, and graphene as substrates for SERS. Materials such as graphene [53,54], TiO_2 [55], and quantum dots [56 - 58] have shown effective substrates for SERS. These substrates involve purely chemical enhancements with no evidence of electromagnetic enhancement and therefore enhancement factors $<10^3$. Studies are going on graphene as a plasmonic material in the infra-red [59].

Broadly, there are three methods used for sample preparation for SERS technique. The first method uses a few microliters of the colloid solution either applied to the sample or mixed with the sample solution and put on the microscope slide for drying. The mixture is then analyzed with the Raman instrument.

Second method includes the deposition process for the metal and then roughens the surface for the generation of optimum surface plasmon. This improves the SERS signal.

In the third technique, SERS substrate is embedded in a sol gel by mixing metal nanoparticles with a photo-reactive chemical. Sol-gel matrix is then exposed to the proper wavelength so that the chemical reacts and forms the nanoparticles *in-situ*.

1.2.5. Applications of SERS

After the discovery, attention towards SERS application has grown exponentially.

Major benefits of SERS technique are the increase in signals' intensity from weak Raman scattering and simplification of these signals in a substantial way [60 - 63]. SERS is a special technique to characterize small numbers of molecules at the plasmonic surfaces. This technique is highly suitable in applications like sensing, imaging, single molecule detection, ultrahigh vacuum and ultrafast science [64 - 67].

Due to its ability for nanoscale analysis of mixture composition SERS technique is useful for studies on the environment, material sciences pharmaceuticals, forensic science, drug and explosives detection, art and archaeological research food quality analysis [68], single algal cell detection [69 - 72] and redox processes at the single molecule level [73]. Some of applications of SERS technique are provided below:

1.2.5.1. Biomarkers

SERS can detect proteins in body fluids because of its ability to sense the presence of low abundance of biomolecules [74]. In one study, pancreatic cancer biomarkers were spotted in the earlier stage using SERS-based immunoassay approach [74]. A SERS-base multiplex protein biomarker detection platform using a microfluidic chip is used in another study to detect several protein biomarkers. This will help to predict the type of disease [75]. This is also used to detect urea and blood plasma label free in human serum [76, 77].

SERS can be effectively used for DNA detection. DNA and RNA detection using SERS technique was first reported using sandwich structures by Mirkin *et al*. Multiplex DNA detection using different Raman active dyes can be achieved using this method [78]. SERS in combination with plasmonic sensing can be applied for quantitative detection of biomolecular interaction [69].

1.2.5.2. Toxic Chemical and Chemical Warfare Agents

Detection of toxic industrial chemicals and chemical warfare agents is very important for human safety and security. SERS is a promising method for the ultrasensitive detection of these chemicals [79]. Explosives such as the half-mustard agent [80] and dinitrobenzenethiol [81] were detected by SERS.

1.2.5.3. Narcotic and Doping Drugs

Amphetamine was successfully detected in XTC tablets using SERS technique [82]. Dihydrocodeine, doxepine, citalopram, trimipramine, carbamazepine, and methadone can be detected in just 1 mg urine or blood sample [83]. In addition, doping drugs in athletics such as clenbuterol, salbutamol, and terbutaline can be successfully detected using SERS and [84].

1.2.5.4. Food Monitoring

SERS is particularly well suited to detect small molecules because of the close proximity of the analyte to the plasmonic structure. Monitoring food for dangerous ingredients such as mycotoxin, pesticides, some colouring agents and antimicrobials can be successfully detected using the technique.

1.2.5.5. Verification and Authentication of Luxury Goods and Currency

Au spheres functionalized with reporter molecules and encased in a silica shell known as SERS nanotags, can be used for the labelling and verification and authentication of jewellery or luxury goods [85]. These nanotags can also help in authentication by embedding them in currency or bank notes [86] during the printing process.

1.3. SURFACE ENHANCED RESONANCE RAMAN SPECTROSCOPY (SERRS)

There is a disadvantage of SERS spectra is that they can be difficult to interpret since the normal Raman spectrum is not necessarily the same as adsorbed on a SERS substrate. The signal enhancement can be so high that Raman bands that are very weak and unnoticeable in Raman spectra can appear in SERS. Trace contaminants may also show additional peaks. Whereas, due to chemical interactions with the metal surface, certain peaks which are observed strongly in conventional Raman might not be present in SERS at all. Because of such problem, Surface-Enhanced Resonance Raman spectroscopy (SERRS) has been developed [87]. It uses both the Surface-Enhancement effect and the Raman Resonance effect, so that SERRS spectra resemble regular Resonance Raman spectra, which makes it much easier to interpretation. Enhancement in Raman signal intensity in SERRS can be as high as 10^{14}.

1.3.1. Advantages of SERRS Over SERS

SERS directly detects the actual molecule under study in a way that is molecularly specific is prone to interference from other species present in the reaction mixture.

In comparison, SERRS is extremely highly sensitive, which makes it a good labelling technique and reduces interference.

There are three other major advantages of SERRS over SERS.

First is the huge additional enhancement in SERRS compared to SERS [88, 89]. For example, the SERS enhancement for pyridine is calculated at 10^{6}, whereas the SERRS enhancement factor for rhodamine is estimated at between 10^{13} and 10^{15}.

Second is the wavelength dependence of the intensity obtained from SERRS. The huge enhancement obtained by SERRS can keep the concentrations of analyte down to well below that required to aggregate the colloid.

Third is that excellent Raman scattering can be obtained from fluorescent molecules. This allows using a very wide range of dyes, both fluorescent and non-fluorescent, as labels.

CONCLUSION

Raman spectroscopy is currently one of the most common spectroscopic techniques for material identification and its physical and chemical environment by analyzing the vibrations by providing a structural fingerprint of molecules. Surface enhanced Raman spectroscopy (SERS) is an excellent technique for characterization in various areas, including environment, pharmaceutical, forensic and food science, etc. It is a highly sensitive technique which can detect different types of analytes, including biomolecules in low concentrations.

CONSENT FOR PUBLICATION

Not applicable.

CONFLICT OF INTEREST

The authors declare no conflict of interest, financial or otherwise.

ACKNOWLEDGEMENTS

Declared none.

REFERENCES

[1]   Smekal A. Zurquantentheorie der dispersion. Naturwissenschaften 1923; 11: 873-5.
        [http://dx.doi.org/10.1007/BF01576902]

[2]   Raman CV, Krishnan KS. A new type of secondary radiation. Nature 1928; 121: 501-2.
        [http://dx.doi.org/10.1038/121501c0]

[3]   Raman CV, Krishnan KS. A new class of spectra due to secondary radiation. Part I. Indian J Phys 1928; 2: 399-419.

[4]   Landsberg G, Mandelstam L. Naturwiss 1928; 16: 557-8.
        [http://dx.doi.org/10.1007/BF01506807]

[5]   Placzek G. Rayleigh Strenung und Raman Effekt, Handbuch der Radiologie II, (in German). Leipzig: AkademischeVerlagsgesellschaft 209.

[6]   Maiman TH. inventor; Hughes Aircraft Co, assignee. Ruby laser systems. United States 1967.

[7]   Rayleigh L. XXXIV. On the transmission of light through an atmosphere containing small particles in suspension, and on the origin of the blue of the sky. Lond Edinb Dublin Philos Mag J Sci 1899; 47: 375-84.
        [http://dx.doi.org/10.1080/14786449908621276]

[8]   Venkatarayudu T. The rule of mutual exclusion. J Chem Phys 1954; 22: 1269.
        [http://dx.doi.org/10.1063/1.1740366]

[9]   Bernath PF. Spectra of atoms and molecules. Oxford university press 2015; p. 304.

[10]   Ferrari AC, Meyer JC, Scardaci V, *et al.* Raman spectrum of graphene and graphene layers. Phys Rev Lett 2006; 97(18): 187401.
        [http://dx.doi.org/10.1103/PhysRevLett.97.187401] [PMID: 17155573]

[11]   Khanna RK. Raman-spectroscopy of oligomeric SiO species isolated in solid methane. J Chem Phys 1981; 74(4): 2108.
        [http://dx.doi.org/10.1063/1.441393]

[12]   Mittal J, Lin KL. The formation of electric circuits with carbon nanotubes and copper using tin solder. Carbon 2011; 49: 4385-91.
        [http://dx.doi.org/10.1016/j.carbon.2011.06.029]

[13]   Mittal J, Lin KL. Bulk thermal conductivity studies of Sn/SnO coated and filled multiwalled carbon nanotubes for thermal interface material. Fuller Nanotube Car N 2017; 25: 301-5.

[14]   Mittal J, Kushwaha N. Over-oxidation of multi-walled carbon nanotubes and formation of fluorescent carbon nanoparticles. Mater Lett 2015; 145: 37-40.
        [http://dx.doi.org/10.1016/j.matlet.2015.01.059]

[15]   Mittal J, Lin KL. Formation of nanojoints between carbon nanotubes and copper nanoparticles. Carbon Lett 2017; 21: 86-92.
        [http://dx.doi.org/10.5714/CL.2017.21.086]

[16]   Pudake RN, Mittal J, Tripathi RM, Tyagi J, Mohanta TK. Biochemical responses of maize seedlings exposed to SnNPs. Micro & Nano Lett 2019; 14: 645-9.
        [http://dx.doi.org/10.1049/mnl.2018.5313]

[17]   Mittal J, Monthioux M, Allouche H, Stephan O, Bacsa W. Room temperature filling of single-wall carbon nanotubes in open air. In AIP Conference Proceedings. 2001; 591: pp. 273-6.
        [http://dx.doi.org/10.1063/1.1426869]

[18]   Kuo-Cheng Ch, Nian-Yi Ch. The biological functions of low-frequency phonons. Sci Sin 1977; 20: 447-57.

[19]   Urabe H, Tominaga Y, Kubota K. Experimental evidence of collective vibrations in DNA double helix (Raman spectroscopy). J Chem Phys 983(78): 5937-9.
        [http://dx.doi.org/10.1063/1.444600]

[20]   Chou KC. Identification of low-frequency modes in protein molecules. Biochem J 1983; 215(3): 465-9.
        [http://dx.doi.org/10.1042/bj2150465] [PMID: 6362659]

[21]   Chou KC. Low-frequency vibrations of DNA molecules. Biochem J 1984; 221(1): 27-31.

[http://dx.doi.org/10.1042/bj2210027] [PMID: 6466317]

[22] Urabe H, Sugawara Y, Ataka M, Rupprecht A. Low-frequency Raman spectra of lysozyme crystals and oriented DNA films: dynamics of crystal water. Biophys J 1998; 74(3): 1533-40.
[http://dx.doi.org/10.1016/S0006-3495(98)77865-8] [PMID: 9512049]

[23] Chou KC. Low-frequency collective motion in biomacromolecules and its biological functions. Biophys Chem 1988; 30(1): 3-48.
[http://dx.doi.org/10.1016/0301-4622(88)85002-6] [PMID: 3046672]

[24] Chou KC. Low-frequency resonance and cooperativity of hemoglobin. Trends Biochem Sci 1989; 14(6): 212-3.
[http://dx.doi.org/10.1016/0968-0004(89)90026-1] [PMID: 2763333]

[25] Jain R, Calderon D, Kierski PR, et al. Raman spectroscopy enables noninvasive biochemical characterization and identification of the stage of healing of a wound. Anal Chem 2014; 86(8): 3764-72.
[http://dx.doi.org/10.1021/ac500513t] [PMID: 24559115]

[26] Taylor PD, Vinn O, Kudryavtsev A, Schopf JW. Raman spectroscopic study of the mineral composition of cirratulid tubes (Annelida, Polychaeta). J Struct Biol 2010; 171(3): 402-5.
[http://dx.doi.org/10.1016/j.jsb.2010.05.010] [PMID: 20566380]

[27] Howell G M, Edwards, M John. Chalmers, Raman Spectroscopy in Archaeology and Art History, Royal Society of Chemistry. 2005.

[28] Vogel B. Raman spectroscopy portends well for standoff explosives detection. IHS Jane's 2008.

[29] Misra AK, Sharma SK, Acosta TE, Porter JN, Bates DE. Single-pulse standoff Raman detection of chemicals from 120 m distance during daytime. Appl Spectrosc 2012; 66(11): 1279-85.
[http://dx.doi.org/10.1366/12-06617] [PMID: 23146183]

[30] Kumar R, Kushwaha N, Mittal J. Superior, rapid and reversible sensing activity of graphene-SnO hybrid film for low concentration of ammonia at room temperature. Sens Actuators B Chem 2017; 244: 243-51.
[http://dx.doi.org/10.1016/j.snb.2016.12.111]

[31] Kumar R, Kumar R, Kushwaha N, Mittal J. Ammonia gas sensing using thin film of MnO_2 nanofibers. IEEE Sens J 2016; 16(12): 4691-5.
[http://dx.doi.org/10.1109/JSEN.2016.2550079]

[32] Kumar R, Mittal J, Kushwaha N, Rao BV, Pandey S, Liu CP. Room temperature carbon monoxide gas sensor using Cu doped OMS-2 nanofibers. Sens Actuators B Chem 2018; 266: 751-60.
[http://dx.doi.org/10.1016/j.snb.2018.03.182]

[33] Kumar R, Jaiswal M, Singh O, Gupta A, Ansari MS, Mittal J. Selective and reversible sensing of low concentration of carbon monoxide gas using Nb doped OMS-2 nanofibers at room temperature. IEEE Sens J 2019; 19(17): 7201-6.
[http://dx.doi.org/10.1109/JSEN.2019.2916485]

[34] Kumar R, Kushwaha N, Mittal J. Ammonia gas sensing activity of Sn nanoparticles film. Sens Lett 2016; 14: 300-3.
[http://dx.doi.org/10.1166/sl.2016.3652]

[35] Yang X, Chang AS, Chen B, Gu C, Bond TC. Multiplexed gas sensing based on Raman spectroscopy in photonic crystal fiber. IEEE Photonics Conference. 447-8.
[http://dx.doi.org/10.1109/IPCon.2012.6358685]

[36] Xu X, Li H, Hasan D, Ruoff RS, Wang AX, Fan DL. Near-field enhanced plasmonic-magnetic bifunctional nanotubes for single cell bioanalysis. Adv Funct Mater 2013; 23: 4332-8.
[http://dx.doi.org/10.1002/adfm.201203822]

[37] Blackie EJ, Le Ru EC, Etchegoin PG. Single-molecule surface-enhanced Raman spectroscopy of

nonresonant molecules. J Am Chem Soc 2009; 131(40): 14466-72.
[http://dx.doi.org/10.1021/ja905319w] [PMID: 19807188]

[38]   Langer J, Jimenez de Aberasturi D, Aizpurua J, *et al.* Present and future of surface-enhanced Raman scattering. ACS Nano 2020; 14(1): 28-117.
[http://dx.doi.org/10.1021/acsnano.9b04224] [PMID: 31478375]

[39]   Fleischmann M, Hendra PJ, McQuillan AJ. Raman spectra of pyridine adsorbed at a silver electrode. Chem Phys Lett 1974; 26: 163-6.
[http://dx.doi.org/10.1016/0009-2614(74)85388-1]

[40]   Jeanmaire DL, Van Duyne RP. Surface Raman spectroelectrochemistry: Part I. Heterocyclic, aromatic, and aliphatic amines adsorbed on the anodized silver electrode. J Electroanal Chem Interfacial Electrochem 1977; 84: 1-20.
[http://dx.doi.org/10.1016/S0022-0728(77)80224-6]

[41]   Albrecht MG, Creighton JA. Anomalously intense Raman spectra of pyridine at a silver electrode. J Am Chem Soc 1977; 99: 5215-7.
[http://dx.doi.org/10.1021/ja00457a071]

[42]   Ritchie RH. Plasma losses by fast electrons in thin films. Phys Rev 1957; 106: 874.
[http://dx.doi.org/10.1103/PhysRev.106.874]

[43]   Stiles PL, Dieringer JA, Shah NC, Van Duyne RP. Surface-enhanced Raman spectroscopy. Annu Rev Anal Chem (Palo Alto, Calif) 2008; 1: 601-26.
[http://dx.doi.org/10.1146/annurev.anchem.1.031207.112814] [PMID: 20636091]

[44]   Camden JP, Dieringer JA, Wang Y, *et al.* Probing the structure of single-molecule surface-enhanced Raman scattering hot spots. J Am Chem Soc 2008; 130(38): 12616-7.
[http://dx.doi.org/10.1021/ja8051427] [PMID: 18761451]

[45]   Jensen L, Aikens CM, Schatz GC. Electronic structure methods for studying surface-enhanced Raman scattering. Chem Soc Rev 2008; 37(5): 1061-73.
[http://dx.doi.org/10.1039/b706023h] [PMID: 18443690]

[46]   Morton SM, Jensen L. Understanding the molecule-surface chemical coupling in SERS. J Am Chem Soc 2009; 131(11): 4090-8.
[http://dx.doi.org/10.1021/ja809143c] [PMID: 19254020]

[47]   Sharma B, Cardinal MF, Kleinman SL, *et al.* High-performance SERS substrates: Advances and challenges. MRS Bull 2013; 38: 615-24.
[http://dx.doi.org/10.1557/mrs.2013.161]

[48]   Sharma B, Frontiera RR, Henry AI, Ringe E, Van Duyne RP. SERS: Materials, applications, and the future. Mater Today 2012; 15: 16-25.
[http://dx.doi.org/10.1016/S1369-7021(12)70017-2]

[49]   Le Ru EC, Blackie E, Meyer M, Etchegoin PG. Surface enhanced Raman scattering enhancement factors: a comprehensive study. J Phys Chem C 2007; 111: 13794-803.
[http://dx.doi.org/10.1021/jp0687908]

[50]   McFarland AD, Young MA, Dieringer JA, Van Duyne RP. Wavelength-scanned surface-enhanced Raman excitation spectroscopy. J Phys Chem B 2005; 109(22): 11279-85.
[http://dx.doi.org/10.1021/jp050508u] [PMID: 16852377]

[51]   Bandarenka HV, Girel KV, Zavatski SA, Panarin A, Terekhov SN. Progress in the development of SERS-active substrates based on metal-coated porous silicon. Materials (Basel) 2018; 11(5): 852.
[http://dx.doi.org/10.3390/ma11050852] [PMID: 29883382]

[52]   Van Duyne RP, Hulteen JC, Treichel DA. Atomic force microscopy and surface-enhanced Raman spectroscopy. I. Ag island films and Ag film over polymer nanosphere surfaces supported on glass. J Chem Phys 1993; 99: 2101-15.
[http://dx.doi.org/10.1063/1.465276]

[53]   Ling X, Xie L, Fang Y, *et al.* Can graphene be used as a substrate for Raman enhancement? Nano Lett 2010; 10(2): 553-61.
[http://dx.doi.org/10.1021/nl903414x] [PMID: 20039694]

[54]   Qiu C, Zhou H, Yang H, Chen M, Guo Y, Sun L. Investigation of n-layer graphenes as substrates for Raman enhancement of crystal violet. J Phys Chem C 2011; 115: 10019-25.
[http://dx.doi.org/10.1021/jp111617c]

[55]   Musumeci A, Gosztola D, Schiller T, *et al.* SERS of semiconducting nanoparticles (TiO_2) hybrid composites). J Am Chem Soc 2009; 131(17): 6040-1.
[http://dx.doi.org/10.1021/ja808277u] [PMID: 19364105]

[56]   Livingstone R, Zhou X, Tamargo MC, Lombardi JR, Quagliano LG, Jean-Mary F. Surface enhanced Raman spectroscopy of pyridine on CdSe/ZnBeSe quantum dots grown by molecular beam epitaxy. J Phys Chem C 2010; 114: 17460-4.
[http://dx.doi.org/10.1021/jp105619m]

[57]   Quagliano LG. Observation of molecules adsorbed on III-V semiconductor quantum dots by surface-enhanced Raman scattering. J Am Chem Soc 2004; 126(23): 7393-8.
[http://dx.doi.org/10.1021/ja031640f] [PMID: 15186179]

[58]   Wang Y, Zhang J, Jia H, *et al.* Mercaptopyridine surface-functionalized CdTe quantum dots with enhanced Raman scattering properties. J Phys Chem C 2008; 112: 996-1000.
[http://dx.doi.org/10.1021/jp077467h]

[59]   Fei Z, Andreev GO, Bao W, *et al.* Infrared nanoscopy of dirac plasmons at the graphene-SiO_2 interface. Nano Lett 2011; 11(11): 4701-5.
[http://dx.doi.org/10.1021/nl202362d] [PMID: 21972938]

[60]   Campion A, Kambhampati P. Surface-enhanced Raman scattering. Chem Soc Rev 1998; 27: 241-50.
[http://dx.doi.org/10.1039/a827241z]

[61]   Dieringer JA, McFarland AD, Shah NC, *et al.* Introductory lecture surface enhanced Raman spectroscopy: new materials, concepts, characterization tools, and applications. Faraday dis 2006; 132: 9-26.
[http://dx.doi.org/10.1039/B513431P]

[62]   Haynes CL, McFarland AD, Van Duyne RP. Surface-enhanced Raman spectroscopy. Anal Chem 2005; 77: 338A-46A.
[http://dx.doi.org/10.1021/ac053456d]

[63]   Pilot R, Signorini R, Durante C, Orian L, Bhamidipati M, Fabris L. A review on surface-enhanced Raman scattering. Biosensors (Basel) 2019; 9(2): 57.
[http://dx.doi.org/10.3390/bios9020057] [PMID: 30999661]

[64]   Doering WE, Nie S. Single-molecule and single-nanoparticle SERS: examining the roles of surface active sites and chemical enhancement. J Phys Chem B 2002; 106: 311-7.
[http://dx.doi.org/10.1021/jp011730b]

[65]   Etchegoin PG, Le Ru EC. A perspective on single molecule SERS: current status and future challenges. Phys Chem Chem Phys 2008; 10(40): 6079-89.
[http://dx.doi.org/10.1039/b809196j] [PMID: 18846295]

[66]   Kneipp K, Kneipp H, Itzkan I, Dasari RR, Feld MS. Ultrasensitive chemical analysis by Raman spectroscopy. Chem Rev 1999; 99(10): 2957-76.
[http://dx.doi.org/10.1021/cr980133r] [PMID: 11749507]

[67]   Moskovits M. Surface-enhanced Raman spectroscopy: a brief retrospective. J Raman Spectrosc 2005; 36: 485-96.
[http://dx.doi.org/10.1002/jrs.1362]

[68]   Andreou C, Mirsafavi R, Moskovits M, Meinhart CD. Detection of low concentrations of ampicillin in

milk. Analyst (Lond) 2015; 140(15): 5003-5.
[http://dx.doi.org/10.1039/C5AN00864F] [PMID: 26087055]

[69]   Xu Z, Jiang J, Wang X, *et al.* Large-area, uniform and low-cost dual-mode plasmonic naked-eye colorimetry and SERS sensor with handheld Raman spectrometer. Nanoscale 2016; 8(11): 6162-72.
[http://dx.doi.org/10.1039/C5NR08357E] [PMID: 26931437]

[70]   Lahr RH, Vikesland PJ. Surface-Enhanced Raman Spectroscopy (SERS). Cellular Imaging of Intracellulary Biosynthesized Gold Nanoparticles ACS Sustainable Chem Eng 2014; 2(7): 1599-608.
[http://dx.doi.org/10.1021/sc500105n]

[71]   Deng YL, Juang YJ. Black silicon SERS substrate: effect of surface morphology on SERS detection and application of single algal cell analysis. Biosens Bioelectron 2014; 53: 37-42.
[http://dx.doi.org/10.1016/j.bios.2013.09.032] [PMID: 24121206]

[72]   Wackerbarth H, Salb C, Gundrum L, *et al.* Detection of explosives based on surface-enhanced Raman spectroscopy. Appl Opt 2010; 49(23): 4362-6.
[http://dx.doi.org/10.1364/AO.49.004362] [PMID: 20697437]

[73]   Cortés E, Etchegoin PG, Lc Ru EC, Fainstein A, Vela ME, Salvarezza RC. Monitoring the electrochemistry of single molecules by surface-enhanced Raman spectroscopy. J Am Chem Soc 2010; 132(51): 18034-7.
[http://dx.doi.org/10.1021/ja108989b] [PMID: 21138263]

[74]   Banaei N, Foley A, Houghton JM, Sun Y, Kim B. Multiplex detection of pancreatic cancer biomarkers using a SERS-based immunoassay. Nanotechnology 2017; 28(45): 455101.
[http://dx.doi.org/10.1088/1361-6528/aa8e8c] [PMID: 28937361]

[75]   Banaei N, Moshfegh J, Mohseni-Kabir A, Houghton JM, Sun Y, Kim B. Machine learning algorithms enhance the specificity of cancer biomarker detection using SERS-based immunoassays in microfluidic chips. RSC Advances 2019; 9: 1859-68.
[http://dx.doi.org/10.1039/C8RA08930B]

[76]   Han YA, Ju J, Yoon Y, Kim SM. Fabrication of cost-effective surface enhanced Raman spectroscopy substrate using glancing angle deposition for the detection of urea in body fluid. J Nanosci Nanotechnol 2014; 14(5): 3797-9.
[http://dx.doi.org/10.1166/jnn.2014.8184] [PMID: 24734638]

[77]   Lin D, Feng S, Huang H, *et al.* Label-free detection of blood plasma using silver nanoparticle-based surface-enhanced Raman spectroscopy for esophageal cancer screening. J biomed nanotech 2014; 10: 478-784.

[78]   Cao YC, Jin R, Mirkin CA. Nanoparticles with Raman spectroscopic fingerprints for DNA and RNA detection. Science 2002; 297(5586): 1536-40.
[http://dx.doi.org/10.1126/science.297.5586.1536] [PMID: 12202825]

[79]   Golightly RS, Doering WE, Natan MJ. Surface-enhanced Raman spectroscopy and homeland security: a perfect match? ACS Nano 2009; 3(10): 2859-69.
[http://dx.doi.org/10.1021/nn9013593] [PMID: 19856975]

[80]   Stuart DA, Biggs KB, Van Duyne RP. Surface-enhanced Raman spectroscopy of half-mustard agent. Analyst (Lond) 2006; 131(4): 568-72.
[http://dx.doi.org/10.1039/b513326b] [PMID: 16568174]

[81]   Sylvia JM, Janni JA, Klein JD, Spencer KM. Surface-enhanced raman detection of 2,4-dinitrotoluene impurity vapor as a marker to locate landmines. Anal Chem 2000; 72(23): 5834-40.
[http://dx.doi.org/10.1021/ac0006573] [PMID: 11128944]

[82]   Sägmüller B, Schwarze B, Brehm G, Schneider S. Application of SERS spectroscopy to the identification of (3,4-methylenedioxy)amphetamine in forensic samples utilizing matrix stabilized silver halides. Analyst (Lond) 2001; 126(11): 2066-71.
[http://dx.doi.org/10.1039/b105321n] [PMID: 11763093]

[83]   Trachta G, Schwarze B, Sägmüller B, Brehm G, Schneider S. Combination of high-performance liquid chromatography and SERS detection applied to the analysis of drugs in human blood and urine. J Mol Struct 2004; 693: 175-85.
[http://dx.doi.org/10.1016/j.molstruc.2004.02.034]

[84]   Izquierdo-Lorenzo I, Sánchez-Cortés S, García-Ramos JV. Adsorption of beta-adrenergic agonists used in sport doping on metal nanoparticles: a detection study based on surface-enhanced Raman scattering. Langmuir 2010; 26(18): 14663-70.
[http://dx.doi.org/10.1021/la102590f] [PMID: 20799745]

[85]   Freeman R, Smith P, Natan M. Alavita Pharmaceuticals Inc, Labelling and authentication of metal objects. United States patent 2005.

[86]   Natan M, Norton S, Freeman R, *et al.* Nanoparticles As Covert Taggants In Currency, Bank Notes, And Related Documents. United States patent 2007.

[87]   McNay G, Eustace D, Smith WE, Faulds K, Graham D. Surface-enhanced Raman scattering (SERS) and surface-enhanced resonance Raman scattering (SERRS): a review of applications. Appl Spectrosc 2011; 65(8): 825-37.
[http://dx.doi.org/10.1366/11-06365] [PMID: 21819771]

[88]   Kneipp K, Wang Y, Kneipp H, *et al.* Single molecule detection using surface-enhanced Raman scattering (SERS). Physical rev let 1997; 78

[89]   Nie S, Emory SR. Probing single molecules and single nanoparticles by surface-enhanced Raman scattering. Science 1997; 275(5303): 1102-6.
[http://dx.doi.org/10.1126/science.275.5303.1102] [PMID: 9027306]

CHAPTER 2

Enhancement Mechanisms and Theory of SERS

Pragya Agar Palod[1,*] and **Manika Khanuja**[2]

[1] *Department of Physics, Shri Vaishnav Vidyapeeth Vishwavidyalaya, Indore - 453111, India*

[2] *Centre for Nanoscience and Nanotechnology, Jamia Millia Islamia, New Delhi - 110025, India*

Abstract: This chapter starts with a brief introduction to basic Raman Spectroscopy, its strengths and bottlenecks leading to ineffective utilization of the technique in its conventional form. Further, the basic principle of Surface Enhanced Raman Spectroscopy (SERS) has been discussed along with its development as a vibrational spectroscopic tool of analytical importance. The milestones related to the development of new methods in SERS have also been covered. The major focus of the chapter is to describe the enhancement mechanisms responsible for magnificent enhancement obtained in SERS as compared to normal Raman Spectroscopy. Out of various mechanisms, the electromagnetic (EM) mechanism has been considered to play the most significant role in enhancement. Localized Surface Plasmon Resonance (LSPR) is chiefly responsible for the whole EM enhancement leading to E^4 enhancement and therefore, it has been discussed in detail. Other mechanisms like chemical and electronic enhancement mechanisms have also been discussed extensively. The dependence of SERS on various factors like substrate, excitation wavelength, size and shape of nanoparticles, etc., has been explained with emphasis on the reported data followed by analysis. Two photon excited SERS has been presented as a special class of SERS. Towards the end, the application of SERS to achieve the ultimate limit of detection by probing single molecules has been emphasised. In this context, Hot spots, the heterogeneous nanoregions causing extremely large enhancements and their various generations have been presented along with preliminary theoretical and experimental results.

Keywords: Analyte, Anti-Stokes, Enhancement Factor, Enhancement Mechanism, Excitation Wavelength, Nanoparticle, Raman Scattering, Resonance, SERS Intensity, SERS Spectrum, Single Molecule *etc.*

2.1. INTRODUCTION

Since the discovery of Raman effect by Dr. Chandrasekhara Venkata Raman and Dr. Kariamanickam Srinivasa Krishnan, in 1928, Raman spectroscopy has been

---

* **Corresponding author Pragya Agar Palod:** Department of Physics, Shri Vaishnav Vidyapeeth Vishwavidyalaya, Indore - 453111, India; Tel: +919691136601; E-mail: agarpragya@gmail.com

considered as a molecular spectroscopic technique of critical importance, as it gains insight into a material's make up or characteristics by utilizing the interaction of light with matter [1]. Irradiation of sample with monochromatic light of suitable energy may force oscillations of electrons, which induces deformation of its electron cloud. This in turn results in change in polarizability. Specific energy transitions of molecular bonds for this change of polarizability, result in to Raman active modes. Therefore it is well established that the information revealed by Raman spectroscopy results, originates from a light scattering process. In a true sense, Raman spectroscopy provides a spectrum characteristic of the molecular fingerprint, which is immensely valuable for identifying a substance. Further, it also provides supplementary information about the lower frequency modes and vibrations that reveal the structure of crystal lattice and molecular backbone.

Raman scattering produces a shift in the frequency of scattered photons. This shift, known as the Raman shift, corresponds to the energy of the molecular vibration ($h\Delta v = hv_M$) with which an excitation photon interacts. As shown in Fig. (**1**), the photons interacting with a molecule in its vibrational ground state produce the scattering signals appearing at the low energy side ($hv_s = hv_L - \Delta v$), known as stokes lines. For photon interaction with a molecule in the first excited vibrational state, the scattering signals emerge at high energy side ($hv_{aS} = hv_L + \Delta v$), known as anti-Stokes lines, where v_L is the frequency of the excitation Laser [2].

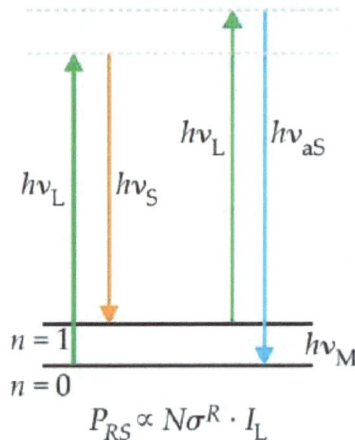

$$P_{RS} \propto N\sigma^R \cdot I_L$$

Fig. (1). Schematic for Raman scattering. Image adapted with reproduced from [2] with permission of American Institute of Physics.

Raman spectroscopy provides a vibrational "fingerprint" of a molecule, allowing us to have insight into (i) how atoms are put together, and (ii) how the molecule interacts with the surrounding molecules. The power of Raman scattering signal (P_{RS}) is a strong function of (i) number of molecules N in the targeted volume, (ii) excitation intensity I_L and (iii) the Raman cross section σ^R, where σ^R is determined by the polarizability derivative of the molecular vibration [2, 3]. Typical Raman cross section being very small, ranges between 10^{-30} and 10^{-25} cm² per molecule. Due to such small cross sections, in Raman spectroscopy, a large number of analyte molecules are required for sufficient conversion rates from excitation Laser to Raman photons. The developmental bottleneck due to the very low detection sensitivity, so far, rendered Raman spectroscopy inherently a weak effect and hence it is barely considered as a tool useful for structural analysis, and not as a feasible technique for ultrasensitive trace detection at the level of single molecules [4, 5].

Further, as organic molecules have been found to have a greater tendency to fluoresce for incident radiation with a shorter wavelength. Therefore monochromatic sources with longer excitation wavelengths were required. It took years for Raman spectroscopy to mature and groom into a standard experimental technique. The first milestone was achieved when the Lasers were discovered in the 1960s. The high intense Laser source and detector technology significantly boosted the achievable detection limit. However, beyond certain points, increasing Laser power to achieve higher sensitivity did not prove to be an intelligent idea, as it enhances the chances of damaging the probed sample. Inherently being a weak scattering process, until 1973, Raman spectroscopy could not earn recognition as a technique of importance either for surface science or for trace analysis [6].

2.1.1. Introduction to SERS

The surface properties have a crucial role in determining many surface chemical processes, including corrosion, electrochemistry, composite formation, catalysis, sintering etc. The development of surface engineering techniques and nanotechnology fuelled the research in the field of composite materials and surfaces with functionalized 'active sites'. Also, the characterization techniques capable of revealing the structure property relationships of such emerging materials gradually came into existence [7]. However, the greatest challenge identified for most of these techniques is; to obtain a good spectral and spatial resolution, that too particularly for low analyte concentrations. The crucial need for the development of *in-situ* techniques capable of operating with ultrahigh sensitivity, surface specificity as well as high spatial, spectral and temporal resolution, was soon recognized by researchers. Surface Enhanced Raman spectroscopy (SERS) is one of the techniques with the capability to achieve an ultrahigh sensitivity down to the single-molecule level using coinage-metal (namely, gold, silver and copper) nanostructures [8]. SERS is a technique based on vibrational spectroscopy. It is a powerful tool for highly sensitive structural identification of low concentration analytes as a result of the amplification of electromagnetic fields caused by the excitation of localized surface plasmons [9]. Fleischmann, Hendra, and McQuillan observed the phenomenon of SERS, for the first time, in 1973 [10]. With this discovery they established that sub-monolayers of small non-resonant organic molecules of pyridine adsorbed onto a roughened surface of silver nanoparticles, exhibit strongly enhanced Raman intensities (with enhancements of the order of 10^6). After this pioneering work, a method for the measurement of the surface enhancement factor was devised by Van Duyne *et al.* [11]. They compared the intensity of the signal of a specific molecule adsorbed on the surface to that of the same molecule in solution and obtained $10^5–10^6$ times enhancement in case of adsorbed pyridine molecules. They further explained that within the electrochemical double layer region, the roughened electrode surface produced an enhancement in the electric field and hence resulted in increased Raman scattering cross section for adsorbed pyridine. The enhancement here refers to the increase in the Raman intensity, overcoming the conventional drawbacks of Raman spectroscopy. The enhancement occurs due to the molecules adsorbed onto the metal surface featuring nanoscale roughness. The results obtained so far, mark the potential of SERS towards the determination of ultralow concentrations of molecules adsorbed on the surface of nanostructures; however, only recent reports have confirmed the progress of SERS towards achieving the anticipated goals [12]. SERS is normally performed on molecules that form chemical bonds of varying strengths and complexities with metallic surfaces. Such surface chemical interactions must themselves be understood in order to interpret the SERS spectrum appropriately [13].

At this point one should understand it clearly that SERS is different from normal Raman spectroscopy in the sense that the presence of metal nanostructures is mandatory for SERS. For understanding SERS, apart from considering the interaction between light and molecules/matter, one must consider the interaction between light and metal nanostructures, as it plays a very detrimental role. SERS being truly surface-selective effect, the metal nanostructures with desired optical properties are of central importance. Additionally, SERS can also be used for detection of the molecular orientation relative to the surface normal. In fact, SERS works on the principle of plasmon-assisted light scattering through the molecules on/near metal nanostructures supporting localized surface plasmon resonance (LSPR). The near-field intensity enhancement is the most prominent effect in SERS. The field intensity redistribution in the close vicinity of plasmonic nanostructures, generates the high local fields. Thus, within the near-field, a broad intensity distribution takes place [14].

The key factors contributing towards establishing SERS as a powerful molecular spectroscopic technique, are wealth of chemical and structural information about the molecules. It is impossible to observe SERS without plasmonic nanostructures. Hence, consideration of the interaction of adsorbed molecules with the underlying metal surface and the electronic properties of the corresponding adsorbate is primarily required for SERS analysis [15]. Various reports claim that the SERS-based analysis forms the basis for ultrasensitive identification and quantitative detection of various biomolecules down to single molecule level ($< 10^{-9}$ M) essential in the fields such as pathological cell diagnostics, biomedicine, environmental surveillance and food safety *etc* [16 - 19].

Though in SERS, the enhancement resulted from various mechanisms; however, the discovery of SMSERS has been marked as a break through and has played a detrimental role in realising the real power of SERS. The major reason for this has been reported the reproducibility of the well-defined substrates [20]. The observation of single-molecule SERS (SMSERS) was achieved for the first time by Kneipp *et al.*, in 1997 by utilizing the extremely large cross sections on the order of 10^{-17} to 10^{-16} cm^2/molecule [21, 22]. This study has proven to be of primary interest because of the requirement of the significant enhancement factor of $> 10^{13}$ for the observation of any signal, had so far been only elusive. The limited availability of the numbers of molecules and substrates, for which successful observations could be carried out, was the main reason for it. SMSERS holds the promise that the signal strength of Raman scattering can challenge the signal strength of fluorescence. With the advances in technology, the potential to control the surface characteristics genuinely, groomed SERS into an interesting analytical tool [20, 23]. Ding *et al.* collected the data from various reports and presented the timeline of the development of various Raman methods originated from SERS right from its discovery till year 2017, as shown in Fig. (**2**) [13]. The various techniques namely, tip-enhanced Raman spectroscopy (TERS), tip-enhanced stimulated Raman spectroscopy (TE-SRS), ultraviolet SERS, shell-isolated nanoparticle enhanced Raman spectroscopy (SHINERS), surface enhanced nonlinear Raman spectroscopy, surface enhanced Raman optical activity (SE-ROA) spectroscopy, near-infrared (NIR) SERS, surface enhanced coherent anti-Stokes Raman spectroscopy (SE-CARS), surface-enhanced femtosecond stimulated Raman spectroscopy (SE-FSRS), and surface-enhanced hyper-Raman spectroscopy (SE-HRS), *etc* are based on the principle of LSPR, the mechanism of electromagnetic field enhancement near the substrate surface.

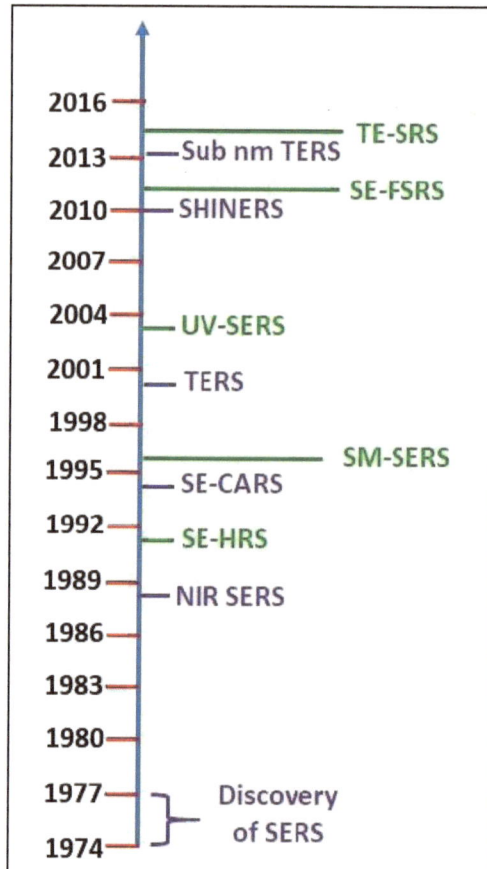

Fig. (2). Milestones related to development of new methods in SERS. Adapted and redrawn with permission from [13].

All these techniques are collectively referred to as plasmon-enhanced Raman spectroscopy (PERS). Fig. (**3**) shows the milestones in terms of number of publications achieved during the development of SERS indicating the tremendous expansion of fundamental understanding as well as application of SERS in various fields [13]. In this figure, SP refers to Surface Plasmon and EM refers to Electromagnetic. The upper and lower parts of the diagram illustrate the number of publications with keyword "surface plasmon" and "SERS" or "surface enhanced Raman", respectively, searched through ISI Web of Science. The increasing number of publications in every decade marks the growing interest of scientific community along with increasing understanding.

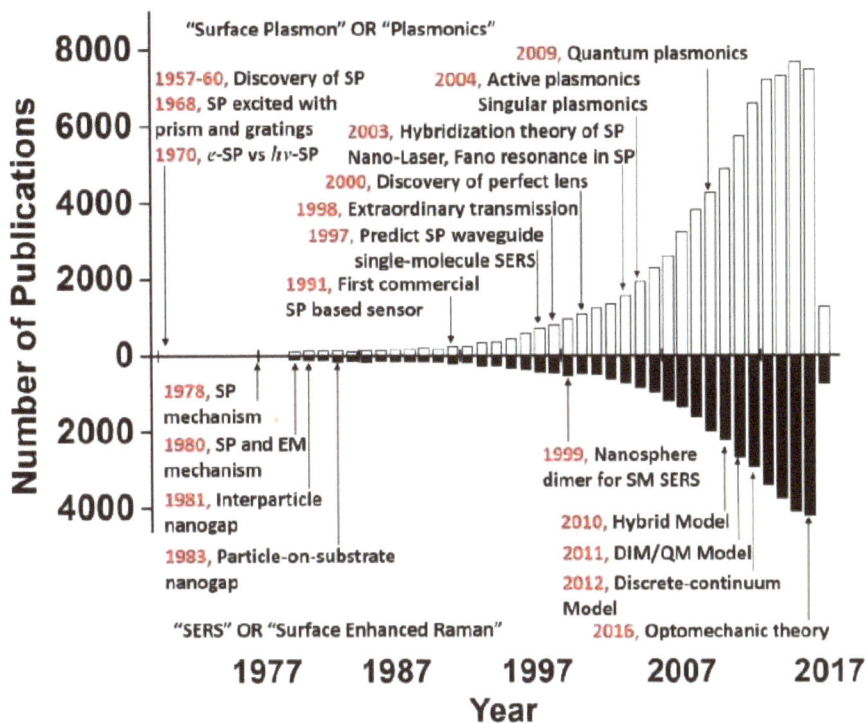

Fig. (3). Milestones achieved during development of SERS in terms of publications with keyword "surface plasmon" and "SERS" or "surface enhanced Raman", respectively, searched through ISI Web of Science. Image adapted with permission from [13].

2.2. ENHANCEMENT MECHANISMS

Enhancement factor in SERS has broadly been accepted to be the outcome of the contributions from: (a) Electromagnetic mechanism (EM), (b) Chemical mechanism (CHEM), (c) Electronic mechanism. Out of these mechanisms, the first mechanism involves enhancements in the intensity of electric field which is a result of plasmon resonance excitation [20]. The second mechanism is the consequence of change in the nature and identity of the adsorbate resulting from the formation of a complex between the molecule (adsorbate) and the metal, which modifies the Raman polarizability tensor of the adsorbate [24]. A key difference between the two mechanisms is that the chemical enhancement is strictly limited to molecules in direct contact with the surface, while in case of electromagnetic mechanism, direct contact between adsorbate and substrate is not a mandate, as far as it is located within a certain sensing volume.

$$\mu_{ind} = \alpha.E \tag{1}$$

As a consequence of excitation of the localized surface plasmon resonance (LSPR) of a nanostructured metal surface, the local electromagnetic field enhances roughly by a factor of 10. Since scaling of Raman scattering varies as E^4; hence the EM enhancement factor is of the order 10^4. CHEM enhancement factor of the order of 10^2 has been reported by the researchers to arise from the excitation of adsorbate localized electronic resonances or metal - to - adsorbate charge - transfer resonances [20]. The above discussed two mechanisms occur due to the square dependence of the intensity of Raman scattering upon the induced dipole moment μ_{ind}, which, in turn, is a function of the Raman polarizability (α) and the magnitude of the incident electromagnetic field (E), as given in equation (1). Though it is experimentally not possible to separate the individual contributions. However, Jensen *et al.* theoretically identified four basic mechanisms responsible for SERS as presented in Fig. (**4**) [25].

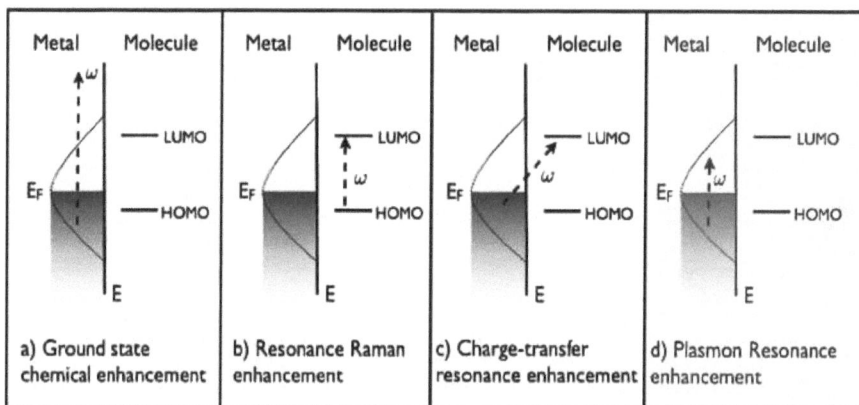

Fig. (4). Schematic presentation of different types of enhancement mechanism in SERS. Image adapted with permission from [25].

(Fig. **4a**) shows the enhancement due to chemical interactions between molecule and nanoparticle, in the ground state, which are not associated with any excitations of the nanoparticle–molecule system. The illustration of the Resonance Raman enhancement has been presented in Fig. (**4b**), for which the excitation wavelength is resonant with a molecular transition. The Charge-transfer (CT) resonance Raman enhancement has been depicted in Fig. (**4c**), with the excitation wavelength being resonant with nanoparticle–molecule CT transitions. Fig. (**4d**) demonstrates the enhancement caused by a very strong local field when the excitation wavelength is resonant with the plasmon excitations in the metal nanoparticle. Out of these mechanisms, the first 3 contributions are grouped together (light peach coloured) because of their chemical origin and are combinedly known as chemical mechanism, while the 4th mechanism (light blue coloured) has a different origin, which is known as the electromagnetic enhancement [25].

2.2.1. Electromagnetic Mechanism (EM)

Electromagnetic enhancement has been accepted as the most dominant and probably the most accurately understood SERS enhancement mechanism. This mechanism is associated only with the photons and metal nanostructures. Various experimental as well as computational studies have been performed to understand the electromagnetic mechanism. The electromagnetic enhancement arises from enhanced optical fields due to excitation of electromagnetic resonances in the metallic structures with roughness features/ particle sizes less than the wavelength of light (on the order of 100 nm) [21, 26, 27]. A more detailed analysis of EM enhancement, along with rigorous expressions has been given by Kerker [28]. Stiles *et al*. have reported that the interaction of an electromagnetic wave and a metal surface, results into the fields at the surface completely different from those observed in the far field. The wave may excite localized surface plasmon resonance (LSPR) on the surface, if the interacting surface is rough enough. The result is that the electromagnetic fields near the surface are amplified. Under the assumption that the intensity of the incident and scattered fields (although at different wavelengths) undergoes enhancement, the possibility for extraordinarily large enhancement of Raman scattering intensity arises. This concept is generally known as the EM enhancement mechanism of SERS. Since its discovery in the 1970s, it has served as a model of extreme importance in developing the understanding of SERS [20]. One of the most desirable feature of SERS is robust and reproducible structures with potential for strong enhancement of electromagnetic field [9]. Significantly enhanced signal strength observed in SERS relative to normal Raman scattering is the outcome of very high local electric fields.

2.2.1.1. Localized Surface Plasmon Resonance (LSPR)

Let us first try to understand the concept of LSPR. It is quite interesting to know that the beautiful colours exhibited by gold and silver nanoparticles upon illumination are the outcome of LSPR.As shown in Fig. (**5**) interaction of a Laser with the surface of a material may set up a collective oscillation of the electrons bound to the metal surface. When this collective oscillation of valence electrons on the surface of a metal (like Au, Ag) is in resonance with the frequency of incident photons, SPR takes place. Surface plasmons are nothing but this collective oscillating mode of electrons and the surface supporting the plasmons is known as plasmonic material. Scattering of light from surface plasmons essentially requires surface roughness or curvature . A large electric field is experienced by the molecule at the surface, if an analyte is present on the roughened metal surface.

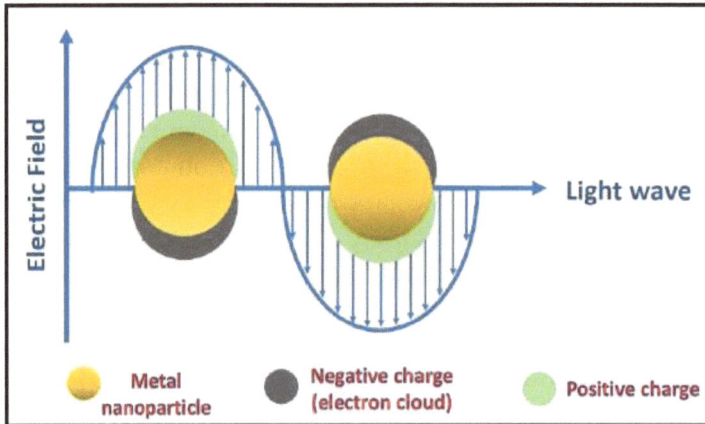

Fig. (5). Schematic representation of LSPR in spherical metal nanoparticle located in a static electric field.

The Raman scattering intensity depends upon the induced polarization caused by the electric field, and hence the Raman signal amplifies by a factor of 10^4 or more [29]. If the feature size of the nanostructure confining the surface plasmons is much smaller than the wavelength of light, they get localized around the nanostructure with a specific frequency, and this phenomenon is known as the LSPR [20, 30, 31]. The optical resonance properties of coinage-metal nanostructures, determines the SERS enhancement of these nanostructures which owing to the excitation of surface plasmon resonance, can significantly enhance the local electromagnetic field [20]. There is no requirement of a direct metal–analyte bond in case of EM enhancement, however it becomes weaker as the separation between the analyte and the surface becomes larger. EM enhancement can be considered to be effective only for the separation of up to ~ 20 A°. Based on the electromagnetic approach surface selection rules have been developed. The extent of the enhancement is a strong function of the surface structure. For two nanoparticles lying close to each other, higher field gradient is produced in the region of close proximity of particles leading to a larger enhancement in that region compared to the regions where particles are not much close. The surface plasmon resonance frequency has a strong dependence on the size, shape and dielectric environment surrounding the particles. Further, for closely spaced particles, the interaction among the surface plasmons of individual particles produces new resonances depending upon the interparticle separation, the incident angle and polarization of the light. Thus, one can conveniently tune the optical properties of individual particles and arrays of interacting particles for the maximum enhancement [29].

2.2.2. Chemical Mechanism (CHEM)

If a bond is formed between the analyte (adsorbate) and the metal surface (substrate), it is the chemical mechanism, which is mainly responsible for the signal enhancement [32]. As a major outcome of this chemisorption, shifting or broadening of the energy levels of the molecule takes place leading to their overlap with the Fermi level of the metal and creation of new electronic states [29, 33]. Electromagnetic enhancement is a long range effect, while chemical enhancement is a short range effect. EM and CHEM, both the mechanisms operate multiplicatively. The contributions of EM and CHEM mechanisms are typically 10^4 or more and 10^2, respectively. Because of less contribution of CHEM enhancement in comparison to EM enhancement, it is difficult to study the CHEM enhancement separately on the system involving EM enhancement. Furthermore, since most of the SERS studies are performed on the roughened surfaces, therefore the simultaneous activity of the two mechanisms also proves to be a problem, as the variation of any experimental parameter in order to probe a system will influence both the mechanisms thereby making the isolation of CHEM enhancement from total enhancement a bit difficult [27, 33]. It is, therefore, important to find a substrate that does not possess the electromagnetic enhancement, retaining the chemical enhancement alone and independently tunable. In order to establish more detailed understanding of CHEM enhancement, theoretical models and calculations have been performed by various researchers. Arya *et al.* put forward a microscopic theory to develop a more insightful understanding of SERS with consideration of both local field as well as chemisorption effects from the adsorbed molecules.

The energy level scheme shown in Fig. (**6**) presents comparison among free molecule, adsorbed molecule and free metal considering silver (Ag) as the substrate and CN^- or CO as the adsorbed molecule. The d band of silver is 4 eV below the Fermi level and it is roughly 4 eV broad. The locations of s-p bands is above the d band and are considered to be 10 eV wide. The positions of shifted and broadened lower level of the adsorbed molecule along with the free molecule levels are also shown in Fig. (**6**). The process of charge transfer between molecule and metal causes a shifting of the lower level of the molecule towards the Fermi energy. Furthermore, the continuum of unoccupied metal states plays the role of intermediate states. When metal absorbs the incident light, these states act as resonant intermediate states and enable efficient charge transfer between the metal and the analyte thereby causing enhanced Raman scattering. Thus, small energy denominators can produce a large enhancement [34]. If the absorbance maximum of an adsorbed molecule changes in such a way so that it approaches closer to the Laser excitation frequency, then the net outcome of this, is, increased cross-section of the molecule. Hence we can consider the CHEM mechanism as a type of resonance Raman effect. Unlike the EM mechanism, CHEM requires essentially a direct interaction between adsorbate and surface, which renders the CHEM enhancement mechanism as a short-range effect, limited to only the first layer of adsorbed molecules [29, 33].

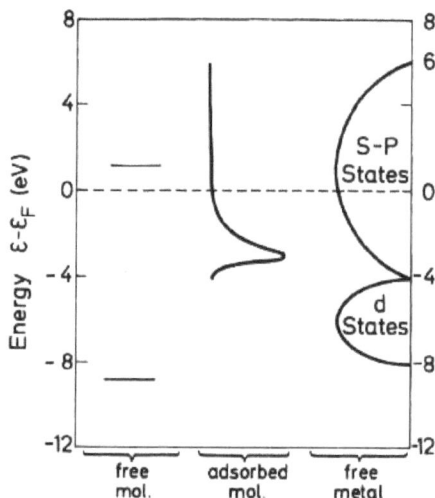

Fig. (6). Energy level scheme to compare free molecule, adsorbed molecule and free metal. Image adapted with permission from [34].

All the theories presented so far, confirm the resonance Raman scattering *via* a charge transfer intermediate state. The adsorbate molecular orbitals are widened into resonances because of interaction with the conduction electrons of the substrate (metal). Resonances with energies lying above (near) the Fermi energy, are partly filled with electrons, while resonances for which energies are well below the Fermi energy, are completely filled. The new possibilities for resonance excitation arise at frequencies much smaller than those of the intrinsic intramolecular excitations of the free molecule because of inclusion of metal states in the process of chemisorption. One such possibility is that the electrons can be excited from the filled adsorbate orbitals to unfilled metal orbitals above the Fermi level (molecule-to-metal charge transfer). Further there exists a possibility for excitation of metal electrons to the partially filled adsorbate affinity level (metal-to-molecule charge transfer). The spectro-electrochemical experiments have been proved to be the most convincing experimental evidence for this picture.

Billmann *et al.* studied and reported SERS intensity *versus* electrode potential plots [35]. The involvement of charge transfer process is a major factor for producing a shift in the electrode potential value corresponding to the SERS intensity maximum with the excitation frequency. Altering the electrode potential leads to changes in the relative energy between the metal's Fermi energy and the adsorbate energy levels, which in turn changes the condition for resonance. When the electrode potential becomes more negative, Fermi energy increases. Therefore metal-to-molecule charge transfer excitations have a red shift, while molecule-to-metal excitations have a blue shift. Further, for adsorbates like N_2, CH_4 and C_2H_6, the SERS effect appears to be faded, the reason being the absence of chemical bond as well as charge transfer band [24, 27, 35]. Surface plasmon scattering has been reported to be an excellent tool for measurement of the electromagnetic properties of nanoparticles. Hence, the surface plasmon scattering properties of single particles before and after chemical treatment can be measured using dark-field microscopy and thus the exact quantitative estimation of the two mechanisms can be carried out [5].

2.2.3. Electronic Mechanism

Electronic mechanism is the name given to a resonance enhancement mechanism from molecules with molecular electronic resonances. *i.e.* if for certain molecules adsorbed onto a metal surface, the frequency of an absorption band and plasmon resonance frequency are close to the frequency of the incident light [36]. The resulting effect is a merger of molecular resonance and surface enhancement, which is popularly known as surface enhanced resonance Raman spectroscopy (SERRS). Stacy and Van Duyne were the first to report this mechanism in 1983 [29, 37, 38]. SERRS presents several unique features, which are absent in Raman scattering, resonance Raman scattering and SERS. It can reveal vibrational as well as electronic information about the adsorbate. Further, it can be used to study a wide concentration range even up to single molecule detection level. Quenching of fluorescence by the surface is another significant attribute, with which good SERRS can be obtained from fluorescing dyes. Rhodamine 6G is a classic example of a strong dye molecule, exhibiting strong fluorescence in the visible range. It prevents the recording of Raman scattering for excitation wavelength in the visible region. However, it gives excellent SERRS with an additional signal 10^6 times greater than for normal Raman scattering 1983 [29, 37, 38]. For practical purposes, one can employ multiple combinations of plasmon resonance frequency, Laser excitation frequency and molar absorptivity of the analyte and it is possible to choose a condition favourable for SERRS to dominate over SERS. Cunningham *et al* reported that a dye can be used as a promising candidate for the wavelengths for which the plasmon resonance and the absorbance maximum differ appreciably. It is because of the ability of a dye to yield a greater enhancement at the molecular resonance frequency than at the plasmon resonance frequency. Fig. (7) demonstrates the absorption spectra for silver colloid (dark) and a dye (light) [39]. In Fig. (7a) , excitation frequency coincides with plasmon absorbance maximum, in (d) it coincides with molecular resonance frequency, (b) and (c) are intermediate frequencies.

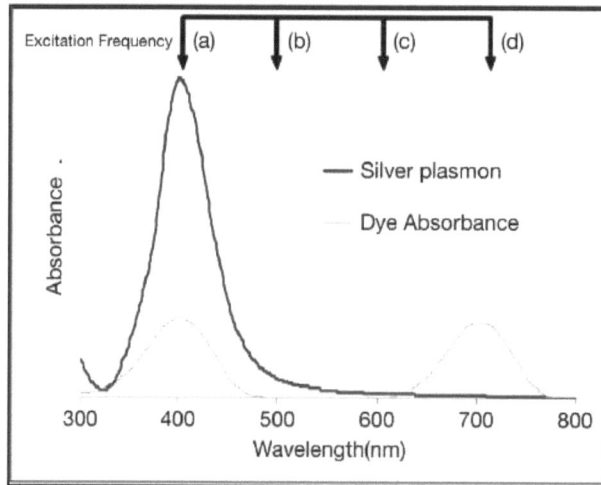

Fig. (7). Absorption profiles of silver colloid and dye with two peaks with molecular resonance (i) coincident and (ii) not coincident with silver plasmon resonance, a-d indicating different Laser excitation frequencies. Image adapted with permission from [39].

Hence, it is possible to study the electronic resonances of molecules adsorbed on a metal surface by analysing the wavelength dependent LSPR shifts. Wavelength dependence of SERS intensity is closely related to the LSPR spectrum. The spectral overlap between the plasmon resonance of nanoparticles and the molecular resonance of the adsorbed species determines the LSPR wavelength shift. The LSPR shift is very small for strong overlaps between LSPR and the molecular resonance wavelength, which in turn, is the result of strong coupling between the molecular resonance and LSPR. A large LSPR shift is observed for the LSPR slightly red shifted with respect to the molecular resonance. On the other hand, excitation at the analyte's electronic resonance frequency causes 10-100 times more enhancement in the Raman scattering intensity in comparison to off resonance excitation, which is referred to as resonance Raman enhancement. Combination of the contributions from the EM and resonance Raman mechanisms, under certain conditions, may be used to achieve SERRS detection at a single-molecule level. The achievable EM and resonant enhancement factors have been reported as 10^{10-11} and 10^4, respectively [40 - 42].

2.3. ENHANCEMENT FACTORS

Enhancement factor (EF) is a figure of central importance in SERS for quantification of the overall signal enhancement. The enhancement factor is defined as the ratio of the Raman signal detected under SERS conditions to the signal obtained under normal conditions, for numbers of active molecules and surface area exposed to the Laser beam, being equal for both [43]. SERS EFs can be experimentally determined by measuring the SERS intensity for the adsorbed molecule on the metal surface (substrate) and the normal Raman intensity of the same, "free" molecule (in the absence of substrate). Further, it requires the normalization of the two intensities to the corresponding number of molecules on the surface (SERS) and conventional Raman, respectively.

$$EF = \frac{I_{SERS}/N_{SERS}}{I_{RAMAN}/N_{RAMAN}} \tag{2}$$

Equation (2) depicts the expression for analytical enhancement factor (EF). Here I and N present the intensity and the number of SERS and Raman active molecules. The number of molecules bound to the enhancing metal substrate is given by N_{SERS} while the number of molecules in the excitation volume is given by N_{RAMAN}. Practically, for any given molecule, the measurement of I_{SERS} and I_{RAMAN} must be carried out independently for a single excitation wavelength. Also, evaluation of the spot size and probe volume must be carried out carefully for determination of the EF analytically. The above expression presents the contributions from all electromagnetic and chemical mechanisms for the system [20, 41, 44, 45].

2.4. E⁴ ENHANCEMENT

The E^4 enhancement approximation predicts that the best spectral location of LSPR coincides with the Laser excitation wavelength, for the maximum electromagnetic enhancement. The maximum enhancement of the incident field intensity at the nanoparticle surface can be obtained using it [8]. This can be understood as follows: As reported, electromagnetic field enhancement in SERS is a two-step process [46, 47]. As shown in Fig. (**8**) in the first step, enhancement of local electromagnetic field takes place around the plasmonic nanoparticles (NPs), at ω_0, the incident frequency (as also discussed in section 2.2.1.1). The role of plasmonic NPs is just like optical antennae, that receive the light from a larger volume and transform the far field in to the near field. The enhancement factor in this step is given by equation (3).

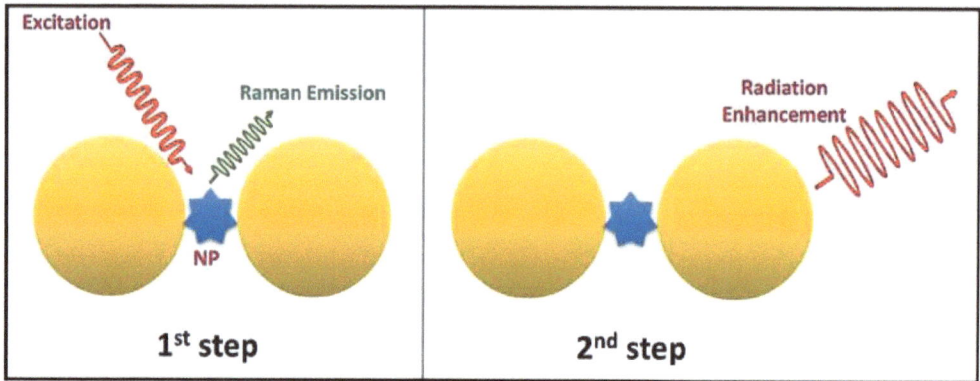

Fig. (8). Schematic presentation to explain E^4 enhancement.

$$G_1(\omega_0, r_m) = \frac{|E_{loc}(\omega_0, r_m)|^2}{|E_0(\omega_0, r_m)|^2} \tag{3}$$

Where $E_{loc}(\omega_0, r_m)$ is the local electric field strength at position r_m of the molecule, ω_0 is the frequency of incident field and E_0 is the incident electric field. The enhancement factor $G_1(\omega_0, r_m)$ is proportional to the square of the electric field. The enhanced local field, in turn, produces a stronger oscillating dipole $p_m(\omega_R, r_m)$ at the Raman scattering frequency ω_R. The dipole radiates, and there is a small but finite probability for the radiated light to be Stokes shifted by the vibrational frequency of the molecule. Since the radiation characteristics of an oscillating dipole significantly depend upon the dielectric properties of its surroundings and the resultant optical resonances [13, 20]. Hence in the second step, which is known as radiation enhancement, the Raman polarizability derivatives of the molecule–NP system result in the enhancement up to 1 to 3 orders of magnitude higher than those of the free molecules. The strong mutual excitation between the induced dipole of molecules (short vertical arrows) and the dipole (and even multipoles) of the NPs (long vertical arrows) induce such enhancement. Now, the plasmonic NPs play the role of transmitting optical antennae and transfer the near field to the far field at the Raman scattered frequency ($\omega_R = \omega_0 - \Delta\omega_{RAMANSHIFT}$). In this step, the enhancement factor $G_2(\omega_R, r_m)$ is proportional to the square of the local electric field (E_{loc}) at ω_R as given in equation (4).

$$G_2(\omega_R, r_m) = \frac{|E_{loc}(\omega_R, r_m)|^2}{|E_0(\omega_R, r_m)|^2} \tag{4}$$

For vibrational modes of adsorbed molecules corresponding to low frequency, the incident and Raman scattered frequency and therefore, the enhancement factors of the two steps (G_1 and G_2) are almost comparable. The overall SERS enhancement is approximately the product of the incident and Raman enhancement processes. Therefore, the SERS enhancement factor is nearly proportional to the fourth power of the local field enhancement as shown in equation (5). Here $\mathbf{E}_{loc}$ and $\mathbf{E}_0$ are the local electric fields in the presence and absence of nanoparticles, respectively. For small Stokes shift, the frequency of the Raman scattered and incident light is considered to be very close [8, 13, 48].

$$G = G_1(\omega_0, r_m) . G_2(\omega_R, r_m)$$

$$= \frac{|E_{loc}(\omega_0, r_m)|^2 . |E_{loc}(\omega_R, r_m)|^2}{|E_0(\omega_0, r_m)|^2 . |E_0(\omega_R, r_m)|^2} \tag{5}$$

$$\approx \frac{|E_{loc}(\omega_R)|^4}{|E_0(\omega_0)|^4}$$

Above equation is known as E^4 approximation for SERS EF. With this understanding one can consider SERS as the light scattered by the plasmonic nanostructure modulated at the frequency of the molecule's vibration [13].

2.5. FACTORS AFFECTING ENHANCEMENT FACTOR

2.5.1. SERS Substrate

The history of SERS is largely the story of the development of SERS substrates, because the substrate, which supports the surface plasmon resonance, is one of the most fundamental requirements for SERS. Since the discovery of SERS, the search for the optimal SERS substrate is an endeavour continuously going on [49, 50]. Specific substrate preparation is the prerequisite for the SERS activity. The chosen metal should have a plasmon in the region close to the frequency of excitation Laser. Because of wide spread application of visible and near infrared frequency Lasers, as excitation sources in Raman spectrometers, silver, gold, and copper are superior choices as substrates, as they have plasmons resonance frequency in this range as shown in Fig. (**9**).

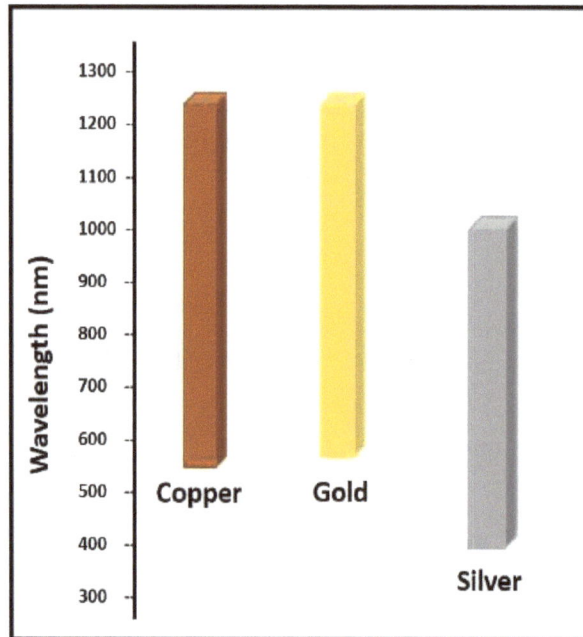

Fig. (9). Wavelength ranges, where silver, gold and copper have been characterized to support SERS.

Out of them all, silver and gold exhibit excellent combination of effective enhancement and chemical stability in air and water. That is why they are the most widely used. The enhancement factor for silver is high for frequencies in the middle of the visible region (around 532 nm) but gold becomes more effective as one moves towards the near IR frequencies. Further, gold is more suitable from the point of view of substrate preparation and storage owing to being chemically more inert as compared to silver. However, analyte adsorption is more efficient with silver in certain cases. Thus, the choice between silver and gold, ultimately depends on the available excitation frequency and the analyte of interest. For the applications, where the chemical stability of the colloid is more important than maximum enhancement, gold is always preferred over other metals like copper, lithium, sodium *etc.*, because it is chemically stable and provides good analyte adsorption. Reportedly, aluminium is effective with UV excitation [15].

A great effort has been made to extend SERS for the surface analysis of various adlayer structures on the transition metals (Pt, Pd, Ru, Rh, Fe, Co, Ni) because of their importance in the fields like electrochemistry, corrosion inhibition and heterogeneous catalysis *etc.* However, in case of transition-metal nanostructures, SERS signal is diminished by two – three orders of magnitude, as compared to coinage - metal nanomaterials. Tian *et al.* devised the strategy named "borrowing

SERS" in order to improve the SERS enhancement factor on transition metal surfaces [51]. According to it, transition metal is coated as an ultrathin overlayer with thickness in the range 2-5 nm on the surface of gold or silver nanostructures. The chemisorption information about the probe molecules residing on the transition-metal overcoating is obtained with an enhancement factor of up to 10^4–10^5 because of the long-range effect generated from the highly SERS-active gold or silver boosts the enhancement mechanism. However, to coat ultrathin layers of target materials on the surfaces of gold or silver nanostructures is no less than a challenge. At the same time it is also important that the coated layer should be free of pinholes, because if analyte is adsorbed directly on the surface of gold or silver nanostructures, it may lead to misinterpretation of the Raman signal [8].

Various techniques have been explored to obtain SERS active substrates with the prime goal to obtain the largest amount of SERS active sites along with the highest EM EF. Few such examples are electrode surfaces roughened by oxidation–reduction cycles, films prepared by physical vapor deposition, assemblies produced by lithography, metal colloids prepared by reduction of metal salt *etc*. One of the most widely studied SERS substrate type is metal nanoparticles prepared as colloidal suspension. In this category, silver colloid is the most popular owing to its high EF for visible radiation, relative ease of synthesis, stability and low cost. Tian *et al*. developed a technique known as shell-isolated nanoparticle-enhanced Raman spectroscopy (SHINERS) [52, 53]. In this technique, a thin dielectric layer over the metal nanoparticle surface effectively screens the metal from a different material interface. For sufficiently thin dielectric spacer layer, molecule bound the surface may experience a different surface chemistry along with maintaining a significant proportion of the field enhancement from the nearby metallic surface. One of the most common shell materials is silica, because silane chemistry favours the functionalization of gold and silver and initiates the shell growth [45]. Another important type of SERS substrates is Tip Enhanced Raman Spectroscopy (TERS). It provides a perfect blend of the high field enhancement of SERS and the imaging capabilities of a scanning probe microscope. It offers extraordinarily high enhancements particularly for biological membranes and cellular systems. Along with high spatial resolution and sensitivity TERS offers single molecule detection potential under specific conditions [54]. Further, these days SERS active substrates are commercially available. This includes Klarite substrate which is nothing but regular array of pyramids coated with nanoscale roughened gold to shape the plasmon and nanotags which comprises stable small gold and silver clusters coated with silica.

2.5.2. Wavelength Dependence

As discussed in section 2.4, the best spectral location of the LSPR for maximum EM enhancement (LSPR λ_{max}) at the surface of nanoparticle, coincides with the Laser excitation wavelength. However, in reality, the situation is different. The theory says that achieving electromagnetic enhancement of the incident as well radiated fields is necessary. In Raman scattering, the wavelengths for these enhancements are different [20]. If excitation corresponds to the plasmon resonance wavelength of the noble metal nanostructures, EM fields at the metal surface are enhanced enormously, which is responsible for the major enhancement (EM mechanism). However, Raman resonance mechanism prevails, when excitation is at an electronic resonance frequency of the analyte (as discussed in the section electronic enhancement).

To have a deeper understanding of the enhancement mechanisms, detailed comparison of SERES profiles with the LSPR spectra of SERS substrates can be a great choice. Wavelength scanned SERES (WS SERES) is the name given to the SERS measurement for several Laser excitation wavelengths. Despite holding the great potential towards improving the understanding of the SERS, reportedly less than 1 percent literature on SERS has focused on WS SERES. One reason for the limitation of this technique is the restriction imposed by the tunability of the excitation Laser and detection system, which actually determine the number of data points. Due to the unavailability of a precisely tunable Laser source with a broad operational range, low data point density and/or limited spectral coverage is the common drawback in most of the reports on SERES. Due to such limitations, direct comparison of the excitation profiles to the spectral location of the LSPR λ_{max} and hence, the full proof generalizations from SERES data are restricted.

Among various publications reported on WS SERES, the work done by McFarland *et al.* has been one of the most popular articles, cited in various reports. This group examined a monolayer of benzenethiol (a non-resonant molecule) on Ag nanoparticle arrays deposited using nanosphere lithography (NSL) using WS SERES studies with varying LSPR [44]. Throughout the visible region of the electromagnetic spectrum, WS SERES data were recorded at different wavelengths for different peak positions of benzenethiol (*viz.* 1575 cm⁻¹, 1081 cm⁻¹, 1009 cm⁻¹). As a representative, Fig. (**10**) shows a characteristic WS excitation profile for the 1575 cm⁻¹ peak position. Gaussian curves fitted to the data for each excitation profile mark that the highest SERS enhancement factor corresponded to the excitation wavelengths, which were shifted higher in energy than the spectral location of the LSPR extinction maximum. The magnitude of the displacement was observed to be approximately half of the Raman Stokes shift.

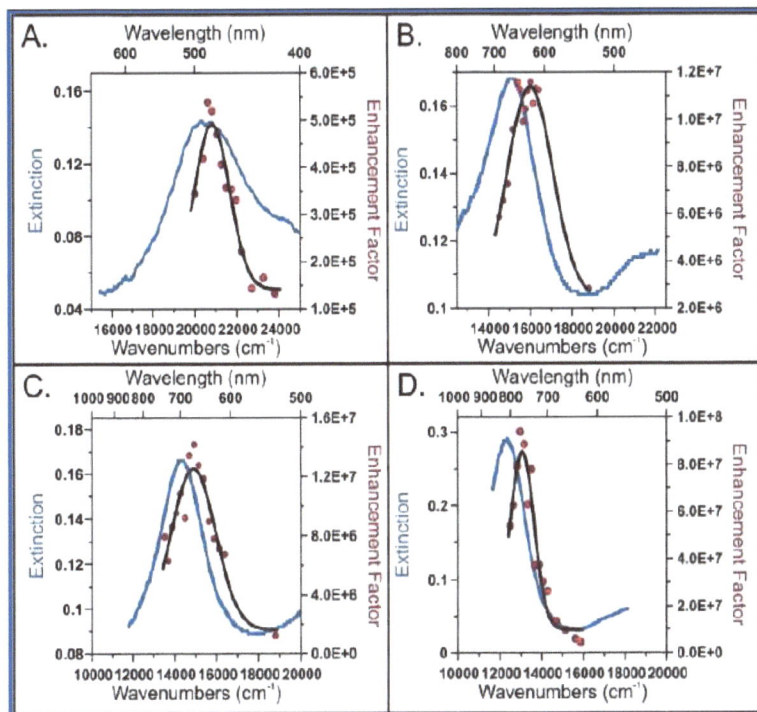

Fig. (10). SERS spectra of benzenethiol with cyclohexane for peak 1575 cm^{-1} for substrate annealed at 300 C for 1 h. (A) LSPR λ_{max}= 489 nm, profile fit maximum at $\lambda_{ex max}$ = 480 nm, (B) LSPR λ_{max}= 663 nm, profile fit maximum at $\lambda_{ex max}$ = 625 nm, (C) LSPR λ_{max} = 699 nm, profile fit maximum at $\lambda_{ex max}$ = 671 nm and (D) LSPR λ_{max}= 810 nm, profile fit maximum at $\lambda_{ex max}$ = 765 nm. Image adapted with permission from [44].

For example, in Fig. (**10**) (A), the measured value of LSPR λ_{max} for the substrate was 489 nm (20450 cm^{-1}). The Gaussian fit indicates that the peak position of the excitation profile, $\lambda_{ex,max}$ to be at 480 nm (20833 cm^{-1}). The calculated value of the peak EF turned out to be 5.5×10^5. This observation is in line with the wavelength dependence of the SERS intensity predicted by the electromagnetic mechanism [44, 55]. Throughout the experiment, for different benzenethiol bands, it was observed that for all the LSPR λ_{max} values, the maximum SERS enhancement invariably occurs, when the substrate LSPR λ_{max} located between λ_{ex} (Laser excitation wavelength) and λ_{vib} (wavelength that is Raman-scattered by the analyte molecules). This signifies a "compromise" location, where both the incident and scattered photons can be resonantly enhanced. This conclusion supports the EM mechanism [13, 44].

2.5.3. Distance Dependence

An important characteristic of the EM mechanism is that the direct contact between the adsorbate and the surface is not mandatory, as far as it is within a certain sensing volume. It has certain benefits in the cases where surface immobilized biomolecules are involved [56]. However, the electromagnetic field decays rapidly as one moves away from the surface. Since the field enhancement around a small metal sphere decays as r^{-3}. Owing to E^4 approximation, the overall distance dependence varies as r^{-12}. Consideration of the increased surface area scaling according to r^2, for shells of molecules at an increased distance from the nanoparticle, yields the distance dependence of the SERS intensity of r^{-10}.

$$I_{SERS} = \left(\frac{a+r}{a}\right)^{-10} \tag{6}$$

where I_{SERS} is the intensity of the Raman signal, a is the average size of the field-enhancing features (radius of curvature of the roughness) on the surface, and r is the distance from the surface to the adsorbate [48]. However, r^{-10} dependence described by equation (6) is purely theoretical. In practice, the factors like the particle shapes, sizes, Interparticle interactions *etc.* lead to deviations from the r^{-10} dependence [48]. Dieringer *et al.* deposited Al_2O_3 multilayers of varying thicknesses (highly uniform and controlled thin films) using atomic layer deposition onto Ag film over nanosphere (AgFON) surfaces and studied dependence of SERS on distance. Al_2O_3 acts as a spacer layer between Ag film and pyridine [57].

Fig. (**11a**) shows the SERS spectra for pyridine adsorbed on AgFON surfaces coated with four different thicknesses of Al_2O_3. A plot of the relative intensity of 1594 cm^{-1} band as a function of Al_2O_3 thickness is shown in Fig. (**11b**). Fitting of the experimental data to equation (6) yields the average size of the enhancing particle a = 12.0 nm. The above results establish that SERS is a long range effect owing to the EM enhancement [57].

2.5.4. Dependence on the Size and Shape of the Nanoparticles

As confirmed by Mie theory, LSPR is highly dependent on size and shape of the particle. Therefore, it is possible to tune the LSPR to a specific wavelength by varying the size and shape of the fabricated nanoparticle. Nanosphere lithography (NSL) is a valuable and cost effective technique for deposition of periodic particle arrays (PPA) of metals. Hynes *et al.* examined the extinction spectra of Ag PPAs with varied shapes and aspect ratios on mica substrates as shown in Fig. (**12**) [58].

In this figure, a and b indicate in-plane width and out-of-plane height of the nanoparticle. For extinction peaks F, G and H, all triangular particles had identical value for a, while a blue shift in the LSPR λ_{max} was observed with increase in the nanoparticle height (b), which results in decrease in aspect ratio (a/b). Similarly for extinction spectra D, E and H particle diameter (a) was varied for more or less similar height (b). Similar effect of blue shift in the LSPR λ_{max} with decrease in aspect ratio was observed. F and C peaks correspond to same sample before and after annealing marking the blue shift for reduction in aspect ratio. As evident from the extinction spectra, λ_{max} can be tuned to the desired values by selection of suitable geometry and aspect ratio of nanoparticles [58].

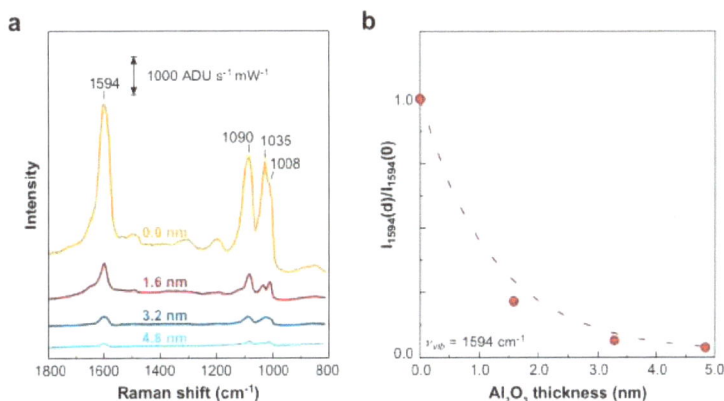

Fig. (11). (a). SERS spectra of pyridine adsorbed onto silver film over nanosphere samples treated with different thicknesses of alumina for excitation wavelength 532 nm. **(b)** Plot of SERS intensity as a function of alumina thickness for the 1594 cm^{-1} band. Dashed curve shows the fit of this data to equation (6). Images adapted with permission from [57].

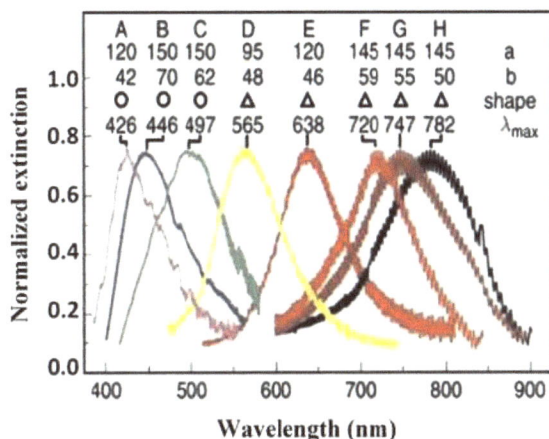

Fig. (12). Modification to the LSPR maximum of periodic particle arrays with changing nanoparticle (a) in-plane width and (b) out-of-plane height. Image adapted with permission from [58].

Here one should think of a quasi-static approach, in which z polarized light of wavelength λ, is incident on a spherical nanoparticle of radius a. In the long wavelength limit (a/λ < 0.1), the assumption of uniform electric field in the vicinity of the nanoparticle, can be considered to be valid, as shown in Fig. (5) [20]. However, analytical solution of the extinction spectrum is possible only for spheres and spheroids, while for all other geometries, it must be approximated [59, 60]. Some of the popular numerical methods are discrete dipole approximation (DDA), the finite difference time domain (FDTD) method, finite element method (FEM), volume integral methods *etc.* Among these methods, DDA is one of the most favoured methods used to describe non-spherical particles. This method divides the nanoparticle into finite number of polarizable elements, known as dipoles. These dipoles interact with each other only through dipole–dipole interactions. These coupled dipoles interact with the applied field. Polarization is induced as a result of the interaction of dipoles with local electric field. This induced polarization is used to determine the extinction and scattering spectra of the particle by applying the Claussius – Mossotti polarizabilities with some radiative reaction correction [30, 61 - 63]. The results calculated using DDA and the finite-difference time domain methods match well with the experiments [61, 64, 65]. The LSPR is very much sensitive to the surrounding dielectric environment and hence it is used for many sensing experiments [30].

2.6. TWO-PHOTON EXCITED SERS

The two photon excited Raman spectroscopy is the name given to the anti-Stokes Raman scattering occurring from pumped vibrational levels. Let us try to explore the Physics behind it. The Raman Stokes and anti-Stokes transitions lead respectively to the population and depopulation of the first excited vibrational level, which is described by rate equation (7) [66].

$$\frac{dN_1}{dt} = (N_0 - N_1)\sigma^{SERS} - \frac{N_1}{\tau_1} \tag{7}$$

Here N_0 and N_1 are the ground and first excited vibrational state populations, respectively. σ^{SERS} represents the SERS scattering cross section and τ_1 is the lifetime of the first excited vibrational state. The first term in equation (7) shows that the excited vibrational level is populated by Stokes scattering and depopulated by anti-Stokes scattering and the second term shows the depopulation of the excited level through the spontaneous decay.

The process has been depicted in Fig. (13) as well. In this figure, v=1 and v=0 represent vibrational excited and ground states. With the assumption of steady state and weak saturation ($\exp\{-h\Delta v/KT\} \leq \sigma^{SERS}I_L\tau_1 \ll 1$), the stokes and anti-stokes power are expressed as follows:

$$P_S^{SERS} = N_0\sigma^{SERS}I_L \tag{8}$$

$$P_{aS}^{SERS} = (N_0\sigma^{SERS}\tau_1 I_L + N_0 e^{\frac{-h\Delta v}{KT}})\sigma^{SERS}I_L \tag{9}$$

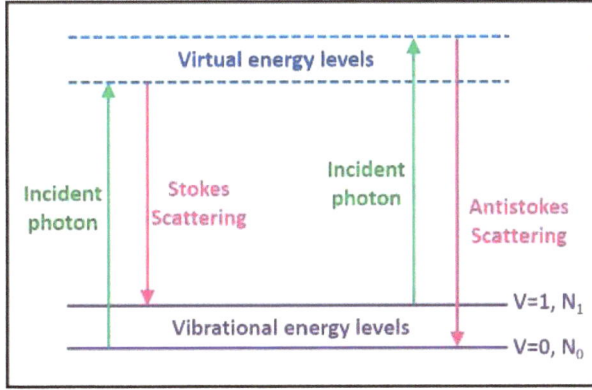

Fig. (13). Schematic representation of population and depopulation of vibrational energy levels.

Theoretically, estimation for the anti-stokes to Stokes SERS signal ratio is given by:

$$\frac{P_{aS}^{SERS}}{P_S^{SERS}} = \sigma^{SERS}\tau_1 I_L + e^{\frac{-h\Delta v}{KT}} \tag{10}$$

Here, the first term depicts the anti-stokes to stokes signals ratio due to vibrational pumping (SERS population), while the second term describes the effect of the Boltzmann /thermal population, Δv being the molecular vibrational frequency. In normal Raman scattering typical cross section is of the order of 10^{-30} cm^2 per molecule. Hence, the anti-stokes scattering can be considered to be negligible. However, in case of SERS occurring at extremely high enhancement levels, spontaneous pumping of excited vibrational state has been observed. In SERS experiments, measurements carried out at relatively low excitation intensities, anti-Stokes to Stokes SERS signal ratios have been found to differ significantly from thermal population. To account for the anti-Stokes to Stokes signal ratios experimentally obtained for rhodamine 6G and other dye molecules, the product of the cross section and vibrational lifetime ($\sigma^{SERS}\tau_1$), must be on the order of 10^{-27} cm^{2}s. Assuming vibrational lifetimes of the order of 10 ps, the estimated scattering cross section should be of the order of 10^{-16} cm^2 per molecule. The enhanced scattering cross section, in case of SERS with vibrational pumping, hence accounts for such enormous enhancement factors around 10^{14}. It can be inferred that the number of molecules involved in the SERS process at the extremely high enhancement level must be very small. The evidence for the pumping effect comes from (i) anti-Stokes to Stokes signal ratios, exceeding those from a Boltzmann distribution (ii) the anti-Stokes signal's quadratic dependence on the excitation intensity, and (iii) presence of a component of corresponding to Raman transition from first excited to second excited state in Stokes signal [66].

Fig. (14) shows Stokes and anti-Stokes SERS spectra of crystal violet attached to isolated and aggregated gold nanospheres. The enhancement factor is too small, in case of isolated particles, to populate the first excited vibrational state appreciably. As evident from Fig. (14b), Due to the low thermal population, the high frequency modes are almost absent on the anti-stokes side. While for nanoparticle aggregates / clusters, presence of strong anti-Stokes signals on high frequency side Fig. (14d) marks the population of the first excited state by the strong Raman process. One must infer that the vibrational levels mainly populated by the SERS process and not by thermal population, give rise to the anti-Stokes scattering. The appearance of strong anti-Stokes SERS signal is referred to as surface enhanced pumped anti-Stokes Raman scattering (SEPARS). In this process, the first excited vibrational levels are populated because of Stokes scattering with an extremely high effective cross section, giving rise to an increase of anti-Stokes signals. Such vibrational pumping leads to the non-resonant SERS enhancement factors on the order of 10^{14} [4, 14, 67].

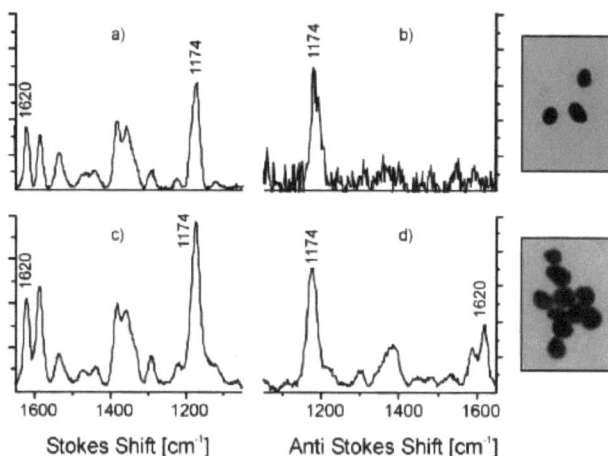

Fig. (14). Stokes and anti-stokes SERS spectra of crystal violet attached to isolated and aggregated gold nanospheres. Image adapted with permission from [4].

Looking more closely at anti-Stokes Raman scattering occurring from pumped vibrational levels, one finds it to be a two-photon Raman process, in a real sense. One photon populates the excited vibrational state, a second photon takes part in the anti-Stokes scattering. Considering the two-photon process, effective two-photon cross section $(\sigma_{aS,nl}^{SERS})$ can be used to describe the power of anti-Stokes signal. Equation (9) can be rewritten as:

$$P_{aS,nl}^{SERS} = N_0 \sigma_{aS,nl}^{SERS} I_L^{\,2} \tag{11}$$

Where,

$$\text{Where } \sigma_{aS,nl}^{SERS} = \sigma_S^{SERS} \sigma_{aS}^{SERS} \tau_1 \tag{12}$$

Fig. (**15**) shows the plots of anti-stokes and stokes SERS signals *vs* excitation Laser intensity. The quadratic and linear fits to anti-stokes and stokes data verify the predicted dependence upon excitation intensity. Under the assumption of a SERS cross section of approximately 10^{-16} cm^2 and a vibrational lifetime on the order of 10 ps, effective two photon cross sections is determined to be about 10^{-43} cm^{4}s. This makes the two-photon excited Raman probe with a cross section more than 7 orders of magnitude higher than the typical cross sections for two-photon excited fluorescence [68]. The vibrational level is exploited as a real intermediate state in the two-photon process. Continuous wave Lasers operating at relatively low excitation intensities can be utilized because of high cross section. Another advantage is that the anti-Stokes spectra are measured at the high-energy side of the excitation laser, which is free from fluorescence [69]. This effect is also known as Surface Enhanced Hyper Raman Scattering (SEHRS), in which two photons of one excitation Laser are involved in generating an incoherent Raman scattering on the vibrational quantum states. Hyper-Raman scattering (HRS) is an optical phenomenon associated with nonlinear properties of the medium [70].

As illustrated in the Jablonski diagram in Fig. (**16**) two photon excited Raman scattering involves three photons, in which a system is excited to a virtual state (shown with dashed lines) by absorbing two photons of the frequency (ω_0) simultaneously and emits one photon at frequency $(2\omega_0 \pm \omega)$, where ω_0 is the incident frequency and ω is the molecular vibration frequency. The stokes and anti-stokes scattering correspond to the transitions $2\omega_0 - \omega$ and $2\omega_0 + \omega$, respectively. Thus in HRS, the Raman signals are shifted relative to the doubled energy of the excitation Laser [70]. In the process of HRS, the symmetry selection rules followed are different from regular one-photon Raman scattering. Hence it is very much possible to probe the so-called "silent modes," *i.e.* the vibrations that are generally absent in Raman and IR absorption spectra. Linear dependence of the power (P_{RS}) of Raman Scattering signals on the excitation intensity (see Introduction) disappears in HRS. The HRS signal is a quadratic function of the excitation intensity. HRS being a two-photon excited process, has cross sections on the order of 10^{-65} cm^{4}s, which are at a scale much smaller than regular one-photon excited Raman scattering cross sections (see Introduction). The resonances between optical fields and surface plasmons, in SERS result in large enhancements of Raman scattering signals of molecules in the vicinity of metal nanostructures. These local optical fields provide the key effect for the observation of SEHRS signals from the molecules located in the nanometer proximity of the nanostructures. Thus, the inherently low, two photon scattering cross sections of HRS may be enhanced to an order of 10^{-45} cm^{4}s in SEHRS. The non-linear (quadratic) dependence of intensity on the excitation field, enables SEHRS to achieve benefits even to a much greater extent from the high local optical fields than SERS.

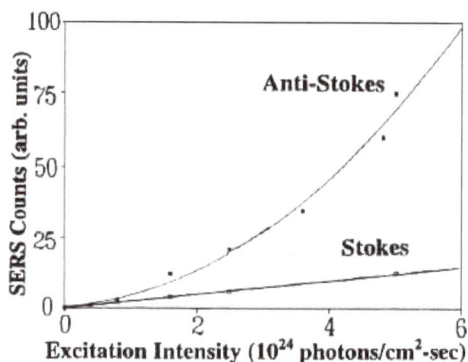

Fig (15). Anti-Stokes and Stokes SERS signals *vs* excitation intensity for the 1174 cm^{-1} Raman band of crystal violet. Image adapted with permission from [66].

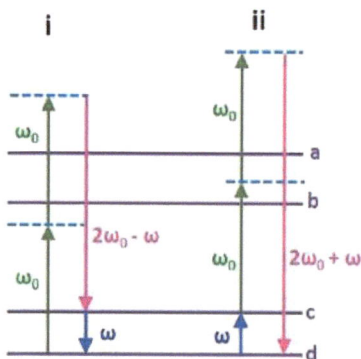

Fig. (16). HRS (i) Stokes and (ii) anti-Stokes jablonski diagram.

The schematic in Fig. (**17a**) shows the HRS and RS processes. Here ν_L, ν_H and ν_S are the frequencies corresponding to Laser excitation, HRS and Raman Stokes scattering, respectively. SEHRS (top) and SERS (bottom) spectra measured from crystal violet adsorbed on silver nanoclusters are shown in Fig. (**17b**) for excitation wavelength 850 nm, while both the spectra measured simultaneously using the first and second diffraction order of the spectrograph, are depicted in the middle spectrum. Despite of low scattering cross section, SEHRS signal appears at levels competitive to SERS. This establishes that enhancement of SEHRS is at least 10^6 times more than that of SERS. This fact is quite in line with the quadratic dependence of the effect on the enhanced local optical fields [71].

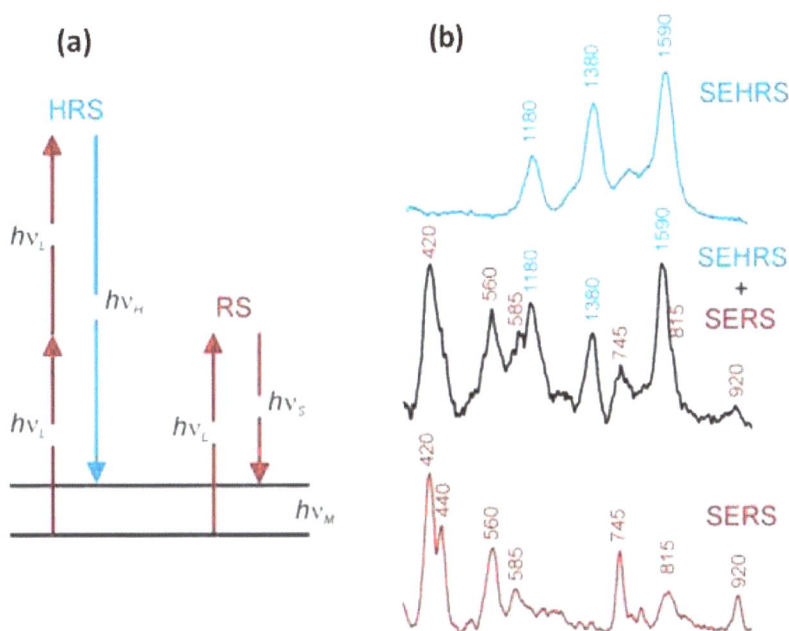

Fig. (17). (a) Schematic to show HRS and RS, (b) SERS and SEHRS spectra of crystal violet on silver nanoclusters, for excitation wavelength 850 nm and power density 10^7 W/cm^2. Images adapted with permission from [4].

One and two photon excited Raman scattering, both can profit from surface enhancement in the presence of metal nanostructures. SEHRS spectrum of a centrosymmetric molecule exhibit significant relative intensity differences because of the surface effect. Also it can contain new vibrational bands, which are conventionally absent in its SERS spectrum. While, the two-photon excited spectrum for a non-centrosymmetric molecule, more or less resembles its SERS spectrum. Hence, structurally sensitive vibrational information complementary to those obtained by SERS, can be obtained using SEHRS. Decrease in the probed volume and the freedom of using longer excitation wavelengths are additional advantages of SEHRS over one photon excited SERS. As SEHRS combines the advantages of two-photon spectroscopy with the structural information of vibrational spectroscopy, it is useful for a number of analytical applications. Since the two processes give different insights into molecular symmetry, the combination of one and two-photon excited SERS is a tool for thorough investigation of different molecules and molecule-nanostructure interactions [72].

2.7. SINGLE MOLECULE DETECTION USING SERS

So far, we have talked about average enhancements for a collection of molecules adsorbed in a random fashion on the surface. However, it is worth noting that this average enhancement is an outcome of the enhancements experienced by all the individual molecules on the surface. For most of the substrates the EF has extreme spatial variations following a long tail distribution. Just like pareto distribution in economics presents the fact that the largest part of the global wealth is controlled by a very small fraction of the population. Similarly, in SERS, this distribution emphasises on the point that the largest part of the signal is contributed by a very tiny number of molecules, out of all the analyte molecules distributed randomly on the surface of the substrate. Putting it differently one can think of a single molecule with the highest enhancement producing the SERS intensity as large as produced jointly by thousands of randomly adsorbed molecules. This indicates that the average EF is nearly 1000 times smaller than the maximum EF [73, 74].

As shown in Fig. (**18**), small enhancement factors of only 10^7–10^8 are required for molecules with relatively high differential Raman scattering cross sections ($d\sigma/d\Omega$) typically on the order of 10^{-27}–10^{-28} cm^{2}sr^{-1} (resonant molecules such as dyes) for their consideration for SMSERS. While, an additional boost nearly of the order 10^3 is required to the enhancement factor, for the non-resonant molecules with "normal" differential Raman scattering cross sections ($d\sigma/d\Omega$) on the order of 10^{-29} – 10^{-30} cm^{2}sr^{-1}. *i.e.* in order to observe single molecule events, they must yield extremely large EFs of the order of 10^9–10^{11} [74]. Hence to achieve the ultimate limit of detection by probing a single molecule using SERS has been of paramount interest among the researchers. Detecting single molecules with high sensitivity and molecular specificity has received great scientific attention due to the widespread applications in many fields such as chemistry, biology, medicine, pharmacology, and environmental science *etc* [9, 21]. Apart from its applications in the fields of scientific interest, surprisingly, SMSERS has emerged as a very novel, powerful and minimally invasive technique in the areas of cultural heritage research.

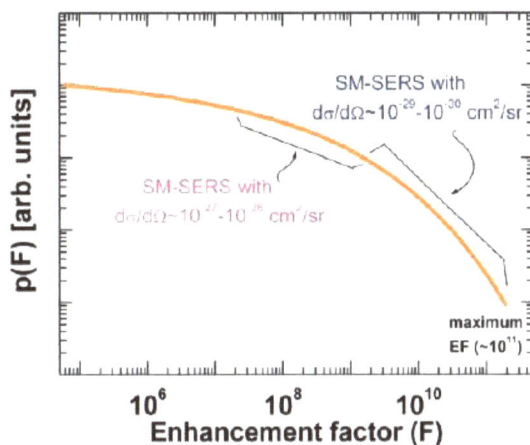

Fig. (18). Probability distribution for SERS enhancement factor, for substrates suitable for single molecule detection using SERS. Image adapted with permission from [74].

With the aim of rejuvenating the reader's interest, authors present a classic example of a water colour painting created by Winslow Homer in 1887 with the title "For to Be a Farmer's Boy". Fig. (**19**) shows the painting as it appears at present. Brosseau *et al.* performed a detailed scientific study on this artwork and digitally reconstructed it, which presents the expected appearance of the painting immediately after its completion [75]. The area of the artwork from the top left corner of the artwork was selected for sampling. The SERS spectra were recorded directly on single particles of red lake pigments from the selected sample area. The colour of the grains was observed to vary among red, yellow and purple. SERS spectra were recorded for the specific pigment grain and compared with the spectra of reference washes of cochineal, burnt carmine and Indian purple (a pigment prepared by W&N by precipitating cochineal with copper sulphate). The high sensitivity of the SERS spectrum of carminic acid to changes in the chemical environment has been reported to be the major reason for the observed variation in the spectra. After careful analysis of the spectral evidences it was concluded that the selected grain was most likely composed of Indian purple. Further, the sample was also found to contain other grains with a mixture of yellow, reddish purple, and bright red [75]. Thus, the basis for the reconstruction of the artwork lies in the molecular identification of the specific colourants present in the specific pigment grains. The fading of pigments clearly indicates the shifting of the colour balance of Homer's creation caused by photochemical degradation of some pigments over the time. The 'colourless' sky depicted in Fig. (**19**), has been replaced by a vibrant autumn sunset in the reconstructed digital artwork [9, 75]. This validates the potential of SERS in the field of art preservation and restoration.

Fig. (19). Water color painting made by Winslow Homer in 1887 from "For to be a farmer's boy" as it appears at present (Source: Art Institute of Chicago). The anticipated appearance of the painting immediately after its completion can be observed in the Digital reconstruction [75].

2.8. HOT SPOTS

In 1997, two reports were published nearly simultaneously by the groups of Nie and Kneipp, that claimed single-molecule detection using SERS. With this, SERS became the first vibrational spectroscopy to achieve the unsurpassed limit of detection for chemical systems, an ideal for analytical metrology [21, 22]. The path breaking discovery of single molecule SERS (SMSERS), which has originated from nanoparticle aggregates, was then taken after by many researchers. Aggregates necessarily have a nanoparticle junction. Here we define a new term "hot spot". A hot spot is a junction or close interaction of two or more plasmonic objects where at least one object has a small radius of curvature on the nm scale. Extremely high optical field intensities are a must in order to measure the Raman signals from single molecules in order to compensate for the extremely small Raman cross sections. A hot spot is capable of concentrating an incident electromagnetic field and effectively amplify the near field between and around the nanostructures [76]. The 'hot spots' are generated because of remarkably heterogeneous nature of the near field around the features such as tips, crevices, edges or junctions *etc*. In comparison to the rest of the analyte molecules, these hot spots bring about much higher SERS enhancement factors. For example, electric field enhancement for silver nanospheres lies somewhere between 10^3 and 10^4, while for silver nanocubes of similar dimensions, it is 10 to 100 times higher. Fig. (**20**) shows TEM images of (A) silver nanocube and (B) nanosphere. It also shows the field enhancement around them (C and D), calculated with DDA for excitation wavelength 514 nm [77].

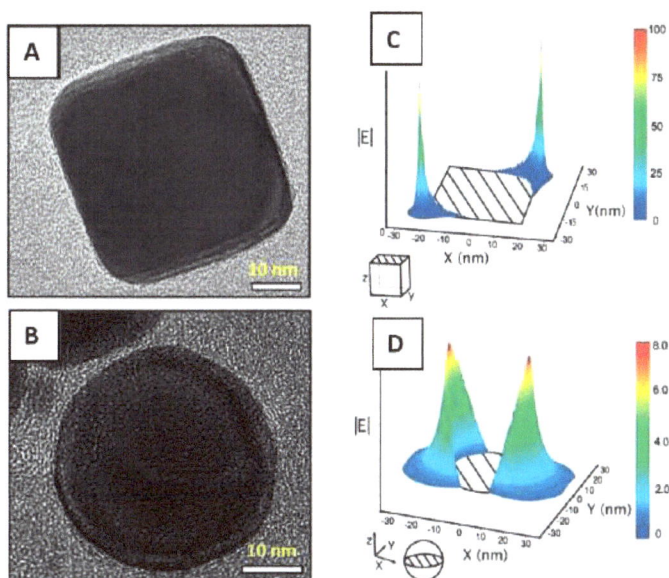

Fig. (20). TEM images of (A) silver nanocube and (B) nanosphere. Field enhancement around (C) a silver nanocube and (D) symmetrical nanosphere (calculated with DDA with 514 nm excitation). Images adapted with permission from [77].

Now-a-days, for the observation of SMSERS, the Hot spots are considered as a prerequisite. Also, it is the hot spots, which determine the limit for the achievable spatial resolution for SERS [43]. Apart from detection, subtle molecular details and dynamics can also be examined with the help of the structural information contained in the SMSER spectrum. Owing to the additional 10^6 resonance enhancement, resonant dye molecules are commonly used to assess SERS substrates for their application in SMSERS. Typically, SMSERS substrates contain solution-phase, solid-supported, TERS systems or some combination of them, all spatially organized in order to have hot spot region in the nanometer gaps between different components. Achieving better control on the particle assembly in order to optimize the enhancement between coupled nanoparticles, has been the major focus of the development of SMSERS substrates [45]. The presence of hot spots on SERS substrates is responsible for a large part of the average enhancement. However, the chances of accidentally finding a molecule in a hot spot, *i.e.*, positioned exactly in the gap between the particles is very small. Contrary to it, to find a molecule somewhere on the surface of one of the two particles, it is much more likely, where it experiences only moderate enhancements. Therefore, both top-down and bottom-up nanofabrication techniques are being explored for systematic generation of hot spots with desired configuration.

2.8.1. Generations of Hot Spots

Based on the structural configuration and dimensional complexity, hot spots can be classified in terms of various generations:

The first generation hotspots: Exhibit moderate SERS activity. They are generated in the assemblies of single nano-objects, such as nanospheres, nanocubes or nanorods suspended in a homogeneous medium. However, assemblies of single nanoparticles with sharp corners and/or with intraparticle gaps (< 2 nm), can be designed intelligently such that they exhibit much higher SERS activity. Nanostars, nanoflowers and other multibranched nanostructures are some typical examples. Fig. (21) depicts microscopic images of first generation hot spots *viz.* silver nanoflower, gold nanostar and silver nanocube experimentally obtained by various groups to observe SERS.

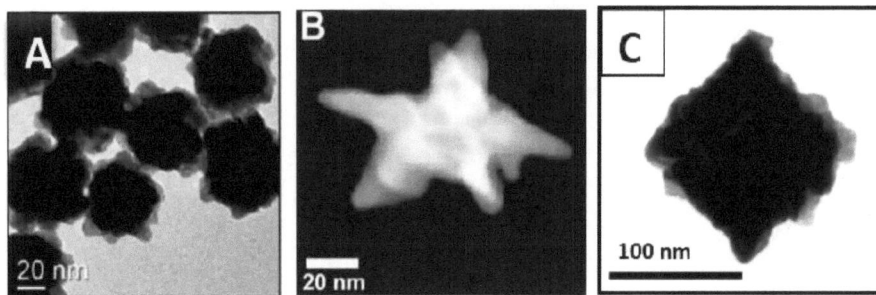

Fig. (21). (A) SEM image of Au nanoflower (Image adapted with permission from [89]), (B) STEM image of Au nanostar (Image adapted with permission from [90]), (C) TEM image of Au nanocube (Image adapted with permission from [91]).

The second generation of SERS hotspots: emerge from coupled nanostructures with controllable interparticle nanogaps or interunit nanogaps in nanopatterned surfaces. Some typical examples include the nanoparticle dimers, oligomers and nanoparticle arrays. Fig. (22) depicts dimers and trimers of gold nanoparticles grown by Sergiienko *et al* [78]. Such hotspots exhibit exceptionally high SERS activity.

Ding *et al.* studied SERS distribution in various gold nanoparticles using finite element simulations [8]. The analysis of the simulation results revealed that the coupled plasmonic nanostructures result in average SERS intensity, typically 2 to 4 orders of magnitude higher than those from single nanostructures and hence, they should be preferred for trace molecule detection. The tiny volume occupied by the second-generation hotspots, must contain extremely strong electromagnetic field, for detection and analysis of trace amounts of molecules, including single molecules, provided the probe molecule is located at the hotspots [8, 13].

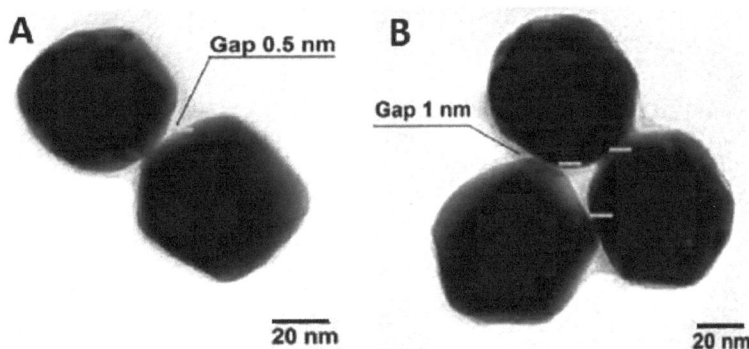

Fig. (22). TEM images of (A) dimers and (B) trimers of gold nanoparticles (adapted with permission from [78]).

The third generation hotspots: Though second generation hotspots produce very intense field concentrations; however, for some of the widely used materials with immense industrial applications (such as silicon, ceramics *etc*) it has been found bit difficult to squeeze into the tiny and narrow regions of the hotspots as formed by coupled nanostructures. For the surface analysis of such materials by SERS, one needs to take account of the EM coupling effect of the probe materials and the SERS active nanostructures. In this case, plasmonic nanostructures are designed so that they can have hotspots right on the surface of the material, which is to be probed. The hotspots generated from hybrid structures consisting of plasmonic nanostructures and other materials (used as substrate) are referred to as third generation hotspots. Third generation hotspots essentially result from the hybridization of the EM field scattered from the plasmonic nanoparticles and an EM field reflected from the substrate material surfaces. The resulting SERS EF depends crucially on the shape and size of the plasmonic nanostructures as well as dielectric properties of the substrate (probe) materials.

Ding *et al.* also studied SERS distribution in various gold nanoparticles on the flat surface of silicon and platinum using finite element simulations. Table **1** demonstrates various parameters related to the hybrid structures. In these table E, <GNP>, G^{NP}_{max} and k refer to electric field, average and maximum SERS enhancement factors at the outer surface of a nanoparticle and wavevector of incident light, respectively. Because of stronger coupling between Au and Pt, average SERS EF is one order of magnitude larger for Pt than for Si. Interestingly, enhancement can also be tuned by using different plasmonic nanostructures. For example, EF is at least 2 orders of magnitude larger for nanocubes than nanospheres. For the case of dimers, hotspots are simultaneously created at particle-substrate and interparticle nanogaps. EF has been found to be three orders of magnitude higher for nanosphere dimers on Si than single nanosphere. SERS EF increases with increase in the refractive index of the dielectric material [8].

Table 1. Parameters related to finite element simulations performed on hybrid nanostructures of various SERS active nanoparticles on Si and Pt.

S. No.	substrate	Particle type	λ_{max} (nm)	$<G_{sub}>$	G^{sub}_{max}	G^{NP}_{max}
1	Si	Sphere	530	94	2.03×10^3	-
2	Pt		535	1.30×10^3	4.32×10^4	-
3	Si	Cube	660	1.48×10^6	1.42×10^7	-
4	Pt		825	7.31×10^5	5.15×10^6	-
5	Si	Nanosphere dimer	600	1.04×10^5	6.85×10^6	6.76×10^7
6	Pt		630	1.02×10^6	9.01×10^7	7.19×10^7
7	Si	Au@Sio2 dimer	580	1.74×10^4	8.37×10^5	6.56×10^6
8	Pt		600	2.07×10^5	1.16×10^7	1.39×10^7

CONCLUSION

Undoubtedly, SERS has now been stablished as a tool of extreme importance for ultrasensitive trace detection. Though, majorly the enhancement mechanisms of SERS have been understood. As discussed, electromagnetic and chemical mechanisms play the most dominant role in the SERS enhancement and have been studied thoroughly. However, still there is a plenty of scope to study other mechanisms in more detailed manner, so far considered to be of less importance. Hopefully, a deeper understanding of these mechanisms in near future, will open up the new dimensions in strengthening the theory and application of SERS in the fields unveiled till date.

CONSENT FOR PUBLICATION

Not applicable.

CONFLICT OF INTEREST

The author declares no conflict of interest, financial or otherwise.

ACKNOWLEDGEMENTS

Declared none.

REFERENCES

[1]   Raman CV, Krishnan KS. A New Type of Secondary Radiation. Nature 1928; 121(3048): 501-2.
[http://dx.doi.org/10.1038/121501c0]

[2]   Kneipp K. Surface-enhanced Raman scattering. Phys Today 2007; 60(11): 40-6.
[http://dx.doi.org/10.1063/1.2812122]

[3]   Kneipp J, Kneipp H, Kneipp K. SERS--a single-molecule and nanoscale tool for bioanalytics. Chem
Soc Rev 2008; 37(5): 1052-60.
[http://dx.doi.org/10.1039/b708459p] [PMID: 18443689]

[4]   Kneipp K, Kneipp H, Kneipp J. Surface-enhanced Raman scattering in local optical fields of silver and
gold nanoaggregates-from single-molecule Raman spectroscopy to ultrasensitive probing in live cells.
Acc Chem Res 2006; 39(7): 443-50.
[http://dx.doi.org/10.1021/ar050107x] [PMID: 16846208]

[5]   Doering WE, Nie S. Single-Molecule and Single-Nanoparticle SERS: Examining the Roles of Surface
Active Sites and Chemical Enhancement. J Phys Chem B 2002; 106(2): 311-7.
[http://dx.doi.org/10.1021/jp011730b]

[6]   Panneerselvam R, Liu GK, Wang YH, *et al.* Surface-enhanced Raman spectroscopy: bottlenecks and
future directions. Chem Commun (Camb) 2017; 54(1): 10-25.
[http://dx.doi.org/10.1039/C7CC05979E] [PMID: 29139483]

[7]   Woodruff DP. Modern Techniques of Surface Science. Cambridge: Cambridge University Press 2016.
[http://dx.doi.org/10.1017/CBO9781139149716]

[8]   Ding SY, Yi J, Li JF, Ren B, Wu DY, Panneerselvam R, *et al.* Nanostructure-based plasmon-enhanced
Raman spectroscopy for surface analysis of materials. Nat Rev Mater 2016; 1: 16021.
[http://dx.doi.org/10.1038/natrevmats.2016.21]

[9]   Sharma B, Frontiera RR, Henry AI, Ringe E, Van Duyne RP. SERS: Materials, applications, and the
future. Mater Today 2012; 15(1-2): 16-25.
[http://dx.doi.org/10.1016/S1369-7021(12)70017-2]

[10]   Fleischmann M, Hendra PJ, McQuillan AJ. Raman spectra of pyridine adsorbed at a silver electrode.
Chem Phys Lett 1974; 26(2): 163-6.
[http://dx.doi.org/10.1016/0009-2614(74)85388-1]

[11]   Jeanmaire DL, Van Duyne RP. Surface raman spectroelectrochemistry: Part I. Heterocyclic, aromatic,
and aliphatic amines adsorbed on the anodized silver electrode. J Electroanal Chem Interfacial
Electrochem 1977; 84(1): 1-20.
[http://dx.doi.org/10.1016/S0022-0728(77)80224-6]

[12]   Schatz GC, Young MA, Van Duyne RP. Electromagnetic Mechanism of SERS. In: Kneipp K,
Moskovits M, Kneipp H, Eds. Surface-Enhanced Raman Scattering: Physics and Applications. Berlin,
Heidelberg: Springer Berlin Heidelberg 2006; pp. 19-45.
[http://dx.doi.org/10.1007/3-540-33567-6_2]

[13]   Ding SY, You EM, Tian ZQ, Moskovits M. Electromagnetic theories of surface-enhanced Raman
spectroscopy. Chem Soc Rev 2017; 46(13): 4042-76.
[http://dx.doi.org/10.1039/C7CS00238F] [PMID: 28660954]

[14]   Kneipp K, Kneipp H. Probing the plasmonic near-field by one- and two-photon excited surface
enhanced Raman scattering. Beilstein J Nanotechnol 2013; 4: 834-42.
[http://dx.doi.org/10.3762/bjnano.4.94] [PMID: 24367752]

[15]   Schlücker S. Surface-enhanced Raman spectroscopy: concepts and chemical applications. Angew
Chem Int Ed Engl 2014; 53(19): 4756-95.
[http://dx.doi.org/10.1002/anie.201205748] [PMID: 24711218]

[16]   O. McAnally M, C. Schatz G, C. Stair P, P. Van Duyne R. Identification of Dimeric Methyl alumina

Surface Species during Atomic Layer Deposition Using Operando Surface-Enhanced Raman Spectroscopy. J Am Chem Soc 2017; 139(6): 2456-63.
[http://dx.doi.org/10.1021/jacs.6b12709] [PMID: 28135417]

[17]　Wang Z, Zong S, Wu L, Zhu D, Cui Y. SERS-Activated Platforms for Immunoassay: Probes, Encoding Methods, and Applications. Chem Rev 2017; 117(12): 7910-63.
[http://dx.doi.org/10.1021/acs.chemrev.7b00027] [PMID: 28534612]

[18]　Xu L, Yan W, Ma W, *et al.* SERS encoded silver pyramids for attomolar detection of multiplexed disease biomarkers. Adv Mater 2015; 27(10): 1706-11.
[http://dx.doi.org/10.1002/adma.201402244] [PMID: 25641772]

[19]　Zhang H, Xu L, Tian Y, Chen M, Liu X, Chen F. Controlled synthesis of hollow Ag@Au nano-urchins with unique synergistic effects for ultrasensitive surface-enhanced Raman spectroscopy. Opt Express 2017; 25(23): 29389-400.
[http://dx.doi.org/10.1364/OE.25.029389]

[20]　Stiles PL, Dieringer JA, Shah NC, Van Duyne RP. Surface-enhanced Raman spectroscopy. Annu Rev Anal Chem (Palo Alto, Calif) 2008; 1(1): 601-26.
[http://dx.doi.org/10.1146/annurev.anchem.1.031207.112814] [PMID: 20636091]

[21]　Kneipp K, Wang Y, Kneipp H, Perelman LT, Itzkan I, Dasari RR, *et al.* Single Molecule Detection Using Surface-Enhanced Raman Scattering (SERS). Phys Rev Lett 1997; 78(9): 1667-70.
[http://dx.doi.org/10.1103/PhysRevLett.78.1667]

[22]　Nie S, Emory SR. Probing Single Molecules and Single Nanoparticles by Surface-Enhanced Raman Scattering. Science 1997; 275(5303): 1102-6.
[http://dx.doi.org/10.1126/science.275.5303.1102] [PMID: 9027306]

[23]　Moskovits M. Surface-enhanced Raman spectroscopy: a brief retrospective. J Raman Spectrosc 2005; 36(607): 485-96.
[http://dx.doi.org/10.1002/jrs.1362]

[24]　Moskovits M. Surface-enhanced spectroscopy. Rev Mod Phys 1985; 57(3): 783-826.
[http://dx.doi.org/10.1103/RevModPhys.57.783]

[25]　Jensen L, Aikens CM, Schatz GC. Electronic structure methods for studying surface-enhanced Raman scattering. Chem Soc Rev 2008; 37(5): 1061-73.
[http://dx.doi.org/10.1039/b706023h] [PMID: 18443690]

[26]　Wokaun A. Surface-Enhanced Electromagnetic Processes. Solid State Phys 1984; 38: 223-94.
[http://dx.doi.org/10.1016/S0081-1947(08)60314-8]

[27]　Campion A. E. Ivanecky J, M. Child C, Foster M. On the Mechanism of Chemical Enhancement in Surface-Enhanced Raman Scattering. J Am Chem Soc 1995; 117(47): 11807-8.
[http://dx.doi.org/10.1021/ja00152a024]

[28]　Kerker M. Estimation of surface-enhanced raman scattering from surface-averaged electromagnetic intensities. J Colloid Interface Sci 1987; 118(2): 417-21.
[http://dx.doi.org/10.1016/0021-9797(87)90477-2]

[29]　Littleford RE, Graham D, Smith E, Khan I. Surface-Enhanced Raman Scattering (SERS), Applications. Reference Module in Chemistry, Molecular Sciences and Chemical Engineering. 2014.

[30]　Kosuda KM, Wiester J, Wustholz KL, Van Duyne RP, Groarke RJ. Nanostructures and Surface-Enhanced Raman Spectroscopy. Reference Module in Materials Science and Materials Engineering. 2016.
[http://dx.doi.org/10.1016/B978-0-12-803581-8.00611-1]

[31]　Zhang Y, He S, Guo W, *et al.* Surface-Plasmon-Driven Hot Electron Photochemistry. Chem Rev 2018; 118(6): 2927-54.
[http://dx.doi.org/10.1021/acs.chemrev.7b00430] [PMID: 29190069]

[32]   Otto A. The 'chemical' (electronic) contribution to surface-enhanced Raman scattering. J Raman Spectrosc 2005; 36(607): 497-509.
[http://dx.doi.org/10.1002/jrs.1355]

[33]   Kambhampati P, Child CM, Foster MC, Campion A. On the chemical mechanism of surface enhanced Raman scattering: Experiment and theory. J Chem Phys 1998; 108(12): 5013-26.
[http://dx.doi.org/10.1063/1.475909]

[34]   Arya K, Zeyher R. Theory of surface-enhanced Raman scattering from molecules adsorbed at metal surfaces. Phys Rev B 1981; 24(4): 1852-65.
[http://dx.doi.org/10.1103/PhysRevB.24.1852]

[35]   Billmann J, Otto A. Electronic surface state contribution to surface enhanced Raman scattering. Solid State Commun 1982; 44(2): 105-7.
[http://dx.doi.org/10.1016/0038-1098(82)90410-0]

[36]   Otto A, Futamata M. Electronic mechanisms of SERS. In: Otto A, Futamata M, Eds. Top. Appl. Phys. 1970; pp. 147-82.

[37]   Shim S, Stuart CM, Mathies RA. Resonance Raman cross-sections and vibronic analysis of rhodamine 6G from broadband stimulated Raman spectroscopy. ChemPhysChem 2008; 9(5): 697-9.
[http://dx.doi.org/10.1002/cphc.200700856] [PMID: 18330856]

[38]   Zrimsek AB, Chiang N, Mattei M, *et al.* Single-Molecule Chemistry with Surface- and Tip-Enhanced Raman Spectroscopy. Chem Rev 2017; 117(11): 7583-613.
[http://dx.doi.org/10.1021/acs.chemrev.6b00552] [PMID: 28610424]

[39]   Cunningham D, Littleford RE, Smith WE, *et al.* Practical control of SERRS enhancement. Faraday Discuss 2006; 132(0): 135-45.
[http://dx.doi.org/10.1039/B506241A] [PMID: 16833113]

[40]   Hildebrandt P, Stockburger M. Surface-enhanced resonance Raman spectroscopy of Rhodamine 6G adsorbed on colloidal silver. J Phys Chem 1984; 88(24): 5935-44.
[http://dx.doi.org/10.1021/j150668a038]

[41]   Duyne RPV. Laser Excitation of Raman Scattering from Adsorbed Molecules on Electrode Surfaces. Chemical and Biochemical Applications of Lasers 1979; pp. 101-84.

[42]   Zhao J. A. Dieringer J, Zhang X, C. Schatz G, P. Van Duyne R. Wavelength-Scanned Surface-Enhanced Resonance Raman Excitation Spectroscopy. J Phys Chem C 2008; 112(49): 19302-10.
[http://dx.doi.org/10.1021/jp807837t]

[43]   C. Maher R. SERS hot spots. In C. Maher R. Raman Spectroscopy for Nanomaterials Characterization 2012; 215-60.

[44]   McFarland AD, Young MA, Dieringer JA, Van Duyne RPD, McFarland AA, Young M, *et al.* Wavelength-scanned surface-enhanced Raman excitation spectroscopy. J Phys Chem B 2005; 109(22): 11279-85.
[http://dx.doi.org/10.1021/jp050508u] [PMID: 16852377]

[45]   Bruzas I, Lum W, Gorunmez Z, Sagle L. Advances in surface-enhanced Raman spectroscopy (SERS) substrates for lipid and protein characterization: sensing and beyond. Analyst (Lond) 2018; 143(17): 3990-4008.
[http://dx.doi.org/10.1039/C8AN00606G] [PMID: 30059080]

[46]   Ding SY, Zhang XM, Ren B, Tian ZQ. Surface-Enhanced Raman Spectroscopy (SERS): General Introduction 2014.
[http://dx.doi.org/10.1002/9780470027318.a9276]

[47]   Yamamoto Y, Ozaki Y, Itoh T. Recent progress and frontiers in the electromagnetic mechanism of surface-enhanced Raman scattering. J Photochem Photobiol Photochem Rev 2014; 21.
[http://dx.doi.org/10.1016/j.jphotochemrev.2014.10.001]

[48]   Spaeth S, Dickey M, T. Carron K. Determination of the Distance Dependence and Experimental Effects for Modified SERS Substrates Based on Self-Assembled Monolayers Formed Using Alkanethiols. J Phys Chem B 1999; 103(18): 3640-6.
[http://dx.doi.org/10.1021/jp984454i]

[49]   Mosier-Boss PA. Review of SERS Substrates for Chemical Sensing. Nanomaterials (Basel) 2017; 7(6): 142.
[http://dx.doi.org/10.3390/nano7060142] [PMID: 28594385]

[50]   Smith WE. Practical understanding and use of surface enhanced Raman scattering/surface enhanced resonance Raman scattering in chemical and biological analysis. Chem Soc Rev 2008; 37(5): 955-64.
[http://dx.doi.org/10.1039/b708841h] [PMID: 18443681]

[51]   Tian ZQ, Ren B, Li JF, Yang ZL. Expanding generality of surface-enhanced Raman spectroscopy with borrowing SERS activity strategy. Chem Commun (Camb) 2007; (34): 3514-34.
[http://dx.doi.org/10.1039/b616986d] [PMID: 18080535]

[52]   Li JF, Zhang YJ, Ding SY, Panneerselvam R, Tian ZQ. Core-Shell Nanoparticle-Enhanced Raman Spectroscopy. Chem Rev 2017; 117(7): 5002-69.
[http://dx.doi.org/10.1021/acs.chemrev.6b00596] [PMID: 28271881]

[53]   Li JF, Zhang YJ, Rudnev AV, et al. Electrochemical shell-isolated nanoparticle-enhanced Raman spectroscopy: correlating structural information and adsorption processes of pyridine at the Au(hkl) single crystal/solution interface. J Am Chem Soc 2015; 137(6): 2400-8.
[http://dx.doi.org/10.1021/ja513263j] [PMID: 25625429]

[54]   Kurouski D. Advances of tip-enhanced Raman spectroscopy (TERS) in electrochemistry, biochemistry, and surface science. Vib Spectrosc 2017; 91: 3-15.
[http://dx.doi.org/10.1016/j.vibspec.2016.06.004]

[55]   Kneipp K, Kneipp H, Itzkan I, Dasari RR, Feld MS. Surface-enhanced Raman scattering and biophysics. J Phys Condens Matter 2002; 14(18): R597-624.
[http://dx.doi.org/10.1088/0953-8984/14/18/202]

[56]   Dick LA, Haes AJ, Van Duyne RP. Distance and Orientation Dependence of Heterogeneous Electron Transfer: A Surface-Enhanced Resonance Raman Scattering Study of Cytochrome c Bound to Carboxylic Acid Terminated Alkanethiols Adsorbed on Silver Electrodes. J Phys Chem B 2000; 104(49): 11752-62.
[http://dx.doi.org/10.1021/jp0029717]

[57]   Dieringer JA, McFarland AD, Shah NC, et al. Surface enhanced Raman spectroscopy: new materials, concepts, characterization tools, and applications. Faraday Discuss 2006; 132(0): 9-26.
[http://dx.doi.org/10.1039/B513431P] [PMID: 16833104]

[58]   Haynes CL, Van Duyne RP. Nanosphere Lithography: A Versatile Nanofabrication Tool for Studies of Size-Dependent Nanoparticle Optics. J Phys Chem B 2001; 105(24): 5599-611.
[http://dx.doi.org/10.1021/jp010657m]

[59]   Link S, El-Sayed MA. Spectral Properties and Relaxation Dynamics of Surface Plasmon Electronic Oscillations in Gold and Silver Nanodots and Nanorods. J Phys Chem B 1999; 103(40): 8410-26.
[http://dx.doi.org/10.1021/jp9917648]

[60]   Duval Malinsky M, Kelly KL, Schatz GC, Van Duyne RP. Nanosphere Lithography: Effect of Substrate on the Localized Surface Plasmon Resonance Spectrum of Silver Nanoparticles. J Phys Chem B 2001; 105(12): 2343-50.
[http://dx.doi.org/10.1021/jp002906x]

[61]   Draine BT, Flatau PJ. Discrete-Dipole Approximation For Scattering Calculations. J Opt Soc Am A Opt Image Sci Vis 1994; 11(4): 1491-9.
[http://dx.doi.org/10.1364/JOSAA.11.001491]

[62]   Purcell EM, Pennypacker CR. Scattering and Absorption of Light by Nonspherical Dielectric Grains.

Astrophys J 1973; 186: 705.
[http://dx.doi.org/10.1086/152538]

[63] Kelly KL, Coronado E, Zhao LL, Schatz GC. The Optical Properties of Metal Nanoparticles: The Influence of Size, Shape, and Dielectric Environment. J Phys Chem B 2003; 107(3): 668-77.
[http://dx.doi.org/10.1021/jp026731y]

[64] Taflove A. Computational electrodynamics: the finite-difference time-domain method Boston. Artech House 1995.

[65] Yang W, Schatz GC, Van Duyne RP. Discrete dipole approximation for calculating extinction and Raman intensities for small particles with arbitrary shapes. J Chem Phys 1995; 103(3): 869-75.
[http://dx.doi.org/10.1063/1.469787]

[66] Kneipp K, Wang Y, Kneipp H, Itzkan I, Dasari RR, Feld MS. Population pumping of excited vibrational states by spontaneous surface-enhanced Raman scattering. Phys Rev Lett 1996; 76(14): 2444-7.
[http://dx.doi.org/10.1103/PhysRevLett.76.2444] [PMID: 10060701]

[67] Kneipp K, Kneipp H, Itzkan I, Dasari RR, Feld MS. Ultrasensitive chemical analysis by Raman spectroscopy. Chem Rev 1999; 99(10): 2957-76.
[http://dx.doi.org/10.1021/cr980133r] [PMID: 11749507]

[68] Mertz J, Xu C, Webb WW. Single-molecule detection by two-photon-excited fluorescence. Opt Lett 1995; 20(24): 2532-4.
[http://dx.doi.org/10.1364/OL.20.002532] [PMID: 19865276]

[69] Kneipp K, Kneipp H. Two-Photon Excited Surface-Enhanced Raman Scattering. In: Kneipp K, Moskovits M, Kneipp H, Eds. Surface-Enhanced Raman Scattering: Physics and Applications. Berlin, Heidelberg: Springer Berlin Heidelberg 2006; pp. 183-96.
[http://dx.doi.org/10.1007/3-540-33567-6_9]

[70] Denisov VN, Mavrin BN, Podobedov VB. Hyper-Raman scattering by vibrational excitations in crystals, glasses and liquids. Phys Rep 1987; 151(1): 1-92.
[http://dx.doi.org/10.1016/0370-1573(87)90053-6]

[71] Kneipp H, Kneipp K, Seifert F. Surface-enhanced hyper-Raman scattering (SEHRS) and surface-enhanced Raman scattering (SERS) by means of mode-locked Ti:sapphire laser excitation. Chem Phys Lett 1993; 212(3-4): 374-8.
[http://dx.doi.org/10.1016/0009-2614(93)89340-N]

[72] Gühlke M, Heiner Z, Kneipp J. Combined near-infrared excited SEHRS and SERS spectra of pH sensors using silver nanostructures. Phys Chem Chem Phys 2015; 17(39): 26093-100.
[http://dx.doi.org/10.1039/C5CP03844H] [PMID: 26377486]

[73] Le Ru EC, Etchegoin PG. Quantifying SERS enhancements. MRS Bull 2013; 38(8): 631-40.
[http://dx.doi.org/10.1557/mrs.2013.158]

[74] Blackie EJ, Le Ru EC, Etchegoin PG. Single-molecule surface-enhanced Raman spectroscopy of nonresonant molecules. J Am Chem Soc 2009; 131(40): 14466-72.
[http://dx.doi.org/10.1021/ja905319w] [PMID: 19807188]

[75] Brosseau CL, Casadio F, Van Duyne RP. Revealing the invisible: using surface-enhanced Raman spectroscopy to identify minute remnants of color in Winslow Homer's colorless skies. J Raman Spectrosc 2011; 42(6): 1305-10.
[http://dx.doi.org/10.1002/jrs.2877]

[76] Kleinman SL, Frontiera RR, Henry AI, Dieringer JA, Van Duyne RP. Creating, characterizing, and controlling chemistry with SERS hot spots. Phys Chem Chem Phys 2013; 15(1): 21-36.
[http://dx.doi.org/10.1039/C2CP42598J] [PMID: 23042160]

[77] Rycenga M, Kim MH, Camargo PHC, Cobley C, Li ZY, Xia Y. Surface-enhanced Raman scattering: comparison of three different molecules on single-crystal nanocubes and nanospheres of silver. J Phys

Chem A 2009; 113(16): 3932-9.
[http://dx.doi.org/10.1021/jp8101817] [PMID: 19175302]

[78]   Sergiienko S, Moor K, Gudun K, Yelemessova Z, Bukasov R. Nanoparticle-nanoparticle *vs.* nanoparticle-substrate hot spot contributions to the SERS signal: studying Raman labelled monomers, dimers and trimers. Phys Chem Chem Phys 2017; 19(6): 4478-87.
[http://dx.doi.org/10.1039/C6CP08254H] [PMID: 28120963]

[79]   Yu X, Cai H, Zhang W, *et al.* Tuning chemical enhancement of SERS by controlling the chemical reduction of graphene oxide nanosheets. ACS Nano 2011; 5(2): 952-8.
[http://dx.doi.org/10.1021/nn102291j] [PMID: 21210657]

[80]   Xia L, Chen M, Zhao X, Zhang Z, Xia J, Xu H, *et al.* Visualized method of chemical enhancement mechanism on SERS and TERS. J Raman Spectrosc 2014; 45(7): 533-40.
[http://dx.doi.org/10.1002/jrs.4504]

[81]   Su JP, Lee YT, Lu SY, Lin JS. Chemical mechanism of surface-enhanced Raman scattering spectrum of pyridine adsorbed on Ag cluster: ab initio molecular dynamics approach. J Comput Chem 2013; 34(32): 2806-15.
[http://dx.doi.org/10.1002/jcc.23464] [PMID: 24166008]

[82]   Seki H. Surface enhanced Raman scattering of pyridine on different silver surfaces. J Chem Phys 1982; 76(9): 4412-8.
[http://dx.doi.org/10.1063/1.443556]

[83]   Qian XM, Nie SM. Single-molecule and single-nanoparticle SERS: from fundamental mechanisms to biomedical applications. Chem Soc Rev 2008; 37(5): 912-20.
[http://dx.doi.org/10.1039/b708839f] [PMID: 18443676]

[84]   Kneipp K, Kneipp H, Itzkan I, Dasari RR, Feld MS, Dresselhaus MS. Nonlinear Raman Probe of Single Molecules Attached to Colloidal Silver and Gold Clusters. In: Shalaev VM, Ed. Optical Properties of Nanostructured Random Media. Berlin, Heidelberg: Springer Berlin Heidelberg 2002; pp. 227-49.
[http://dx.doi.org/10.1007/3-540-44948-5_11]

[85]   Kneipp J, Kneipp H, Kneipp K. Two-photon vibrational spectroscopy for biosciences based on surface-enhanced hyper-Raman scattering. Proc Natl Acad Sci USA 2006; 103(46): 17149-53.
[http://dx.doi.org/10.1073/pnas.0608262103] [PMID: 17088534]

[86]   Kneipp J, Kneipp H, Wittig B, Kneipp K. One- and two-photon excited optical ph probing for cells using surface-enhanced Raman and hyper-Raman nanosensors. Nano Lett 2007; 7(9): 2819-23.
[http://dx.doi.org/10.1021/nl071418z] [PMID: 17696561]

[87]   Kim NH, Hwang W, Baek K, *et al.* Smart SERS Hot Spots: Single Molecules Can Be Positioned in a Plasmonic Nanojunction Using Host-Guest Chemistry. J Am Chem Soc 2018; 140(13): 4705-11.
[http://dx.doi.org/10.1021/jacs.8b01501] [PMID: 29485275]

[88]   Bumbrah GS, Sharma RM. Raman spectroscopy – Basic principle, instrumentation and selected applications for the characterization of drugs of abuse. Egypt J Forensic Sci 2016; 6(3): 209-15.
[http://dx.doi.org/10.1016/j.ejfs.2015.06.001]

[89]   Xie J, Zhang Q, Lee JY, Wang DI. The synthesis of SERS-active gold nanoflower tags for *in vivo* applications. ACS Nano 2008; 2(12): 2473-80.
[http://dx.doi.org/10.1021/nn800442q] [PMID: 19206281]

[90]   Rodríguez-Lorenzo L, Álvarez-Puebla RA, Pastoriza-Santos I, *et al.* Zeptomol detection through controlled ultrasensitive surface-enhanced Raman scattering. J Am Chem Soc 2009; 131(13): 4616-8.
[http://dx.doi.org/10.1021/ja809418t] [PMID: 19292448]

[91]   Chang H, Lee YY, Lee HE, *et al.* Size-controllable and uniform gold bumpy nanocubes for single-particle-level surface-enhanced Raman scattering sensitivity. Phys Chem Chem Phys 2019; 21(18): 9044-51.
[http://dx.doi.org/10.1039/C9CP00138G] [PMID: 30916087]

CHAPTER 3

Plasmonic and Non-plasmonic Resonance Materials in SERS

Harsimran Singh Bindra[1] and **Ranu Nayak**[2,*]

[1] *School of Biotechnology, S.K. University of Agricultural Sciences and Technology of Jammu, Jammu and Kashmir, India*

[2] *Amity Institute of Nanotechnology, Amity University Uttar Pradesh, Noida, India*

Abstract: This chapter explores the recent advances in implementing diverse plasmonic and non-plasmonic nanomaterials actively for Surface Enhanced Raman Scattering (SERS). This chapter, chiefly presents a detailed outline on the potential applications of SERS probe either as Raman reporter molecule (RRM), conjugates of RRM with plasmonic nanomaterials, silica encapsulated conjugate probe of RRM with plasmonic nanomaterials. Overall, this chapter collectively emphasizes addressing the performance mechanism along with pros and cons of different materials.

Keywords: Biomedical Application, Gold, Hybrid Plasmonic Nanoparticle, Label-free, Labeled SERS Probe, Metal Nanoparticle, Non-destructive, Non-plasmonic SERS, Plasmonic SERS, Plasmonic Substrate, Raman Dye, Raman Reporter, RRM, SERS, Signal Enhancement, Silicon, Silver, SiO_2 Shell, Surface Enhanced Raman, ZnO.

3.1. INTRODUCTION

A major requirement in sensing and diagnostic application is to amplify the receiving signal without much loss in signal to noise (S/N) ratio. There are different sensitive detection methods, for instance, electrochemical detection, fluorescence-based detection, field effect transistor (FET) based electronic detection, *etc.* Most of these techniques have some benefits and limitations. The Discovery of the plasmonic effect due to the interaction of photons with phonons across the exterior of a metal/semiconductor in nanoscale has led to the invention of surface enhanced Raman scattering (SERS) effect.

---

[*] **Corresponding author Ranu Nayak:** Amity Institute of Nanotechnology, Amity University Uttar Pradesh, Noida, India; Tel: +919818936883; E-mail: rnayak@amity.edu;ranunayakbose@gmail.com

An enhancement in Raman signal (related to energy transfer between rotational energy levels) is observed when a target molecule/analyte gets adsorbed on a plasmonic or non-plasmonic SERS active substrate. One of the reasons for enhancement in Raman signal on the SERS active substrate is due to the initiation of plasmonic resonating oscillations that arise due to interaction of the free surface conduction electrons on a metal surface with the phonon (atomic) vibration of the target molecule/analyte. Expectedly, a charge transfer occurs between the oscillating electrons and the adsorbed atoms of the target molecule that causes a substantial enhancement in the local electromagnetic (*em*) field and in turn to an extremely large extent that eventually induces dipoles that are detected as enhanced Raman signal intensity up to 10^{11} times. Another reason for signal enhancement up to 10^2- 10^3 can be justified on the account of chemical variation arising from the resonance/charge transfer between the adsorbed analyte and metal surface. Undoubtedly, the contribution of *em* enhancement to SERS is much stronger than the chemical enhancement. A major benefit of SERS based recognition is label-free detection that makes it a less troublesome approach [1]. Since Raman spectra serves as a unique signal, it acts as a fingerprint for selective and highly sensitive detection of analytes including small, large or complex molecules [2 - 4]. Moreover, unlike fluorescence based detection, the occurrence of photo-bleaching is not evidenced [5]. In addition, the strength of the Raman signal can easily be adjusted by varying the dimensions of the nanosensing surface. There are a variety of plasmonic as well as nano-plasmonic SERS solid substrates and photonic nanostructures/nanoparticles that have been developed and explored for chemical and biological based SERS detection. In biological applications, SERS has been successfully applied for versatile applications like - cellular imaging [6 - 8], *in-vivo* tissue detection and imaging [9], portable paper-based immunosensors [10, 11], bead assays [12], *etc*. In all these applications, several Raman Reporter Molecules (RRM) have been conjugated with the SERS active materials to significantly enhance the obtained Raman signals. These are called *Labeled SERS probes*. In order to further improve the performance, the Labeled SERS probes are also encapsulated with nano-dimension layers of insulating materials. For a better understanding of the reader, the proceeding sections of this chapter give a detailed outline of different types of RRM, plasmonic and non-plasmonic nanomaterials.

3.2. RAMAN REPORTER MOLECULES (RRM)

Raman scattering is known to be a weak phenomenon, however, Raman signal obtained from a reporter molecule (also called Raman molecule) can be significantly enhanced as high as 10^9 orders, by adsorbing these molecules onto specific SERS active materials/substrates like nanotextured metal thin film on an insulating substrate or metal nanoparticles of noble metals like Au, Ag [13 - 15].

Typically, there are several RRM that have been reported in works of literature for a variety of applications. For instance, Raman dyes such as Malachite green isothiocyanate, Brilliant blue, Cresyl violet, Rhodamine 6G, Cy3, Cy5 and Methylene blue [16] offer strong signal output. This property of reporter molecules marks their active candidature in cellular imaging. In addition, thiolate molecules like 4-mercaptobenzoic acid, 5,5'-dithiobis(2-nitrobenzoic acid), 4-mercaptopyridine, 4-methoxythiophenol, 4-aminothiophenol are mostly considered for detection using Au substrate as they tend to develop conjugation bridge with the metal substrate *via* carboxylate groups [17, 18]. A disadvantage of using 4-mercaptobenzoic acid and its derivatives as Raman molecule is its limited use in multiplexed assay. This is because the characteristic bands from the phenyl ring exhibit a very small wavenumber shift as a function of the substituent [19]. This causes an overlap of spectral signals that may be difficult to resolve. Larger wavenumber differences with minimum overlapping spectra are seen from conjugated polyenes as RRMs [20]. Many small molecules with characteristic Raman spectral signals have been used *e.g.*, 1,2-bis(4-pyridyl)-ethylene (BPE) [21]. However, small molecules also suffer from overlapping spectra. Olefins and alkynes with aryl substituents (arenethiol being one of the substituent) on both sides have also been used as Raman reporters [22]. Arenethiol substituent enhances chemisorption onto the surface of gold.

Achieving insignificant spectral overlap becomes more and more grim with increased demand for desired reporters. This becomes further complicated with a practical approach incurring the use of large reporter molecules such as Raman dyes which apparently offer complicated spectra. To the best of our knowledge, there is no specific approach reported which highlights the selection criteria of reporter molecules. Typically, to differentiate multiple Raman reporters in a multiplexed signal, it is desirable that the reporter molecules employed should have minimum signal overlapping, so as to secure their relevant spectral features in the obtained multiplexed signal. However, reports are available which outlines a selection of multiple reporters using mathematical tool- "Correlation value" which optimizes the selection scheme [23].

It is clear from the literatures that most commonly, the reporter molecules with $-NH_2$ or $-SH$ moieties are used because of their special binding affinity with the SERS active plasmonic particles like gold (Au) and silver(Ag). However, these SERS probe, *i.e.*, Au/Ag functionalized with RRMs often suffer from leaching and aggregation of the reporter molecule in liquid. Often this is resolved by encapsulating the entire SERS probe in a shell of silica (SiO_2) that not only improves the hydrophilicity of the probe, but also prevents leaching of the reporter molecule and its aggregation in liquid. The SiO_2 shell also provides a platform for further functionalization with biomolecules [24]. A schematic representation of different SERS probes is shown in Fig. (**1**).

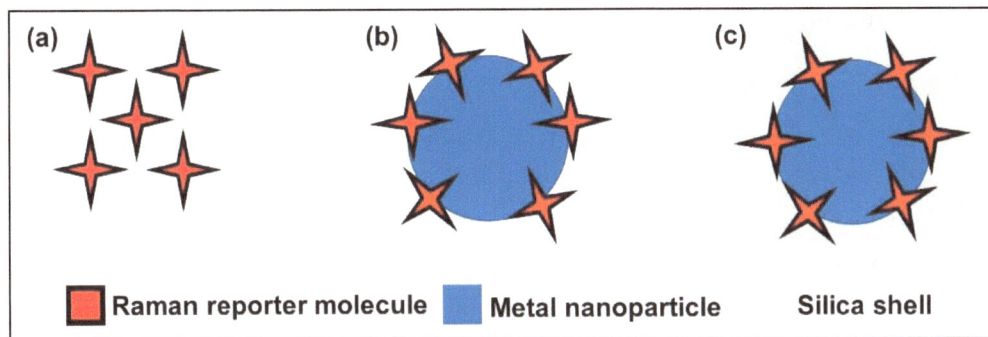

Fig. (1). Schematic representation of three different SERS probe (a) RRM, (b) plasmonic metal nanoparticles in conjugation withRRMs, and (c) encapsulated metal nanoparticle/RRM with SiO_2 shell.

3.3. MATERIALS USED IN SERS

The very first SERS activity was reported in 1974 where enhanced Raman spectra of pyridine on rough silver film was observed; however, at that time it was misinterpreted [25]. Later, Duyne and Albrecht revealed SERS as a new phenomenon in 1977 [26, 27]. In this attempt, Duyne and Albrecht evidenced extraordinary enhancement of Raman signals from molecules that were found to be in the proximity of metallic nanostructures. In the emerging field of photonics and nanoscience, this unique observation acted as an exploratory pathway and accelerated the development of SERS substrate targeting the detection of a wide range of chemical and biological analytes.

3.3.1. Plasmonic SERS Nanomaterials

During the beginning of SERS exploration, SERS active substrates were employed either in colloidal form or in the form of one-dimensional (1D) nanostructures (nanowires) of metals like Au, Ag, or Copper (Cu). In today's date, both nanoparticle colloidal suspensions as well as nanostructures grown on solid substrates are being used as SERS active materials/substrates.

Undoubtedly, it was realized that the morphology (shape, size and orientation) of the plasmonic nanoparticles played a significant role in the enhancement of the Raman signal. Electromagnetic enhancement is supposed to originate from the enhanced localized *em* field in plasmonic metal nanoparticles/nanostructures. For instance, if the metal nanoparticles like Au, Ag, Cu and Aluminum (Al) have dimensions analogous smaller than the wavelength of incident light, then the localized surface plasmon resonance (LSPR) can be projected with a resonance frequency as [28].

$$\omega_{LSPR} = \omega_p/(\sqrt{1 + 2\varepsilon_{diel}}) \tag{1}$$

where, and ω_p is the plasma frequency of bulk material (metal) and ε_{diel} refers to dielectric constant (both real and an imaginary part) of the surrounding medium.

Material selectivity for permitting strong plasmon resonance and SERS depends on the dielectric function (real part). The imaginary part of dielectric constant contributes to damping at the interested wavelength range (covering visible and near infrared range). Cu and Al have high imaginary part thus, damping is stronger, whereas Au and Ag have very low damping but strong plasmon. Multiplexed detection of cancer biomarkers has been reported using SERS-based molecular sentinel (MS) nanoprobes composed of a Raman-labelled DNA hairpin-loop probe and a Ag nanoparticle. In the absence of a complementary target DNA, a strong SERS signal was observed due to the proximity of Raman label with the Ag nanoparticle. On proximity with a complimentary target, hybridization occurred, thus separating the Raman label and Ag nanoparticle, hence quenching the SERS signal [29].

Ag nanoparticles are amongst the strongest plasmonic response; however, Ag is not very environmentally stable; rather, Au nanoparticles are more stable and biocompatible. Overall, Ag and Au are typically applied SERS materials. In general, Ag nanoparticles are known to demonstrate a much higher enhancement factor as well as the potential to lift SERS signals up to several higher orders in contrast to similar sized Au nanoparticles in the visible wavelength region. Nie and Kneipp *et al.* have obtained Raman spectroscopic results of a single Raman molecule using nanostructured Ag as substrate [30, 31]. The obtained Raman signals were found to enhance upto 10^{14}. Nevertheless, Au being highly biocompatible, allows it to get widely explored as SERS material for several bioanalysis. In this aspect, the shapes and dimensions are modified to further enhance the SERS signal.

Au plasmonic substrate in the form of periodic nanocup array with extra hot-spots of gold nanoparticles has been reported to have a higher SERS enhancement factor than the 2D nanoparticle array. The closely spaced hot-spots of 3D gold nanoparticles with separation less than 1 nm tend to strongly confine the *em* field and hence the Raman signal [32].

Apart from Au and Ag nanoparticles, alkali metals like Lithium (Li), Sodium (Na), and Potassium (K) have also been studied for SERS application [33]. Moreover, they offer dynamic chemical properties. Hence, consideration of such materials would not be promising for biomedical applications. Nevertheless, these materials can be explored for other SERS applications if their activities could be optimally controlled.

Other than nanoparticles, 1D (wired) configuration of metal nanostructures have also been explored to support circulating surface Plasmon polaritons which make them to emerge as unique plasmonic nanomaterials [34].Uniform Ag nanowires (diameter ~ 30-60 nm; length ~ 50 μm) with ultra-low plasmon damping during the propagation have been employed in the study of surface plasmon polariton waveguide [34].

Au nanorod configuration demonstrates two plasmonic peaks - transverse band obtained at lower wavelength (~ 520 nm – 540 nm) with a position coincident with Au nanoparticles which is assigned to electron oscillations occurring perpendicular to the long axis. The second peak is a longitudinal band prevailing in the longer wavelength(~ 750 nm) which designates to electron oscillation occurring in pathway along the long axis [35]. The second plasmonic band founds to be sensitive with respect to the aspect ratio of the Au nanorods morphologically tunable between the visible and the near IR wavelength region. This beneficial property of Au nanorods has attracted its applicability as sensitive SERS substrates in biological applications like photo-thermal heating therapy, cell imaging, *in-vivo* disease screening *etc* [36, 37].

Inspired by different shapes of nanomaterials, metallic colloids with morphologies, like nanoprisms [13, 38], nanocubes [39, 40], nanostars [41, 42], nanocages [43] and nanosheets [44, 45] have also been investigated as SERS substrates. Since these shapes offer fine boundaries or twigs of metallic nanostructures, this allows accumulation of charge and cause higher *em* field enhancement compared to spherical nanoparticles. In addition, plasmonic hybridization occurring across distinct tips and cores further leads to an increased local *em* electric field. Of-late, nanostar configuration has gathered more attention in contrast to nanorod configuration due to their sharp apex. The sharp apex can offer scope for boosting the *em* field over the surface of the particle by several orders of magnitude. Recent research has also indicated that symmetric arrangement of nanostar exhibit superior and reproducible SERS response with almost four fold signal enhancement than asymmetric ones [46].

Bimetallic alloy based hybrid plasmonic nanoparticles such as Gold -Palladium (Au-Pd), Gold-Platinum (Au-Pt) and Gold- Silver (Au-Ag) has also been explored

[47 - 51]. For instance, Sun *et al.* reported eco-friendly realization of Au-Pd binary metallic nanoparticles, demonstrating excellent SERS activity and competent catalytic potency [48].

Fan *et al.* designed complex plasmonic nanocapsules having a tri- layered 1D nanostructure core comprising of Silver/Nickel/Silver (Ag/Ni/Ag) encapsulated with a thin layer of silica loaded with monodispersed Ag nanoparticles acting as hotspots [52, 53]. Ag nanoparticles tremendously enhanced the surface area to ~ 1200 μm^2 thus, demonstrating ultrasensitive SERS detection for both chemical and biomolecules. Presence of Ni in these structures introduced a magnetic effect that was successfully manoeuvred to a single live mammalian cell allowing further membrane composition analysed *via* SERS spectroscopy [28]. Magnetic cores can be readily used either for separation or for medical imaging, while plasmonic Ag or Au nanoparticles can be considered as the SERS active substrates.

SiO_2 coated Au / Ag nanoparticles SERS tags as well as SiO_2 encapsulated Raman tags loaded with RRMs have been explored by several scientific group [54 - 57]. Overall, the core-shell structured SERS and Raman tags exhibit brilliant stability which can make them to further act as potential candidate in biomolecule detection.

Interestingly, nanoscale breaches in metallic core-shell structures [also called nanomatryoshkas (NM)] have shown to create plenty of hot spots for SERS as a result of extremely enhanced *em* fields in between the gaps. These nanogaps are made using dielectric layers such as DNA [58, 59], SiO_2 [60] and small molecule Raman reporters such as 4-methylbenzenethiol and 1,4-benzenedithiol [61]. With smaller breaches plasmonic coupling increases but when breach size reaches sub-nm dimension, quantum effects become significant and play a progressively vital role and tend to reduce the *em* enhancement. Therefore, leading to reduction in SERS effect. In addition to breach size, shell thickness could be a prominent factor related to SERS. Hu *et al.* Synthesized thirteen Au NM sof varying dimensions ranging from 41 nm - 100 nm [62]. The realized structures comprise of uniform Au core~ 15nm dimension with anano -gap~1.2 nm formed using thiolate DNA. Further the nano-gaps were filled with RRM. The acquired SERS performance was found to increase when the thickness of Au shell raised to ~ 61nm but on further increase in shell size, a reduction in SERS performance has been observed [62].

The size-dependent SERS signal variation occurs in such a way that with expansion in thickness of shell, the internal as well as external energy modes come closer, thus experiencing enhanced intensity of anti-bonding plasmons.

Whereas, after increasing the resonance of the metallic shell higher than the dimension of the core, the energy dissipates results in reduction in the plasmonic resonance. Thus, in order to perceive optimal SERS signal enhancement, the thickness of the shell should be matching with that of the core's size.

3.3.2. Non-plasmonic SERS Materials

Label-free and non-destructive property is a key merit of surface-enhanced Raman Scattering (SERS), this enables it to act as an efficient technique to monitor the catalysis reaction. However, this technique offers wide scope for identifying the surface-interface information covering surface morphology, configuration, orientation and molecular bonding [25, 63 - 65]. However, during real-time processes, the laser irradiation occurring in the plasmonic system also provides a suitable window for additional side reactions which are either photo or thermo induced [65 - 67]. These side reactions might cause obstacles in analysing the areas *i.e.*, "hot-spots" offering high local surface plasmon resonance effect [68]. In order to avoid these side reactions, practices have been made either by selecting plasmonic metal SERS substrates offering minimal or no scope for occurrence of side reactions or counting on employing non-plasmonic materials. Among the two available options, non-plasmonic materials has demonstrated better scope for interference free SERS monitoring [68, 69]. In this attempt, the non-plasmonic materials employed for SERS monitoring are typically semiconductor materials. The selectivity of these materials is based on the fact that irrespective of *em* enhancement as shown by coinage (SERS active metal) metals, the semiconductor materials exhibit significant chemical enhancement which is apparently considered to be more efficient in limiting the probable occurrence of interference during SERS based catalytic monitoring [70 - 72]. Despite of these properties, semiconductor assisted SERS based catalytic reaction monitoring hasn't been practised on a wider level due the low sensitivity of the SERS active semiconductor. Fortunately, the available literature results contribute actively in paving way that suitably outlines the practical applicability of non-plasmonic material for SERS based activities on a wider scale.

At the initial stage, semiconductor materials showed weak enhancement factors due to limiting plasmon resonances active in their conduction band. But later with improved strategies and the introduction of nano-scaled semiconductor materials, these materials have demonstrated increased enhancement factors $\sim 10^3 - 10^7$ [73]. This has allowed them to gain strong attention and allowed them to emerge as potential candidate for SERS active monitoring alternative to already available metallic SERS active materials [74]. Developing and studying semiconductor-based SERS platforms has been found critical either in case of further increasing the prevailing enhancement factors for real-time applications or else to understand and disclose the mechanistic behaviour of weak chemical enhancement [73]. With a scope of electronic band structure tunability in semiconductors (SC's), a thorough study based on analysis of chemical enhancement mechanism in SERS field can be made believable using semiconductor-based substrates. In addition, from an application point of view, certain semiconductor materials are available which can aid in enhanced SERS effect and comprises of suitably wide optical band gap > 3eV in comparison to metal-based SERS active materials.

Among various SC's studied, metal oxides (MO's) are another materials that have been extensively explored as SERS-active materials. As an early proof of concept, crystalline nickel oxide (NiO) and titanium dioxide (TiO_2) based SERS active materials were reported initially [75, 76]. Pristine TiO_2 was found to exhibit small SERS enhancement [77]. However, later on the semiconductor was found to show enhancement in the SERS intensity on the incorporation of either SERS active metal particles like Ag or Au nanoparticles [77, 78]. Similarly, NiO nanoflakes decorated with Ag nanoparticles showed an improved SERS activity compared to its pristine configuration [79]. Ag nanoparticle decorated NiO nanoflake contributed to SERS active detection of polychlorinated biphenyls upto lowest concentration of 10^{-6} M [79].

Other than NiO and TiO_2, nanostructured zinc oxide (ZnO) is another material which is being explored for SERS monitoring. The selectivity of ZnO is accounted on the basis of its high refractive index that facilitates significant optical confinement which further contributes to increased SERS output [72, 80 - 82]. Besides, its property for enhancing charge-transfer *via* metal to the analyte, it also acts as a benefactor towards improving the SERS effect [72]. In addition, the nanostructured configuration of ZnO offers high surface area enabling maximized on sight absorption of analyte molecule [83]. In this regard Wang *et al.* reported the formation of amorphous ZnO nanocages which demonstrated remarkable SERS activity as a result of presence of plentiful metastable electronic states [84]. These metastable states facilitated the interfacial charge thereby amplified the molecular polarization. Moreover, the nanocage architecture provided a suitable surface area towards analyte molecule adsorption, leading to a notable enhancement factor ~ 6.6×10^5. The results acquired strongly emphasized on formation of strong molecule-semiconductor bonds, which overall marks the suitability of amorphous ZnO nanocage as an effective SERS platform. Further, Lombardi *et al.* reported the SERS behaviour of 4-aminothiophenol using Au/ZnO nanoparticles having average particle size of ~20 nm [72]. Here the author reported an increased Raman signal which was evidenced as a result of ZnO assisted improved charge-transfer from the Au to 4-aminothiophenol RRM. As an extension of this study, the authors presented another report outlining the study of Raman intensity of two unlike molecules namely 4-mercaptopyridine and 4-mercaptobenzoic acid adsorbed on the surface of same metal oxide crystal of size ranging between 18 -31 nm [85]. In this report, the author evidenced an increased Raman intensity prevailing due to formation of high charge-transfer complex developed between molecules adsorbed and metal oxide crystal having particle size ~28 nm. Besides ZnO nanostructures, ZnO quantum dots (QD's) have also been explored for acquiring enhanced SERS excitation [69]. Considering the enhanced materialistic properties of ZnO QD's, Rupa *et al.* reported unique label-free *in-vitro* SERS diagnosis of cancer by bringing ZnO based probes down to quantum scale [69]. The realized semiconductor probes demonstrated an increased SERS performance with enhancement factor raised to ~10^6 at multiple excitation wavelength with detection limit of upto nanomolar concentration. ZnO QD's has been reported to offer compatibility with complementary metal oxide semiconductor technology for development of small sensors [86]. Besides employing quantum scale configuration of ZnO for enhanced SERS applications, it has also been explored as a potential candidate for developing SERS active platforms. The ZnO based SERS platform has been extended to various nanostructures either in the form of nanorods, nanowires, nanocones, *etc.* [87, 88]. Considering this strategy, Shi *et al.* studied the variation in relative energy between highly occupied molecular orbital (HOMO) and lowest unoccupied molecular orbital (LUMO) of probe molecule and valance-conduction band edges of ZnO nanorod on SERS response [87]. Here, authors used two different probe molecules, namely 4-aminothiophenol (4-AT) and 4-mercaptopyridine (4-MP), having different molecular orbital energetics but similar π-electron densities as well as –SH groups that assist in strong binding to the semiconductor surface. It was found that 4-MP molecules demonstrated stronger Raman signals than 4-AT. The difference between the Raman signal detection between two molecules was based on the thermodynamic feasibility of the corresponding charge-transfer processes occurring between the valance band-edge of ZnO and LUMO of the molecule. Further, on exciting 4-MP loaded ZnO nanorod substrate with lower frequency, Raman signals were observed at pumping energy above 2 eV. These results presented a clear outline that existence of thermodynamically promoted charge-transfer process is essential to obtain high SERS effects with magnified molecular polarizability. Later on, using 4-MP/ZnO nanowire and nanocones configuration Yoon *et al.* reported appreciable Raman enhancement ~ 10^3 [88].

Copper oxide (CuO) is another material which has been explored for SERS activities [89, 90]. Guo *et al.* synthesized Cu_2O mesoporous spheres grown in form of a 3D cube shaped superstructure *via* recrystallization-induced self-assembly method [91]. The authors experienced the establishment of a large number of defects which resulted in the development of copper vacancies across the surface. The presence of surface vacancy defects (present in range ~0.5-0.9 eV beneath the conduction band edge of Cu_2O) promotes photo-induced electron transfer between Cu_2O and R6G probe molecule yielding an enhanced Raman scattering with enhancement factor ~8×10^5 and detection sensitivity up to nM concentration. In a recent study, the realization of composite substrate of CuO nanowires and Cu_2O has been performed for SERS active detection of 4-methylbenzenethiol [92]. In addition to SERS activity analysis, the authors also analysed the self-cleaning property of the SERS substrate *via* photocatalytic degradation in visible light range. As per the results obtained, optical illumination improved the substrate's reusability keeping 85% of SERS activity secured.

In comparison to metal oxide nanoparticles, pure noble metal nanoparticles has shown better performance. However, they also demonstrate weak stability due to air oxidation, other than this, thermal energy is also considered as prominent reason for decrease in their sensitivity with time. In order to rectify this issue, conjugates of noble metal nanoparticles (like Au) and MO's (like CuO, ZnO, TiO_2) has been developed. These conjugates demonstrate improved stability in contrast to single noble metals owing to a synergistic effect of noble metal and metal oxide. Particularly, in this case, the occurrence of nanostructured noble metal is in the vicinity of metal oxide which could improve the SERS activity up to certain level. For instance, TiO_2 nanograss decorated with Ag nanoparticles has shown better performance in the detection of R6G as well as 4-ATP molecules [93]. Likewise, Ag nanoparticle decorated ZnO hierarchical structures have been explored for SERS detection of organic pollutants like R6G, 4-chlorophenol, 2,4-dichlorophenoxyacetic acid [94].

Since decades, multiple novel strategies have been reported for improving the electromagnetic wave-molecule interaction behaviour of inorganic semiconductor micro and nanostructures for their better SERS performance. For instance, Alessandri *et al.* reported substantial improvement in Raman scattering enhancements for the colloidal assembly of recyclable SiO_2/TiO_2 core-shell microsphere resonators when methylene blue RRM are adsorbed on the exterior surface [95]. The synthesis of core shell microspheres was performed using a conformable coating of silica with a thin layer of amorphous TiO_2 shell using atomic layer deposition (ALD) followed by annealing at 700°C to form a crystalline anatase phase. Here in this study, the authors postulated their observation on the basis of a theoretical study reported by Ausman and Schatz [96]. The report by Ausman and Schatz counts on the presence of effective total internal reflection results in a momentary electromagnetic fields development which is further essential for generating local surface hot spots. The developed electromagnetic fields overlap with the Raman scattered radiation and result in an increased enhancement factor closely matching with the values classically detected for noble metals.

The search for improving inorganic SC's as alternatives to SERS active metallic materials, has gained a new direction with exploration carbon materials as potential candidates for SERS active performances. In this attempt, the focus has been made improving the electronic states of the material by employing π-conjugated delocalized electronic systems. As a proof of concept, various organic π-conjugated systems offering diverse molecular structures (like graphene) has been designed demonstrating diverse electrical as well as optical properties. In lieu to Raman enhancement, the first report based on graphene SERS active platform was presented by Liu *et al.* [97]. In this study, the author reported the SERS active testing of monolayered graphene (deposited on Si-SiO$_2$ substrate) after depositing common probe molecules namely, Rhodamine 6G, Crystal Violet, Protopphyrin IX, Phthalocyanine using solution-soaking and vacuum evaporation techniques. The SERS effect of graphene platform was compared with the SERS response of Si substrate (used as reference), as per the observed results, the graphene monolayer demonstrated significant Raman intensity with enhancement factor ~2-17 and detection limit ranging between $10^{-8} - 10^{-10}$ M. Interestingly, Raman intensities were found to be decreasing when the number of graphene layers was increased from monolayer to multilayer, this has been assured by additionally testing graphite sheet as SERS active platform where Raman signals were found to be either weak or absent. Thus, it can be ensured that this behaviour is found to be active due to differences in electronic structures of graphene with several layers which encourage doping effect when probe molecules are absorbed [97]. Employing graphene as SERS active material, detection of probe molecules like Protopphyrin IX and Rhodamine 6G has been performed efficiently [98]. Other than graphene, boron nitride (BN) has also been explored for SERS based detection of Rhodamine 6G [99]. In comparison to bulk BN, atomically thin BN film has shown better Raman results. Besides, BN offers high thermal stability (800°C), this allows it to effectively act as a reusable coating over noble metal layers for Raman enhancement [99 - 101].

MXenes is another type of material which has also been explored for SERS activity [102, 103]. It is generally composed of transition metal carbides and nitrides bearing general formula as $M_{n+1}X_nT_x$ where M, X and T_x refers to early transition metal, carbon or nitrogen and surface terminated functional group respectively. Similar to BN, MXenes are prominently used for SERS active detection of R6G molecule [104]. Moreover, they also actively contribute in detection of probe molecules like crystal violet, acid blue and methylene blue [104].

3.4. SOLID SERS SUBSTRATES

SERS activity has been extensively explored on self-organized nanoporous alumina (NPA) substrates sensitized with thin layers of Ag thin film or Ag nanoparticles along with porphyrin as a SERS-reporting molecule. Among the non-lithographic fabrication techniques for SERS substrates, self-organized NPA template with ordered hexagonal pores has attracted great attention on the account of self-organized pore profile with long-range ordering and tunability over dimension and various configuration shapes. A significant SERS enhancement was obtained on the Ag/NPA substrate on synchronizing laser excitation wavelength to the non-radiative surface plasmon resonance modes [116]. A major observation was an additional improvement in SERS signal on depositing analyte on the reverse side of Ag/NPA than on the silvered side [116]. Au layer has also been coated on NPA template to get SERS signal enhancement by almost 10^7 by controlling the thickness of the top Au layer and the bottom NPA template. Plasmonic nanopore arrays also provided a high degree of structural reproducibility, with high-throughput large-area capability for SERS detection that might generate advanced biological applications [117]. Some other examples of the NPA templates based SERS nanostructures include Ag nanowires of < 10 nm coupling gap distance [118], Ag tipped and SiO_2 nanorod clusters [9].

Another simple and low-cost method for fabricating a SERS active platform was reported by growing a double-layer stacked nanoporous Au films on SiO_2 substrate. Molecules like Rhodamine 6G, ascorbic acid, and 4-mercaptobenzoic acid could be detected with limit of detection down to 10^{-13}M (for Rhodamine 6G) could be achieved [119]. Recyclable SERS substrate using thin layer of Au-coated high density ZnO nanorods on Si substrate has also been reported to be highly efficient and low-cost. Recyclability was obtained by simple UV ray assisted cleaning cycles. ZnO being a high refractive index material assisted to concentrated the light and Au is already known as an excellent SERS active material [120].

Recently, a unique strategy has been introduced by Huang *et al.* where graphene quantum dots has been employed as SERS active material [121]. Here, the authors performed the realization of quantum dots (size ~2 nm) directly on Si-SiO_2 substrate using a quasi-equilibrium plasma-enhanced chemical vapor deposition (CVD)for the detection of rhodamine 6G. The platform showed enhancement factor output maximum up to 2.37×10^3 with detection sensitivity down to 10^{-9} M which is comparatively higher than the conventional graphene sheet SERS active platform. Typically for performing substrate assisted SERS analysis, the consideration of two-dimension SERS active material chiefly depends on their high surface area as well as their ability to get deposited on multiple substrates whether hard or flexible [104]. Among Si, glass and paper substrates, the flakes of Ti_2NT_x (MXenes) deposited on paper has shown high SERS performance (enhancement factor $\sim 10^{12}$) while detecting femtomolar concentration of rhodamine 6G [104].

Demerel *et al.* realized 3D nanostructured organic films (20 microns) of α,ω-Diperfluorohexyl-quarterthiophene (DFH-4T) on Si (100) substrate using physical vapor deposition (deposition rate >40 nm; source substrate distance ~7 cm) [122]. The testing of films for SERS activity was performed using methylene blue as probe molecule. Interestingly, in absence of additional plasmonic layer, DFH-4T film showed a unique Raman signal enhancement upto 3.4×10^3. In addition to this, the authors also realized organic metal (Au nanoparticles) hybrid active film comprising of nanostructured DFH-4T films on Si (100) substrate. The hybrid platform aided Raman signal enhancement of up to 10^{10} with analyte detection limit $<10^{-21}$ M.

Silicon (Si) is considered as the most favoured candidate material for developing multifunctional SERS analytical platforms [73]. Investigations of Si-based SERS substrates chiefly comprise Si nanostructures either in form of nanowires, nanocones *etc.* For instance, Cao *et al.* reported the formation of Si nanocones bearing height ~ 25 microns and apex tip ~5 nm [123]. Here, the authors observed that Raman scattering enhancement is dependent on the location as well as energy of the exciting laser falling on the surface of nanocone. Using such configuration, the highest enhancement value has been recorded ~ 10^3 at an excitation value ~ 785 nm [123]. Later on, Khorasaninejad *et al.* reported that Raman enhancement can be augmented as a function of morphology by observing maximum enhancement at excitation value ~833 nm for Si nanowires bearing height ~ 1.1 micron and 115 nm diameter [124, 125]. Besides, they also observed that by employing uniform arranged dense arrays of Si nanowires constructive coupling could be obtained. Wang *et al.* reported the formation of Si and germanium (Ge) nanowires using electroless etching [126]. In this work, the authors obtained enhanced Raman scattering for probe molecules like Rhodamine 6G, p-amino thiophenol. The report emphasizes the reason that enhanced Raman response is evidenced due to charge transfer efficiently occurring across the hydrogenated surfaces, whereas the occurrence of SiO_x capping layers leads to creation of an energy blockade that overpowers this effect. Typically, in these situations, doping plays a vital role in controlling the charge transfer direction [73]. Employing Si (n-type) results in the injection of electrons *via* semiconductor-to-molecule pathway, whereas employing Si (p-type) can result in acceptance of electrons from the adsorbed molecule [73]. In addition to 1D Si nanostructures, Si nanoparticles also offer manifold opportunities to experience enhanced SERS activity. The first report highlighting Si nanoparticle assisted Raman response was reported by Brueck *et al.* [127]. Later on, Liu *et al.* reported the development of Si electrodes [128]. Here authors demonstrated the existence of two major contributions: resonance Raman scattering and morphology-dependent resonance for small size particles (<100 nm) and large size particles (>100 nm), respectively. It is found that these two effects could perform better together thereby enhancing first-order Raman band. Moreover, the two effects have also been found efficient for acquiring information of the host substrate based on the analysis of the higher-order modes resulting from multiphonon scattering. Following this phenomenon, Rodriguez *et al.* developed polydisperse Si nanoparticles with an average size ranging from 200 nm to 400 nm [129]. Employing para-aminobenzoic acid as probe molecule, the authors experienced significant Raman enhancement which they reported as function morphology-dependent resonances.

According to literature history, the fabrication of Si nanoparticles is considered a critical task as the development of these particles involves using tools like standard nanolithography, whereas wet chemical methods are often characterized by high polydispersion [130]. This issue has been suitably resolved by practicing the laser ablation technique. Time dependent (Femtosecond) laser ablation of Si wafers tends to develop Si nanoparticles as well as facilitates their simultaneous allocation onto glass substrate [73]. Si nanoparticles prepared using laser ablation offered size in range of 100 nm to 200 nm [131]. The formed nanoparticles showed Raman enhancement dependent on the particle size and the maximum value was received with particle diameter ~155 nm. Followed by Si, silicon dioxide (SiO_2) is another favoured material which is widely explored for application in Raman spectroscopy. Generally, for SERS application, thin films comprising of microsphere configuration of SiO_2 have been practiced [132]. The SiO_2 microspheres introduce a uniform rise in the Raman signal that can't be compensated by a simple shift to higher objectives. Another advantage of these nanospheres is their facile deposition on the desired surface which overall amplifies the Raman response. Another uniqueness of this arrangement is the eased detachment of nanospheres from the substrate by simply washing with either water, ethanol *etc.* under mild conditions. For instance, using SiO_2 microsphere, the detection limit of methylene blue was enhanced by three order magnitude of molar concentration *i.e.*, 10^{-4} to 10^{-7} M [133]. On the other hand, the detection limit can be further improved up to 10^{-9} M when SiO_2 microspheres are coupled to SERS active substrate [133].

In context to the above-mentioned data, it is clear that the development of plasmonic and non-plasmonic materials / substrates is being done over a decade. Besides, in the last two years, a significant surge in understanding the insights of designing plasmonic and non-plasmonic nanostructures have been observed which are critically evaluated by Barbillon [105]. However, for readers concern and knowledge update, recent advances (2020-till date) in development of plasmonic and non-plasmonic SERS active materials /substrates has been shown in Table **1**.

Table 1. Recent advances in the development of plasmonic and non-plasmonic SERS active materials/substrates.

S. No.	SERS Active Material	Type	Key Aspects and Merit	Ref.
1.	SiO_2 @ Metal (W,Mo,Ti, Nb) carbide @Si	Plasmonic	• Active in detection of R6G; LoD ~10^{-8} M	[106]
2.	Au nano sunflowers	Plasmonic	• Label-free SERS based detection of damaged DNA *via* electro stimulus in apoptotic cancer cells • Development of highly dense surface charge and hot spot for capturing DNA is observed through corresponding SERS spectrum • SERS spectra indicated destruction of base as well as skeleton of cancer cell DNA (harmless to normal cells) at electro stimulus ~ 1.2 V ; 5 mins	[107]
3.	Ag colloids @ CaF_2 glass	Plasmonic	• Diagnostic applicability in detection of breast cancer cells through multivariate analysis (MVA) of SER spectrum • MVA based SERS analysis allowed the variation in healthy and infected sample with sensitivity ~90%; specificity ~ 89% and accuracy ~ 89%	[108]
4.	Au @ TiN_xcore shell nanostructures	Plasmonic	• High thermal and chemical stability • Significant field enhancement at the surface allowing suitable detection of R6G with concentration ~10^{-4} M.	[109]
5.	GO@Ag@TiO_2 @FTO/glass	-	• Improved signal reproducibility, longer-term stability, and higher detection sensitivity compared to pristine TiO_2@FTO/glass • Suitable for detection of CV (10^{-9}M), R6G (10^{-8} M), MG (10^{-7} M) upto 10 cycles of reusability	[109]
6.	GO-AuNR composite plasmonic paper	Plasmonic	• SERS results indicate evaluation of EF maximum ~ 10^7 with LoD for R6G~ 0.1 nM • Based on obtained EF and LoD values, the plasmonic paper shows potential to detect Mitoxantrone upto 5 µM.	[110]
7.	3-Dimensional graphene oxide quantum sensor	Non-Plasmonic	• The molecular sensitivity recorded outline minimum LoD upto ~10^{-15} M. • Real-time application indicate detection of disease biomarker and environmental contaminant Bisphenol-A upto concentration down to femtomolar	[111]

(Table 1) cont.....

S. No.	SERS Active Material	Type	Key Aspects and Merit	Ref.
8.	MoS$_2$ Nanoflowers	Non-Plasmonic	• Enhanced solar induced photodegradation of methyl orange, methyl blue, R6G, Oxytetracyclinehydrochloride • Significant SERS active sensing (involving chemical and electromagnetic charge transfer) of rhodamine B with LoD ~10^{-7} M	[111]
9.	Porous carbon nano-wires	Non-Plasmonic	• Strong charge transfer resonance allowed signal enhancement upto 10^{6}M • Absence of hot spots and no oxidation ensured high repeatability, durability and biocompatibility	[112]
10.	(Q structured) TiO$_2$	Non-Plasmonic	• Presence of oxygen vacancy strongly promotes SERS enhancement • Dimensional reduction of TiO$_2$ to quantum regime allowed EF elevation ~ 10^7 • Suitable selectivity for sensing of breast and cervical cancer cells lines	[113]
11.	MoTe$_2$ films	Non-Plasmonic	• Surface analyte (β-sitosterol) complex formation results in occurrence of chemical enhancement with EF ~ 10^4 • Emergence of low valued signal signs the confirmation of film homogeneity of 7-layered MoTe$_2$ film • Minimal signal loss upto 50 days ensured MoTe$_2$ film as suitable SERS platform for biosensing	[114]
12.	PSA antigen @ Graphene (monolayer) @ Si	Non-Plasmonic	• PSA aptamer shows strong bonding with graphene *via* π-π stacking interactions • Shifting in vibrational frequency of graphene indicates specific binding between PSA and aptamer • With the minimum detection limit ~ 0.01ng/ml, the entire analysis process takes ETA~ 30 min	[115]

CONCLUDING REMARKS

SERS is an extremely sensitive non-invasive technique capable of providing molecular fingerprints of Raman reporter molecules. Reproducible and scalable SERS enhancement is the need for futuristic materials research. Electromagnetic enhancement and chemical enhancement both mechanisms have led to the development of several plasmonic and non-plasmonic SERS nanomaterials. RRMs are further tagged with the available SERS nanomaterials to further enhance SERS signal. Silica shell has also offered higher stability to the unstable SERS materials and Raman molecules. Signal enhancement is also strongly dependent on the shape, size, symmetry and surface morphology of the SERS tag/probe. A significant interest has been gathering across the development of

SERS active nanoprobes typically for biomedical applications. Overall, the development of anatomically uniform SERS nanoprobes with potency to exhibit stable SERS signal intensity for quantitative detection of biomolecules and live cells is highly desired.

CONSENT FOR PUBLICATION

Not applicable.

CONFLICT OF INTEREST

The authors declare no conflict of interest, financial or otherwise.

ACKNOWLEDGEMENTS

Authors (RN) acknowledge Amity University, Uttar Pradesh Noida, India.

REFERENCES

[1]   Abbas A, Tian L, Morrissey JJ, Kharasch ED, Singamaneni S. Hot spot-localized artificial antibodies for label-free plasmonic biosensing. Adv Funct Mater 2013; 23(14): 1789-97.
[http://dx.doi.org/10.1002/adfm.201202370] [PMID: 24013481]

[2]   Bhunia SK, Zeiri L, Manna J, Nandi S, Jelinek R. Carbon-Dot/Silver-Nanoparticle Flexible SERS-Active Films. ACS Appl Mater Interfaces 2016; 8(38): 25637-43.
[http://dx.doi.org/10.1021/acsami.6b10945] [PMID: 27585236]

[3]   Dinish US, Balasundaram G, Chang YT, Olivo M. Actively targeted *in vivo* multiplex detection of intrinsic cancer biomarkers using biocompatible SERS nanotags. Sci Rep 2014; 4: 4075.
[http://dx.doi.org/10.1038/srep04075] [PMID: 24518045]

[4]   Huefner A, Kuan WL, Barker RA, Mahajan S. Intracellular SERS nanoprobes for distinction of different neuronal cell types. Nano Lett 2013; 13(6): 2463-70.
[http://dx.doi.org/10.1021/nl400448n] [PMID: 23638825]

[5]   Lim DK, Jeon KS, Kim HM, Nam JM, Suh YD. Nanogap-engineerable Raman-active nanodumbbells for single-molecule detection. Nat Mater 2010; 9(1): 60-7.
[http://dx.doi.org/10.1038/nmat2596] [PMID: 20010829]

[6]   Gao Jie, Sanchez-Purra Maria Huang, Hao Wang S. Synthesis of different-sized gold nanostars for Raman bioimaging and photothermal therapy in cancer nanotheranostics. Sci China Chem 2017; 60: 1219-29.
[http://dx.doi.org/10.1007/s11426-017-9088-x]

[7]   Bodelón G, Montes-García V, Fernández-López C, Pastoriza-Santos I, Pérez-Juste J, Liz-Marzán LM. Au@pNIPAM SERRS Tags for Multiplex Immunophenotyping Cellular Receptors and Imaging Tumor Cells. Small 2015; 11(33): 4149-57.
[http://dx.doi.org/10.1002/smll.201500269] [PMID: 25939486]

[8]   Jokerst JV, Miao Z, Zavaleta C, Cheng Z, Gambhir SS. Affibody-functionalized gold-silica nanoparticles for Raman molecular imaging of the epidermal growth factor receptor. Small 2011; 7(5): 625-33.
[http://dx.doi.org/10.1002/smll.201002291] [PMID: 21302357]

[9]   Schierhorn M, Lee SJ, Boettcher SW, *et al.* Metal-silica hybrid nanostructures for surface-enhanced raman spectroscopy. Adv Mater 2006; 18: 2829-32.

[http://dx.doi.org/10.1002/adma.200601254]

[10]   Wang R, Kim K, Choi N, *et al.* Highly sensitive detection of high-risk bacterial pathogens using SERS-based lateral flow assay strips. Sens Actuators B Chem 2018; 270: 72-9.
[http://dx.doi.org/10.1016/j.snb.2018.04.162]

[11]   He S, Chua J, Tan EKM, *et al.* Optimizing the SERS enhancement of a facile gold nanostar immobilized paper-based SERS substrate. RSC Advances 2017; 7: 16264-72.
[http://dx.doi.org/10.1039/C6RA28450G]

[12]   Kang H, Jeong S, Koh Y, *et al.* Direct identification of on-bead peptides using surface-enhanced Raman spectroscopic barcoding system for high-throughput bioanalysis. Sci Rep 2015; 5: 10144.
[http://dx.doi.org/10.1038/srep10144] [PMID: 26017924]

[13]   Yang Y, Zhong XL, Zhang Q, *et al.* The role of etching in the formation of Ag nanoplates with straight, curved and wavy edges and comparison of their SERS properties. Small 2014; 10(7): 1430-7.
[http://dx.doi.org/10.1002/smll.201302877] [PMID: 24339345]

[14]   Zeman EJ, Schatz GC. An accurate electromagnetic theory study of surface enhancement factors for silver, gold, copper, lithium, sodium, aluminum, gallium, indium, zinc, and cadmium. J Phys Chem 1987; 91: 634-43.
[http://dx.doi.org/10.1021/j100287a028]

[15]   Reguera J, Langer J, Jiménez de Aberasturi D, Liz-Marzán LM. Anisotropic metal nanoparticles for surface enhanced Raman scattering. Chem Soc Rev 2017; 46(13): 3866-85.
[http://dx.doi.org/10.1039/C7CS00158D] [PMID: 28447698]

[16]   Cao YWC, Jin R, Mirkin CA. Nanoparticles with Raman spectroscopic fingerprints for DNA and RNA detection. Science (80-) 2002; 297: 1536-40.
[http://dx.doi.org/10.1126/science.297.5586.1536]

[17]   Xia X, Li W, Zhang Y, Xia Y. Silica-coated dimers of silver nanospheres as surface-enhanced Raman scattering tags for imaging cancer cells. Interface Focus 2013; 3(3): 20120092.
[http://dx.doi.org/10.1098/rsfs.2012.0092] [PMID: 24427538]

[18]   Osinkina L, Lohmüller T, Jäckel F, *et al.* Synthesis of gold nanostar arrays as reliable, large-scale, homogeneous substrates for surface-enhanced Raman scattering imaging and spectroscopy. J Phys Chem C 2013; 117: 22198-202.
[http://dx.doi.org/10.1021/jp312149d]

[19]   Gellner M, Kömpe K, Schlücker S. Multiplexing with SERS labels using mixed SAMs of Raman reporter molecules. Anal Bioanal Chem 2009; 394(7): 1839-44.
[http://dx.doi.org/10.1007/s00216-009-2868-8] [PMID: 19543719]

[20]   Bianco A, Del Zoppo M, Zerbi G. Experimental CC stretching phonon dispersion curves and electron phonon coupling in polyene derivatives. J Chem Phys 2004; 120(3): 1450-7.
[http://dx.doi.org/10.1063/1.1633760] [PMID: 15268270]

[21]   Kearns H, Shand NC, Smith WE, Faulds K, Graham D. 1064 nm SERS of NIR active hollow gold nanotags. Phys Chem Chem Phys 2015; 17(3): 1980-6.
[http://dx.doi.org/10.1039/C4CP04281F] [PMID: 25475892]

[22]   Schütz M, Müller CI, Salehi M, Lambert C, Schlücker S. Design and synthesis of Raman reporter molecules for tissue imaging by immuno-SERS microscopy. J Biophotonics 2011; 4(6): 453-63.
[http://dx.doi.org/10.1002/jbio.201000116] [PMID: 21298811]

[23]   Sánchez-Purrà M, Roig-Solvas B, Rodriguez-Quijada C, Leonardo BM, Hamad-Schifferli K. Reporter Selection for Nanotags in Multiplexed Surface Enhanced Raman Spectroscopy Assays. ACS Omega 2018; 3(9): 10733-42.
[http://dx.doi.org/10.1021/acsomega.8b01499] [PMID: 30320250]

[24]   Li M, Cushing SK, Zhang J, *et al.* Shape-dependent surface-enhanced Raman scattering in gold-Raman probe-silica sandwiched nanoparticles for biocompatible applications. Nanotechnology 2012;

23(11): 115501.
[http://dx.doi.org/10.1088/0957-4484/23/11/115501] [PMID: 22383452]

[25]   Fleischmann M, Hendra PJ, McQuillan AJ. Raman spectra of pyridine adsorbed at a silver electrode. Chem Phys Lett 1974; 26: 163-6.
[http://dx.doi.org/10.1016/0009-2614(74)85388-1]

[26]   Jeanmaire DL, Van Duyne RP. Surface raman spectroelectrochemistry. Part I. Heterocyclic, aromatic, and aliphatic amines adsorbed on the anodized silver electrode. J Electroanal Chem 1977; 84: 1-20.
[http://dx.doi.org/10.1016/S0022-0728(77)80224-6]

[27]   Albrecht MG, Creighton JA. Anomalously Intense Raman Spectra of Pyridine at a Silver Electrode. J Am Chem Soc 1977; 99: 5215-7.
[http://dx.doi.org/10.1021/ja00457a071]

[28]   Mühlschlegel P, Eisler HJ, Martin OJF, *et al.* Applied physics: Resonant optical antennas. Science (80-) 2005; 308: 1607-9.

[29]   Wang HN, Vo-Dinh T. Multiplex detection of breast cancer biomarkers using plasmonic molecular sentinel nanoprobes. Nanotechnology 2009; 20(6): 065101.
[http://dx.doi.org/10.1088/0957-4484/20/6/065101] [PMID: 19417369]

[30]   Kneipp K, Wang Y, Kneipp H, *et al.* Single molecule detection using surface-enhanced raman scattering (SERS). Phys Rev Lett 1997; 78: 1667-70.
[http://dx.doi.org/10.1103/PhysRevLett.78.1667]

[31]   Nie S, Emory SR. Probing single molecules and single nanoparticles by surface-enhanced Raman scattering. Science (80-) 1997; 275: 1102-6.
[http://dx.doi.org/10.1126/science.275.5303.1102]

[32]   Seo S, Chang TW, Liu GL. 3D Plasmon Coupling Assisted Sers on Nanoparticle-Nanocup Array Hybrids. Sci Rep 2018; 8(1): 3002.
[http://dx.doi.org/10.1038/s41598-018-19256-7] [PMID: 29445092]

[33]   Sharma B, Frontiera RR, Henry AI, *et al.* SERS: Materials, applications, and the future. Mater Today 2012; 15: 16-25.
[http://dx.doi.org/10.1016/S1369-7021(12)70017-2]

[34]   Xia Y, Yang P, Sun Y, *et al.* One-dimensional nanostructures: Synthesis, characterization, and applications. Adv Mater 2003; 15: 353-89.
[http://dx.doi.org/10.1002/adma.200390087]

[35]   Link S, El-Sayed MA. Spectral Properties and Relaxation Dynamics of Surface Plasmon Electronic Oscillations in Gold and Silver Nanodots and Nanorods. J Phys Chem B 1999; 103: 8410-26.
[http://dx.doi.org/10.1021/jp9917648]

[36]   von Maltzahn G, Centrone A, Park JH, *et al.* SERS-coded cold nanorods as a multifunctional platform for densely multiplexed near-infrared imaging and photothermal heating. Adv Mater 2009; 21(31): 3175-80.
[http://dx.doi.org/10.1002/adma.200803464] [PMID: 20174478]

[37]   Wang Y, Yan B, Chen L. SERS tags: novel optical nanoprobes for bioanalysis. Chem Rev 2013; 113(3): 1391-428.
[http://dx.doi.org/10.1021/cr300120g] [PMID: 23273312]

[38]   Ciou SH, Cao YW, Huang HC, *et al.* SERS enhancement factors studies of silver nanoprism and spherical nanoparticle colloids in the presence of bromide ions. J Phys Chem C 2009; 113: 9520-5.
[http://dx.doi.org/10.1021/jp809687v]

[39]   Sun Y, Xia Y. Shape-controlled synthesis of gold and silver nanoparticles. Science (80-) 2002; 298: 2176-9.
[http://dx.doi.org/10.1126/science.1077229]

[40]   McLellan JM, Li ZY, Siekkinen AR, Xia Y. The SERS activity of a supported Ag nanocube strongly depends on its orientation relative to laser polarization. Nano Lett 2007; 7(4): 1013-7.
[http://dx.doi.org/10.1021/nl070157q] [PMID: 17375965]

[41]   Khoury CG, Vo-Dinh T. Gold nanostars for surface-enhanced Raman scattering: synthesis, characterization and optimization. J Phys Chem C 2008; 2008(112): 18849-59.
[http://dx.doi.org/10.1021/jp8054747] [PMID: 23977403]

[42]   Barbosa S, Agrawal A, Rodríguez-Lorenzo L, *et al.* Tuning size and sensing properties in colloidal gold nanostars. Langmuir 2010; 26(18): 14943-50.
[http://dx.doi.org/10.1021/la102559e] [PMID: 20804155]

[43]   Tian L, Gandra N, Singamaneni S. Monitoring controlled release of payload from gold nanocages using surface enhanced Raman scattering. ACS Nano 2013; 7(5): 4252-60.
[http://dx.doi.org/10.1021/nn400728t] [PMID: 23577650]

[44]   Yuan H, Liu Y, Fales AM, Li YL, Liu J, Vo-Dinh T. Quantitative surface-enhanced resonant Raman scattering multiplexing of biocompatible gold nanostars for *in vitro* and *ex vivo* detection. Anal Chem 2013; 85(1): 208-12.
[http://dx.doi.org/10.1021/ac302510g] [PMID: 23194068]

[45]   Xu P, Zhang B, Mack NH, *et al.* Synthesis of homogeneous silver nanosheet assemblies for surface enhanced Raman scattering applications. J Mater Chem 2010; 20: 7222.
[http://dx.doi.org/10.1039/c0jm01322f]

[46]   Niu W, Chua YAA, Zhang W, Huang H, Lu X. Highly Symmetric Gold Nanostars: Crystallographic Control and Surface-Enhanced Raman Scattering Property. J Am Chem Soc 2015; 137(33): 10460-3.
[http://dx.doi.org/10.1021/jacs.5b05321] [PMID: 26259023]

[47]   Amendola V, Scaramuzza S, Agnoli S, Polizzi S, Meneghetti M. Strong dependence of surface plasmon resonance and surface enhanced Raman scattering on the composition of Au-Fe nanoalloys. Nanoscale 2014; 6(3): 1423-33.
[http://dx.doi.org/10.1039/C3NR04995G] [PMID: 24309909]

[48]   Sun D, Zhang G, Jiang X, *et al.* Biogenic flower-shaped Au–Pd nanoparticles: synthesis, SERS detection and catalysis towards benzyl alcohol oxidation. J Mater Chem A Mater Energy Sustain 2014; 2: 1767-73.
[http://dx.doi.org/10.1039/C3TA13922K]

[49]   Lee YW, Kim NH, Lee KY, *et al.* Synthesis and characterization of flower-shaped porous Au-Pd alloy nanoparticles. J Phys Chem C 2008; 112: 6717-22.
[http://dx.doi.org/10.1021/jp710933d]

[50]   Li J-M, Yang Y, Qin D. Hollow nanocubes made of Ag–Au alloys for SERS detection with sensitivity of 10 −8 M for melamine. J Mater Chem C Mater Opt Electron Devices 2014; 2: 9934-40.
[http://dx.doi.org/10.1039/C4TC02004A]

[51]   Bao ZY, Lei DY, Jiang R, *et al.* Bifunctional Au@Pt core-shell nanostructures for *in situ* monitoring of catalytic reactions by surface-enhanced Raman scattering spectroscopy. Nanoscale 2014; 6(15): 9063-70.
[http://dx.doi.org/10.1039/C4NR00770K] [PMID: 24976250]

[52]   Xu X, Kim K, Li H, Fan DL. Ordered arrays of Raman nanosensors for ultrasensitive and location predictable biochemical detection. Adv Mater 2012; 24(40): 5457-63.
[http://dx.doi.org/10.1002/adma.201201820] [PMID: 22887635]

[53]   Xu X, Li H, Hasan D, *et al.* Near-field enhanced plasmonic-magnetic bifunctional nanotubes for single cell bioanalysis. Adv Funct Mater 2013; 23: 4332-8.
[http://dx.doi.org/10.1002/adfm.201203822]

[54]   Kong X, Yu Q, Zhang X, *et al.* Synthesis and application of surface enhanced Raman scattering (SERS) tags of Ag@SiO 2 core/shell nanoparticles in protein detection. J Mater Chem 2012; 22:

7767-74.
[http://dx.doi.org/10.1039/c2jm16397g]

[55]   Doering WE, Nie S. Spectroscopic tags using dye-embedded nanoparticles and surface-enhanced Raman scattering. Anal Chem 2003; 75(22): 6171-6.
[http://dx.doi.org/10.1021/ac034672u] [PMID: 14615997]

[56]   Mulvaney SP, Musick MD, Keating CD, *et al.* Glass-coated, analyte-tagged nanoparticles: A new tagging system based on detection with surface-enhanced Raman scattering. Langmuir 2003; 19: 4784-90.
[http://dx.doi.org/10.1021/la026706j]

[57]   Kong X, Yu Q, Lv Z, Du X. Tandem assays of protein and glucose with functionalized core/shell particles based on magnetic separation and surface-enhanced Raman scattering. Small 2013; 9(19): 3259-64.
[http://dx.doi.org/10.1002/smll.201203248] [PMID: 23585333]

[58]   Lim DK, Jeon KS, Hwang JH, *et al.* Highly uniform and reproducible surface-enhanced Raman scattering from DNA-tailorable nanoparticles with 1-nm interior gap. Nat Nanotechnol 2011; 6(7): 452-60.
[http://dx.doi.org/10.1038/nnano.2011.79] [PMID: 21623360]

[59]   Kang JW, So PTC, Dasari RR, Lim DK. High resolution live cell Raman imaging using subcellular organelle-targeting SERS-sensitive gold nanoparticles with highly narrow intra-nanogap. Nano Lett 2015; 15(3): 1766-72.
[http://dx.doi.org/10.1021/nl504444w] [PMID: 25646716]

[60]   Lin L, Zapata M, Xiong M, *et al.* Nanooptics of Plasmonic Nanomatryoshkas: Shrinking the Size of a Core-Shell Junction to Subnanometer. Nano Lett 2015; 15(10): 6419-28.
[http://dx.doi.org/10.1021/acs.nanolett.5b02931] [PMID: 26375710]

[61]   Lin L, Gu H, Ye J. Plasmonic multi-shell nanomatryoshka particles as highly tunable SERS tags with built-in reporters. Chem Commun (Camb) 2015; 51(100): 17740-3.
[http://dx.doi.org/10.1039/C5CC06599B] [PMID: 26490180]

[62]   Hu C, Shen J, Yan J, *et al.* Highly narrow nanogap-containing Au@Au core-shell SERS nanoparticles: size-dependent Raman enhancement and applications in cancer cell imaging. Nanoscale 2016; 8(4): 2090-6.
[http://dx.doi.org/10.1039/C5NR06919J] [PMID: 26701141]

[63]   Camden JP, Dieringer JA, Wang Y, *et al.* Probing the structure of single-molecule surface-enhanced Raman scattering hot spots. J Am Chem Soc 2008; 130(38): 12616-7.
[http://dx.doi.org/10.1021/ja8051427] [PMID: 18761451]

[64]   Chen G, Wang Y, Yang M, *et al.* Measuring ensemble-averaged surface-enhanced Raman scattering in the hotspots of colloidal nanoparticle dimers and trimers. J Am Chem Soc 2010; 132(11): 3644-5.
[http://dx.doi.org/10.1021/ja9090885] [PMID: 20196540]

[65]   Li X, Chen G, Yang L, *et al.* Multifunctional Au-Coated TiO2 Nanotube Arrays as Recyclable SERS Substrates for Multifold Organic Pollutants Detection. Adv Funct Mater 2010; 20: 2815-24.
[http://dx.doi.org/10.1002/adfm.201000792]

[66]   Joseph V, Engelbrekt C, Zhang J, Gernert U, Ulstrup J, Kneipp J. Characterizing the kinetics of nanoparticle-catalyzed reactions by surface-enhanced Raman scattering. Angew Chem Int Ed Engl 2012; 51(30): 7592-6.
[http://dx.doi.org/10.1002/anie.201203526] [PMID: 22806949]

[67]   Zhou Q, Li X, Fan Q, Zhang X, Zheng J. Charge transfer between metal nanoparticles interconnected with a functionalized molecule probed by surface-enhanced Raman spectroscopy. Angew Chem Int Ed 2006; 45(24): 3970-3.
[http://dx.doi.org/10.1002/anie.200504419] [PMID: 16683285]

[68] Qi D, Yan X, Wang L, Zhang J. Plasmon-free SERS self-monitoring of catalysis reaction on Au nanoclusters/TiO2 photonic microarray. Chem Commun (Camb) 2015; 51(42): 8813-6.
[http://dx.doi.org/10.1039/C5CC02468D] [PMID: 25920346]

[69] Haldavnekar R, Venkatakrishnan K, Tan B. Non plasmonic semiconductor quantum SERS probe as a pathway for *in vitro* cancer detection. Nat Commun 2018; 9(1): 3065.
[http://dx.doi.org/10.1038/s41467-018-05237-x] [PMID: 30076296]

[70] Musumeci A, Gosztola D, Schiller T, *et al.* SERS of semiconducting nanoparticles (TiO$_{(2)}$ hybrid composites). J Am Chem Soc 2009; 131(17): 6040-1.
[http://dx.doi.org/10.1021/ja808277u] [PMID: 19364105]

[71] Wang X, Shi W, She G, Mu L. Surface-Enhanced Raman Scattering (SERS) on transition metal and semiconductor nanostructures. Phys Chem Chem Phys 2012; 14(17): 5891-901.
[http://dx.doi.org/10.1039/c2cp40080d] [PMID: 22362151]

[72] Wang Y, Ruan W, Zhang J, *et al.* Direct observation of surface-enhanced Raman scattering in ZnO nanocrystals. J Raman Spectrosc 2009; 40: 1072-7.
[http://dx.doi.org/10.1002/jrs.2241]

[73] Alessandri I, Lombardi JR. Enhanced Raman Scattering with Dielectrics. Chem Rev 2016; 116(24): 14921-81.
[http://dx.doi.org/10.1021/acs.chemrev.6b00365] [PMID: 27739670]

[74] Demirel G, Usta H, Yilmaz M, *et al.* Surface-enhanced Raman spectroscopy (SERS): An adventure from plasmonic metals to organic semiconductors as SERS platforms. J Mater Chem C Mater Opt Electron Devices 2018; 6: 5314-35.
[http://dx.doi.org/10.1039/C8TC01168K]

[75] Yamada H, Yamamoto Y, Tani N. Surface-enhanced raman scattering (SERS) of adsorbed molecules on smooth surfaces of metals and a metal oxide. Chem Phys Lett 1982; 86: 397-400.
[http://dx.doi.org/10.1016/0009-2614(82)83531-8]

[76] Yamada H, Yamamoto Y. Surface enhanced Raman scattering (SERS) of chemisorbed species on various kinds of metals and semiconductors. Surf Sci 1983; 134: 71-90.
[http://dx.doi.org/10.1016/0039-6028(83)90312-6]

[77] Prakash J, Sun S, Swart HC, *et al.* Noble metals-TiO$_2$ nanocomposites: From fundamental mechanisms to photocatalysis, surface enhanced Raman scattering and antibacterial applications. Appl Mater Today 2018; 11: 82-135.
[http://dx.doi.org/10.1016/j.apmt.2018.02.002]

[78] Song W, Wang Y, Zhao B. Surface-Enhanced Raman Scattering of 4-Mercaptopyridine on the Surface of TiO$_2$ Nanofibers Coated with Ag Nanoparticles. J Phys Chem C 2007; 111: 12786-91.
[http://dx.doi.org/10.1021/jp073728b]

[79] Zhou Q, Meng G, Huang Q, *et al.* Ag-nanoparticles-decorated NiO-nanoflakes grafted Ni-nanorod arrays stuck out of porous AAO as effective SERS substrates. Phys Chem Chem Phys 2014; 16(8): 3686-92.
[http://dx.doi.org/10.1039/c3cp54119c] [PMID: 24419246]

[80] Khan MA, Hogan TP, Shanker B. Gold-coated zinc oxide nanowire-based substrate for surface-enhanced Raman spectroscopy. J Raman Spectrosc 2009; 40: 1539-45.
[http://dx.doi.org/10.1002/jrs.2296]

[81] Chen L, Luo L, Chen Z, *et al.* ZnO/Au composite nanoarrays as substrates for surface-enhanced Raman scattering detection. J Phys Chem C 2010; 114: 93-100.
[http://dx.doi.org/10.1021/jp908423v]

[82] Cheng C, Yan B, Wong SM, *et al.* Fabrication and SERS performance of silver-nanoparticle-decorated Si/ZnO nanotrees in ordered arrays. ACS Appl Mater Interfaces 2010; 2(7): 1824-8.
[http://dx.doi.org/10.1021/am100270b] [PMID: 20515071]

[83]   Gao M, Xing G, Yang J, *et al.* Zinc oxide nanotubes decorated with silver nanoparticles as an ultrasensitive substrate for surface-enhanced Raman scattering. Mikrochim Acta 2012; 179: 315-21.
[http://dx.doi.org/10.1007/s00604-012-0898-y]

[84]   Wang X, Shi W, Jin Z, *et al.* Remarkable SERS Activity Observed from Amorphous ZnO Nanocages. Angew Chem Int Ed Engl 2017; 56(33): 9851-5.
[http://dx.doi.org/10.1002/anie.201705187] [PMID: 28651039]

[85]   Sun Z, Zhao B, Lombardi JR. ZnO nanoparticle size-dependent excitation of surface Raman signal from adsorbed molecules: Observation of a charge-transfer resonance. Appl Phys Lett 2007; 91 Epub ahead of print.
[http://dx.doi.org/10.1063/1.2817529]

[86]   Sandhu A. Nanowire networks for macroelectronic devices. Nat Nanotechnol 2009; 1 Epub ahead of print.
[http://dx.doi.org/10.1038/nnano.2009.120]

[87]   Wang X, She G, Xu H, *et al.* The surface-enhanced Raman scattering from ZnO nanorod arrays and its application for chemosensors. Sens Actuators B Chem 2014; 193: 745-51.
[http://dx.doi.org/10.1016/j.snb.2013.11.097]

[88]   Shin HY, Shim EL, Choi YJ, Park JH, Yoon S. Giant enhancement of the Raman response due to one-dimensional ZnO nanostructures. Nanoscale 2014; 6(24): 14622-6.
[http://dx.doi.org/10.1039/C4NR04527K] [PMID: 25355156]

[89]   Fu S-Y, Hsu Y-K, Chen M-H, Chuang CJ, Chen YC, Lin YG. Silver-decorated hierarchical cuprous oxide micro/nanospheres as highly effective surface-enhanced Raman scattering substrates. Opt Express 2014; 22(12): 14617-24.
[http://dx.doi.org/10.1364/OE.22.014617] [PMID: 24977557]

[90]   Wang Y, Hu H, Jing S, *et al.* Enhanced Raman scattering as a probe for 4-mercaptopyridine surface-modified copper oxide nanocrystals. Anal Sci 2007; 23(7): 787-91.
[http://dx.doi.org/10.2116/analsci.23.787] [PMID: 17625318]

[91]   Lin J, Shang Y, Li X, *et al.* Ultrasensitive SERS Detection by Defect Engineering on Single Cu_2O Superstructure Particle. Adv Mater 2017; 29: 1604797.
[http://dx.doi.org/10.1002/adma.201604797]

[92]   Xu K, Yan H, Tan CF, *et al.* Hedgehog Inspired CuO Nanowires/Cu_2O Composites for Broadband Visible-Light-Driven Recyclable Surface Enhanced Raman Scattering. Adv Opt Mater 2018; 6: 1701167.
[http://dx.doi.org/10.1002/adom.201701167]

[93]   Xu SC, Zhang YX, Luo YY, *et al.* Ag-decorated TiO_2 nanograss for 3D SERS-active substrate with visible light self-cleaning and reactivation. Analyst (Lond) 2013; 138(16): 4519-25.
[http://dx.doi.org/10.1039/c3an00750b] [PMID: 23774192]

[94]   Shaik UP, Hamad S, Mohiddon MA, *et al.* Morphologically manipulated Ag/ZnO nanostructures as surface enhanced Raman scattering probes for explosives detection. J Appl Phys 2016; 119: 7. Epub ahead of print.
[http://dx.doi.org/10.1063/1.4943034]

[95]   Alessandri I. Enhancing Raman scattering without plasmons: unprecedented sensitivity achieved by TiO_2 shell-based resonators. J Am Chem Soc 2013; 135(15): 5541-4.
[http://dx.doi.org/10.1021/ja401666p] [PMID: 23560442]

[96]   Ausman LK, Schatz GC. Whispering-gallery mode resonators: Surface enhanced Raman scattering without plasmons. J Chem Phys 2008; 129(5): 054704. Epub ahead of print.
[http://dx.doi.org/10.1063/1.2961012] [PMID: 18698918]

[97]   Ling X, Xie L, Fang Y, *et al.* Can graphene be used as a substrate for Raman enhancement? Nano Lett 2010; 10(2): 553-61.

[http://dx.doi.org/10.1021/nl903414x] [PMID: 20039694]

[98] Xie L, Ling X, Fang Y, Zhang J, Liu Z. Graphene as a substrate to suppress fluorescence in resonance Raman spectroscopy. J Am Chem Soc 2009; 131(29): 9890-1.
[http://dx.doi.org/10.1021/ja9037593] [PMID: 19572745]

[99] Cai Q, Li LH, Yu Y, *et al.* Boron nitride nanosheets as improved and reusable substrates for gold nanoparticles enabled surface enhanced Raman spectroscopy. Phys Chem Chem Phys 2015; 17(12): 7761-6.
[http://dx.doi.org/10.1039/C5CP00532A] [PMID: 25714659]

[100] Wang J, Ma F, Liang W, *et al.* Optical, photonic and optoelectronic properties of graphene, h-NB and their hybrid materials. Nanophotonics 2017. Epub ahead of print.
[http://dx.doi.org/10.1515/nanoph-2017-0015]

[101] Cai Q, Mateti S, Yang W, *et al.* Boron Nitride Nanosheets Improve Sensitivity and Reusability of Surface-Enhanced Raman Spectroscopy. Angew Chem Int Ed Engl 2016; 55(29): 8405-9. Epub ahead of print.
[http://dx.doi.org/10.1002/anie.201600517] [PMID: 27112577]

[102] Naguib M, Kurtoglu M, Presser V, *et al.* Two-dimensional nanocrystals produced by exfoliation of Ti3 AlC2. Adv Mater 2011; 23(37): 4248-53.
[http://dx.doi.org/10.1002/adma.201102306] [PMID: 21861270]

[103] Huang K, Li Z, Lin J, Han G, Huang P. Two-dimensional transition metal carbides and nitrides (MXenes) for biomedical applications. Chem Soc Rev 2018; 47(14): 5109-24.
[http://dx.doi.org/10.1039/C7CS00838D] [PMID: 29667670]

[104] Sarycheva A, Makaryan T, Maleski K, *et al.* Two-Dimensional Titanium Carbide (MXene) as Surface-Enhanced Raman Scattering Substrate. J Phys Chem C 2017; 121: 19983-8.
[http://dx.doi.org/10.1021/acs.jpcc.7b08180]

[105] Barbillon G. Latest Novelties on Plasmonic and Non-Plasmonic Nanomaterials for SERS Sensing. Nanomaterials (Basel) 2020; 10(6): 1200.
[http://dx.doi.org/10.3390/nano10061200] [PMID: 32575470]

[106] Lan L, Fan X, Gao Y, *et al.* Plasmonic metal carbide SERS chips. J Mater Chem C Mater Opt Electron Devices 2020; 8: 14523-30.
[http://dx.doi.org/10.1039/D0TC03512B]

[107] Qi G, Wang D, Li C, Ma K, Zhang Y, Jin Y. Plasmonic SERS Au Nanosunflowers for Sensitive and Label-Free Diagnosis of DNA Base Damage in Stimulus-Induced Cell Apoptosis. Anal Chem 2020; 92(17): 11755-62.
[http://dx.doi.org/10.1021/acs.analchem.0c01799] [PMID: 32786448]

[108] Ştiufiuc GF, Toma V, Buse M, *et al.* Solid Plasmonic Substrates for Breast Cancer Detection by Means of SERS Analysis of Blood Plasma. Nanomaterials (Basel) 2020; 10(6): 1212.
[http://dx.doi.org/10.3390/nano10061212] [PMID: 32575924]

[109] Wang L, Wang S, Mei L, *et al.* Nanostructures Composed of Dual Plasmonic Materials Exhibiting High Thermal Stability and SERS Enhancement. Part Part Syst Charact 2021; 2000321.
[http://dx.doi.org/10.1002/ppsc.202000321]

[110] Ponlamuangdee K, Hornyak GL, Bora T, *et al.* Graphene oxide/gold nanorod plasmonic paper-a simple and cost-effective SERS substrate for anticancer drug analysis. New J Chem 2020; 44: 14087-94.
[http://dx.doi.org/10.1039/D0NJ02448A]

[111] Ganesh S, Venkatakrishnan K, Tan B. Tailoring carbon for single molecule detection – Broad spectrum 3D quantum sensor. Sens Actuators B Chem 2020; 317: 128216.
[http://dx.doi.org/10.1016/j.snb.2020.128216]

[112] Chen N, Xiao TH, Luo Z, *et al.* Porous carbon nanowire array for surface-enhanced Raman

spectroscopy. Nat Commun 2020; 11(1): 4772.
[http://dx.doi.org/10.1038/s41467-020-18590-7] [PMID: 32973145]

[113] Keshavarz M, Kassanos P, Tan B, *et al.* Metal-oxide surface-enhanced Raman biosensor template towards point-of-care EGFR detection and cancer diagnostics. Nanoscale Horiz 2020; 5: 294-307.
[http://dx.doi.org/10.1039/C9NH00590K]

[114] Fraser JP, Postnikov P, Miliutina E, *et al.* Application of a 2D Molybdenum Telluride in SERS Detection of Biorelevant Molecules. ACS Appl Mater Interfaces 2020; 12(42): 47774-83.
[http://dx.doi.org/10.1021/acsami.0c11231] [PMID: 32985181]

[115] Liu S, Huo Y, Bai J, *et al.* Rapid and sensitive detection of prostate-specific antigen *via* label-free frequency shift Raman of sensing graphene. Biosens Bioelectron 2020; 158: 112184.
[http://dx.doi.org/10.1016/j.bios.2020.112184] [PMID: 32275212]

[116] Terekhov SN, Kachan SM, Panarin AY, Mojzes P. Surface-enhanced Raman scattering on silvered porous alumina templates: role of multipolar surface plasmon resonant modes. Phys Chem Chem Phys 2015; 17(47): 31780-9.
[http://dx.doi.org/10.1039/C5CP04197J] [PMID: 26563558]

[117] Choi D, Choi Y, Hong S, Kang T, Lee LP. Self-organized hexagonal-nanopore SERS array. Small 2010; 6(16): 1741-4.
[http://dx.doi.org/10.1002/smll.200901937] [PMID: 20333691]

[118] Wang HH, Liu CY, Wu S. Highly raman-enhancing substrates based on silver nanoparticle arrays with tunable sub-10 nm gaps. Adv Mater 2006; 18: 491-5.
[http://dx.doi.org/10.1002/adma.200501875]

[119] Tang L, Liu Y, Liu G, *et al.* A Novel SERS Substrate Platform: Spatially Stacking Plasmonic Hotspots Films. Nanoscale Res Lett 2019; 14(1): 94.
[http://dx.doi.org/10.1186/s11671-019-2928-8] [PMID: 30868395]

[120] Sinha G, Depero LE, Alessandri I. Recyclable SERS substrates based on Au-coated ZnO nanorods. ACS Appl Mater Interfaces 2011; 3(7): 2557-63.
[http://dx.doi.org/10.1021/am200396n] [PMID: 21634790]

[121] Liu D, Chen X, Hu Y, *et al.* Raman enhancement on ultra-clean graphene quantum dots produced by quasi-equilibrium plasma-enhanced chemical vapor deposition. Nat Commun 2018; 9(1): 193. Epub ahead of print.
[http://dx.doi.org/10.1038/s41467-017-02627-5] [PMID: 29335471]

[122] Yilmaz M, Babur E, Ozdemir M, *et al.* Nanostructured organic semiconductor films for molecular detection with surface-enhanced Raman spectroscopy. Nat Mater 2017; 16(9): 918-24.
[http://dx.doi.org/10.1038/nmat4957] [PMID: 28783157]

[123] Cao L, Nabet B, Spanier JE. Enhanced Raman scattering from individual semiconductor nanocones and nanowires. Phys Rev Lett 2006; 96(15): 157402. Epub ahead of print.
[http://dx.doi.org/10.1103/PhysRevLett.96.157402] [PMID: 16712194]

[124] Khorasaninejad M, Walia J, Saini SS. Enhanced first-order Raman scattering from arrays of vertical silicon nanowires. Nanotechnology 2012; 23(27): 275706. Epub ahead of print.
[http://dx.doi.org/10.1088/0957-4484/23/27/275706] [PMID: 22710724]

[125] Khorasaninejad M, Dhindsa N, Walia J, *et al.* Highly enhanced Raman scattering from coupled vertical silicon nanowire arrays. Appl Phys Lett 2012; 101: 22. Epub ahead of print.
[http://dx.doi.org/10.1063/1.4764057]

[126] Wang X, Shi W, She G, Mu L. Using Si and Ge nanostructures as substrates for surface-enhanced Raman scattering based on photoinduced charge transfer mechanism. J Am Chem Soc 2011; 133(41): 16518-23.
[http://dx.doi.org/10.1021/ja2057874] [PMID: 21939241]

[127] Zaidi SH, Chu AS, Brueck SRJ. Optical properties of nanoscale, one-dimensional silicon grating

structures. J Appl Phys 1996; 80: 6997-7008.
[http://dx.doi.org/10.1063/1.363774]

[128] Liu FM, Ren B, Wu JH, *et al.* Enhanced-Raman scattering from silicon nanoparticle substrates. Chem Phys Lett 2003; 382: 502-7.
[http://dx.doi.org/10.1016/j.cplett.2003.10.102]

[129] Rodriguez I, Shi L, Lu X, Korgel BA, Alvarez-Puebla RA, Meseguer F. Silicon nanoparticles as Raman scattering enhancers. Nanoscale 2014; 6(11): 5666-70.
[http://dx.doi.org/10.1039/C4NR00593G] [PMID: 24764023]

[130] Fenollosa R, Meseguer F, Tymczenko M. Silicon Colloids: From Microcavities to Photonic Sponges. Adv Mater 2008; 20: 95-8.
[http://dx.doi.org/10.1002/adma.200701589]

[131] Kuznetsov AI, Kiyan R, Chichkov BN. Laser fabrication of 2D and 3D metal nanoparticle structures and arrays. Opt Express 2010; 18(20): 21198-203.
[http://dx.doi.org/10.1364/OE.18.021198] [PMID: 20941016]

[132] Dmitriev PA, Baranov DG, Milichko VA, *et al.* Resonant Raman scattering from silicon nanoparticles enhanced by magnetic response. Nanoscale 2016; 8(18): 9721-6.
[http://dx.doi.org/10.1039/C5NR07965A] [PMID: 27113352]

[133] Alessandri I, Bontempi N, Depero LE. Colloidal lenses as universal Raman scattering enhancers. RSC Advances 2014; 4: 38152-8.
[http://dx.doi.org/10.1039/C4RA07198K]

CHAPTER 4

Application of Raman Spectroscopy in Amyloid Research

Sandip Dolui[1], Kaushik Bera[1], Animesh Mondal[1], Krishnendu Khamaru[1] and **Nakul C Maiti[1,*]**

[1] Structural Biology and Bioinformatics Division, Indian Institute of Chemical Biology, Council of Scientific and Industrial Research, 4, Raja S.C. Mullick Road, Kolkata 700032, India

Abstract: Raman scattering spectroscopy was discovered in 1928 by CV Raman and KS Krishnan. The technique has developed enormously and it is becoming useful in many ways for studying biochemical events and structural intricacy of biological macromolecules such as RNA, DNA, protein and their assemblies. The focus of this review is on the recent application of Raman spectroscopy to research achievements of protein aggregation and fibrillation of several proteins and peptides. Particularly we analyzed the protein secondary structure of different assembly structures captured in the fibril formation pathway with a particular focus on oligomeric intermediate which is believed now to be most cytotoxic. This intermediate structure attains characteristic morphological features while the constituent protein may/may not differ much in their secondary and tertiary structure in the native physiological conditions. Conformation states of proteins in the oligomeric state obtained by Raman spectroscopic analysis, particularly aid in comprehending the structure of the oligomer and overall mechanisms of fibrillation and amyloid formation. It has been established that the backbone amide band and side-chain vibrations of amino acid residues present in protein molecules largely affected the fibril formation pathway and it follows a concerted reaction pathway, *i.e.* the protein molecule transform into β- sheet rich amyloid fibril *via* formation of an oligomeric intermediate. However, the intriguing and interesting fact is that the proteins maintain/attain some helical pattern in the oligomeric step. Raman analyses established the distribution of residues in both helical and β-domain, possesses similarity with molten globule like structure. However, in the fibrillar state, the protein backbone attains anti-parallel β-sheet structure and several side-chain residues may be exposed on the surface of the protein and it is evidenced in the Raman spectra of the fibrils. The review particularly focuses on the aggregation and amyloid-like fibril formation of hen egg-white lysozyme (HEWL) and discusses different aspects of fibril formation mechanisms based on Raman spectroscopic data analysis.

*** Corresponding author Nakul C Maiti:** Structural Biology and Bioinformatics Division, Indian Institute of Chemical Biology, Council of Scientific and Industrial Research, 4, Raja S.C. Mullick Road, Kolkata 700032, India; Tel: 03324995940; E-mail: ncmaiti@iicb.res.in

Keywords: Aggregates, Amyloid, Diseases, Human, Oligomer, Protein, Raman, Spectroscopy Lysozyme.

4.1. INTRODUCTION

4.1.1. Basics of Raman Spectroscopy

Spectroscopic techniques such as the absorption, emission, or scattering of electromagnetic radiation are methods based on the interaction of light with atoms or molecules. A molecule is defined as two or more bonded atoms that are involved in various types of motion and intermolecular interactions, and so contain different forms of energy. It can interact with incoming light, electromagnetic radiation such as visible light (photon), and a transition in its energy state (excited state) thus may occur. Most of the scattered photons from the molecule remain with the same energy and this is called elastic (Rayleigh) scattering. Although the interaction of the molecule with the photon, an exchange of energy equivalent to a quantum of vibration (related to the molecule) energy may also occur between the two (Fig. **1**). This causes differences in the energy of some scattered light by vibrational (also for rotational) energy differences of the molecule. This kind of inelastic scattering is referred as Raman scattering and the phenomenon is known as the Raman effect [1] (Fig. **2**). The energy (frequency) difference is defined as the Raman shifts and is measured in a unit of cm^{-1}. The Raman scattering is, however, a relatively very weak process with a probability that one in ~10^7 photons scatters in-elastically. It was discovered by Sir C.V. Raman in 1928 [1]. The molecules are excited with any single wavelength of light, from deep UV to near infrared, to produce a significant Raman scattering.

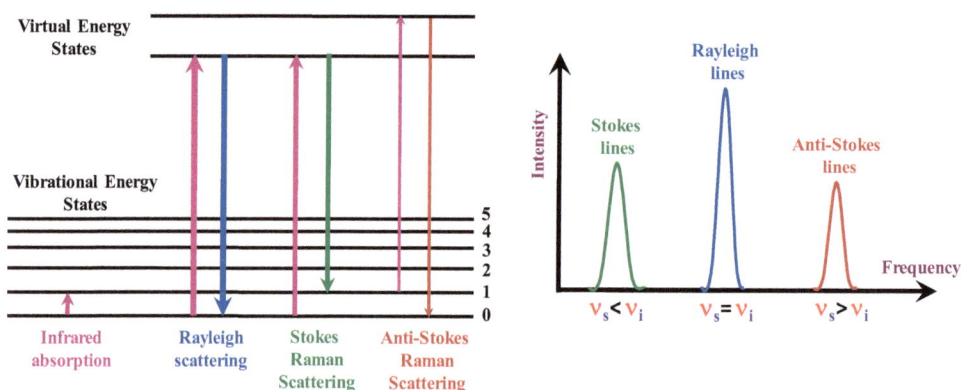

Fig. (1). A schematic diagram of the energy transitions involved in Infra-red absorption, Rayleigh scattering and Raman scattering (Stokes & anti-Stokes). The numbers (0, 1, 2...) represents a vibrational quantum number of different vibrational energy levels under the electronic ground state. B. Schematic spectral representation of Rayleigh lines & Raman lines (Stokes & anti-Stokes).

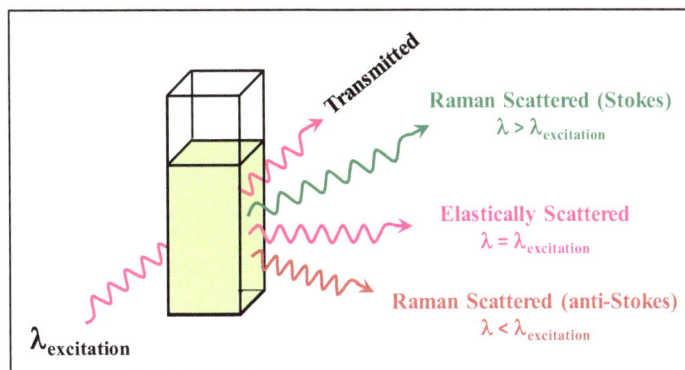

Fig. (2). Elastic and Raman (Stokes & anti-Stokes) scattering processes in a solution.

Thus the Raman scattering is a two photon process involving the interaction of light with molecule (Fig. **2**). The molecule in a monochromatic light source, such as under a laser beam; it first absorbs a photon and then transits to a virtual state followed by the emission of a photon that has exchanged energy with the molecular vibration [2 - 5]. In case the energy of the emitted photon does not alter and remains the same as the incoming photon, it is referred to as Rayleigh scattering, which is millions of times higher in intensity than that of Raman scattering. If the energy of the emitted photon is weaker than the incoming photon, it suggests that the molecule in the ground state obtained energy from the photon and transited to the excited state. This denotes the Stokes Raman scattering, which is measured in a conventional Raman experiment. On the other side, if the incoming photon gains energy from the molecule that is already in an excited vibrational state, it will radiate a stronger photon while the excited molecule falls back to the ground state; this process is called anti-Stokes Raman scattering. The intensity of anti-Stokes Raman scattering is weaker comparative to Stokes Raman scattering due to the smaller population in the excited vibrational state compared to the populations in the ground state according to the Boltzmann distribution law.

The Raman shifts, the Stokes and anti-Stokes features are a direct measure of the vibrational energies of the molecule. Thus the measurement of frequency (wavelength) of an inelastically scattered photons in the Raman spectrum are associated with different mode (energy) of molecular vibrations and the Raman intensity (scattering probability) depends on the polarizability of the oscillating bond (molecule). The molecular configuration and bonding pattern, therefore, are well reflected in the Raman spectrum. It can provide information related to the structure and chemical composition of a molecule's in their different state of existence. Protein molecules are also very sensitive and produce wonderful Raman signature [6, 7] that provides inherent molecular details related to its structure and function.

4.1.2. Raman Spectra of Proteins and Peptides

The polypeptide backbone of the protein molecule is mainly made up of 20 natural amino acids connected to each other by amide/peptide (CONH) linkages. The molecular bonds within the protein such as, C=O, C=C, N-H C-C, S–S, C–S, S–H are rich in π-electron and thus, have larger polarizability and produce good Raman signals [4, 6 - 17]. The localized vibrations of these bonds generally produce intense Raman bands (Fig. **3**). Aliphatic and other non-romantic side chains produce weak Raman bands.

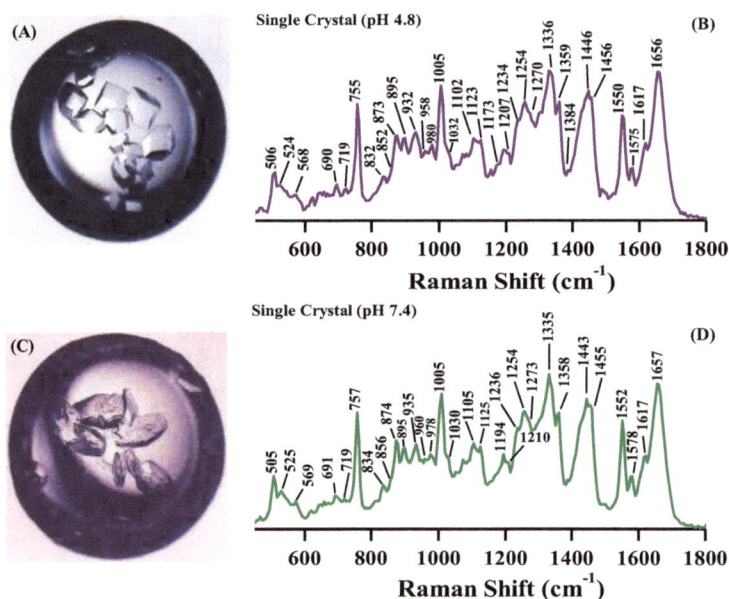

Fig. (3). Microscopic images of hen egg white lysozyme (HEWL) crystals grown in two different pHs (pH 4.8 and pH 7.4) and their Raman spectra. A. Single crystal of lysozyme pH 4.8, produced in the droplets of concentration 45mg/ml in hanging drops methods. (B) Raman spectra of lysozyme crystal obtained by 532 nm laser excitation. (C) single crystal of lysozyme at pH 7.4 and (D) Raman spectra of the crystal. Raman experimental condition was same as in panel B.

Most of the Raman bands for a protein molecule are in the spectral window of (400-2000) cm^{-1} (Fig. **3**) [18 - 23]. Localized groups such as OH, NH, SH produce Raman frequency in the high energy region (2500-4000) cm^{-1}. However, based on the origin, the bands in the Raman spectra of protein may be conveniently divided into the polypeptide backbone modes (amide mode) and modes associated with the side chain residues. The amide A and amide B bands appear at high frequency region [24]. Both the bands are associated with the NH stretching and appear at~ 3500cm⁻¹and 3100 cm⁻¹ [6, 15]. These two bands are very weak in the dispersive Raman spectra of the protein. Amide I mode appear in the frequency range 1640cm⁻¹ to 1700 cm⁻¹ [7, 8, 10, 25 - 30]. It arises mainly from the stretching vibration of C=O; however, a small amount of out-of-phase C-N stretching also contributes to this band. It is very sensitive to the backbone conformation and is less affected by the side chains of the amino acid residues. The amide II vibrations produce Raman signature at ~1550 cm⁻¹; it consists of an out-of-phase combination of C-N stretching and N-H bending motions. Amide III band originates from C-N stretching coupled with N-H bending vibrations and appears at(1200-1340) cm⁻¹ [26, 27, 31, 32].

The Raman band associated with the amide I mode is mostly used to estimate the secondary structure of proteins [16, 33 - 35]. The C=O bond can form a hydrogen bond with the CONH groups of another peptide bond at different positions in the same chain or another polypeptide backbone. In a α-helical protein the hydrogen bonds in the polypeptide backbone are created between the C=O group and the NH on the same chain and Raman band characteristics band appear at (1650-1662) cm⁻¹. The band appears at 1670 cm⁻¹in the β- sheet structure and the hydrogen bond is created with the C=O and NH groups from nearby chains which are arranged either in parallel or anti-parallel modes. In case of loose β-sheet and disordered structure, it is observed at about 1640 cm⁻¹. Similarly, amide III bands in the frequency range of (1200-1350) cm⁻¹ also provide complementary information about the protein backbone conformations. The amide III mode of vibration that appears at (1230-1235) cm⁻¹ is assigned to the β-sheet conformation, band near (1240-1250) cm⁻¹for random coil geometry and (1260-1300)cm⁻¹for α-helical conformation [8, 9, 34, 36, 37]. Asher's group established important protein conformation based on Amide III and amide II bands of Raman spectra obtained upon UV laser excitation [32].

The Raman region between (900-1000) cm⁻¹ also contain typical protein conformation bands [26, 38]. The presence of backbone helical signature appears at (930-940) cm⁻¹, β-stand signature at (980-998) cm⁻¹and disordered segments at 960 cm⁻¹ [10, 38]. The Cα-H bending mode appears at 1380-1400 cm⁻¹. The band intensity is significantly enhanced in ultraviolet resonance Raman spectra and it is also very much sensitive to protein backbone conformation. The β-sheet conformation gives the band at ~1396 cm⁻¹ and it shifts to ~1387 cm⁻¹ for disordered (PPII) conformation. However, in the α-helical conformation, the Cα-H bending mixes with and N-H vibration and is coupled; consequently, the band intensity becomes weaker. The C-H stretching band at 1450 cm⁻¹ was assigned to CH₂-, CH₃- deformation and CH₂- scissoring motions for overall protein backbone and side chain deformations. Localized groups such as NH, OH, SH give frequency in the high energy region (2500-4000) cm^{-1}. Due to intra/intermolecular H-bond formation, the Raman bands associated with NH, OH groups often become broad. The vibrations linked to these bonds are also used to gather information related to protein tertiary structure and functional activity involving the OH group.

Raman marker bands denoting aromatic side chains of residues in proteins are relatively strong [9, 37, 39 - 42]. The Raman frequency and intensity of the Trp and Tyr bands have been rigorously studied by normal mode calculation and experiments. The microenvironment and surroundings of phenyl/indole rings, hydrophobicity and hydrogen bonding pattern also contribute to the position and intensity of these bands. A prominent peak at 755 cm^{-1} denotes in-phase symmetric breathings vibration of benzene and pyrrole (indole ring) of the Trp residue. The tryptophan bands near 1340 and 1360 cm^{-1} are regarded as the two-component bands of a Fermi doublet [43]. These two bands are linked to the fermi resonances between the N-C stretching band at 1340 cm^{-1} and a combination band of out-of-plane deformation modes involving the pyrrole and benzene ring of tryptophan residue. The environmental surroundings of Trp residues could be assumed by considering the intensity ratio(I_{1360}/I_{1340}) of the doublet bands [40]. The accepted notion is that the intensity ratio (I_{1360}/I_{1340}) greater than unity indicates a net hydrophobic microenvironment of the indole ring and thus, in a more hydrophobic environment, the relative intensity of the 1360 cm^{-1} increases compared to 1340 cm^{-1}. Similarly the tyrosine residues exhibited vibrations bands at ~832, ~855 ~1207 and ~1617 cm^{-1} in the protein samples. The previous investigations demonstrated that the relative Raman intensity ratio (I_{850}/I_{830}) is strongly dependent upon the hydrogen-bonding nature of the tyrosyl OH group [44 - 47]. When it acts as a strong H-bond donor the ratio becomes ~0.4. Conversely, if the phenoxyl oxygen acts as the acceptor of a strong H-bond the ratio becomes 2.5. If it acts as both a donor and an acceptor of hydrogen bonds, the ratio becomes about 1.25.

The majority of proteins contain free sulfhydryl (S-H) and disulfide(-S-S-) bonds. The free S-H groups, found in proteins give very weak band in the visible Raman spectrum [6, 48]. However, the S-H groups in proteins can be detected with high sensitive Raman spectroscopy in the frequency interval between 2500 and 2700 cm^{-1}. The position and intensity of the band also vary depending on its involvement in H-bond formation [6]. It is observed that the frequency of S-H group gets lowered by 25-60 cm^{-1} if it acts as a strong H-bond donor. Raman markers of disulfide bonds occur in the low frequency region between 700 and 450 cm^{-1} [15, 37, 48 - 50]. The protein disulfide (-S-S-) stretching vibration band is very much depends on its geometry. The relative intensity of this band could be used as a qualitative measure of the S-S bond and native tertiary structure of the protein. Raman spectra of protein HEWL is shown in Fig. (**3**) and the band assignment was based on previous investigations [7, 19, 30, 47, 51, 52]. The list of bands is summarized in Table **1**. Four disulfide bridges (S-S bonds) provide large stability to the protein tertiary structure. Three of the disulfide bonds, Cys6-Cys127, Cys30-Cys115 and Cys64-Cys80, preferred gauche-gauche-gauche (ggg) conformation and the Raman band appeared at 506 cm^{-1} in the Raman spectra of lysozyme crystal.

Table 1. Raman vibrational bands (cm^{-1}) of lysozyme (HEWL) as monomer, oligomer and fibril states.

Crystal	Monomer	Oligomer	Fiber	Modes of Raman vibration
505	-	-	-	S-S
525	-	-	-	S-S
691	-	-	-	C-S, Met
757	762	762	762	Trp
834,856	834,855	834,854	834,849	Tyr
874	879	878	883	Trp
895,935	900,932	903,932	900,935	C_α-C stretching (α-helix)
960,978	958	957	987	Skeletal β- strand, C_α-C stretching
1005	1005	1005	1007	Phe
1030	1028	1029	1033	Phe
1105,1125	1075,1107,1128	1082,1105,1127	1082,1105,1127	CH_2 symmetric rock+ Cα-C stretching
1210	1202	1200	1208	Tyr, Phe
1236	1235	1234	1231	Type II β-turn, strand, Amide III
1254	1250	1245,1258	1244,1257	Poly-l-proline, Amide III
1273	1260	-	-	α-helix, Amide III
1335,1358	1337,1358	1338,1358	1337,1359	Trp, C_α-H deformation,
1443,1455	1449	1452	1442,1456	CH_2,CH_3& CH deformation and scissoring
1552	1551	1553	1552	Trp
1578	1580	1578	1582	Trp
1617	1617	1618	1618	Tyr aromatic vibration
1657	1658	1659	-	Amide I, α- helix
-	-	-	1673	Amide I, β-sheet

4.1.3. Raman Techniques in Bio-Analysis

Raman spectroscopy was discovered in 1928, however it was not applied in biology for many years. With the discovery of lasers in the 1960s and in subsequent years, availability of several high quality optical sources including UV and visible lasers revolutionise the Raman spectroscopic field [53 - 58]. In addition, with remarkable advancement and innovations in optical and spectroscopic technologies such as CCD detector, holographic notch filter, fiber optical probe, grating technology Raman spectroscopy slowly entered different fields including molecular biology [2, 4, 30, 53, 59 - 66]. The recent development of Raman spectrometershave dramatically increased the instrument sensitivity. The combination of Raman spectroscopy with atomic force microscopy (AFM) or scanning electron microscopy (SEM) has further improved the method to probe the structural aspects of proteins and their assembly structure such as amyloid of fibrilar aggregates [61, 67 - 73]. With the advent of confocal microscopic system Raman spectroscopy emerged as a novel tool for investigating protein aggregation and associated event of amyloidosis and cell imaging with great special resolution [74 - 77]. Surface-enhanced Raman spectroscopy (SERS) is also becoming a powerful analytical method for investigating macroscopic detail of the amyloid aggregates [78 - 81]. Based on SERS methods are developed that can probe very small quantity of samples with high spatial and spectral resolution [82]. For instance scanning near-field microscopy (SNOM) and tip-enhanced Raman spectroscopy (TERS) can provide nanometre scale spatial [78, 83]. This advantage makes it an essential and unique technique in the field of biology and materials science to explore the molecular basis of structure and function. Using TERS and SNOM even information about the amino acid composition on the surface amyloid fibrils can be inferred.

Raman microscopic based methods are capable to probe small protein samples and chemical mapping, by mainly combining Raman spectrometer with an optical microscope. Most Raman microscopes including the confocal systems in general integrate a dispersive Raman spectrometer along with optical microscope which uses optical fiber for transmission of the Raman signal coming from optical microscopic stage to the analyzer. However, standalone open configured systems are also provide great spatial arrangements. The microscopic arrangement in the confocal Raman setup is either upright or inverted alignment to collect the dispersive Raman spectra of protein samples. The machines are often equipped with more than one excitation laser sources spanning from UV-visible to near infrared regions .To reduce photo degradation of the sample near-IR laser is preferred as an excitation source to record the protein samples. In this range a 785 nm laser line is preferred for protein samples to overcome background fluorescence. However, shorter wavelength laser lines such as at 633 or 532 nm are useful for samples that scatter very weakly and also their use is very common for samples which produce very weak fluorescence. To obtain a good quality of spectra of aggregated protein samples a Raman microscopy-based method known as drop-coating deposition Raman (DCDR) is developed [35, 84 - 86]. Earlier investigation was performed to study amyloid fibril formation of α-synuclein [35, 63], lysozyme [42] and amyloid β [36] peptides and the structural changes during protein aggregation.

Different techniques such as visible dispersive Raman and ultraviolet resonance Raman (UVRR) spectroscopy are applied for studies protein structures, protein-protein interaction and their activities. Both the techniques are also used to investigate native amyloid protein aggregates in different conditions. Based on selective laser excitation resonance and engaging deep UV resonance Raman (DUVRR) spectroscopy, Ashers group elucidate fibril core structural organization and provided the psi (Ψ) dihedral angle of the protein backbone [4, 22, 31, 32, 87, 88]. Thus, the resonance Raman method can provide the information of the changes in protein secondary and tertiary structure and their surrounding during the protein aggregation [10, 45, 89]. The amide II band remains very weak and overlapped with bands from other localized group vibrations. However, the band becomes very prominent in deep UV resonance Raman spectrum [87]. Coupling this UV-resonance experiment with the H/D exchange experiments the Asher group could able to quantify the amount of disordered and very compact hydrophobic structural component which is inaccessible to solvent the in amyloid fibrils [4, 30, 88, 90, 91]. In addition to the determination of the global structural changes from analysis of the backbone vibration, recent applications of site-specific approaches, such as isotopic labelling of specific amino acid has greatly expanded the application of Raman spectroscopy to probe conformational changes with residue specific resolution [92].

Depending on the sample condition and purposes different Raman technique are designed [11, 33, 73, 78, 79, 89, 92 - 97]. However, Raman spectra data recorded using these methods contain a large amount of information and methods are developed to extract and analyze the important information present in the acquired spectra. Recent multivarient analysis method uses the Raman spectrum more efficiently and provides relevant information directly from the sample spectrum with/without the blank corrections, resolution enhancement, or subjective interpretations. In antibody aggregation studies, in small molecule binding to enzyme crystal common methods include making difference spectra and principal component analysis (PCA) [11, 51]. Also often carried out partial least squares (PLS) and multivariate curve resolution (MCR) analysis [94, 98 - 101]. Raman data analysis also includes aspects spectral variance analysis and spectral deconvolution/curve fitting and quantitative regression analysis [63, 102]. Pre-processing methods such as baseline correction, smoothing, and normalization are often required to derive quantitative information. Linear discriminant analysis (LDA) is also required depending on the requirement.

4.2. AMYLOID PROTEIN AGGREGATES AND THEIR RAMAN SIGNATURE

4.2.1. Amyloid Protein Aggregates

Protein is the key biological macromolecule that maintains normal cellular functions and provides most structural component of the cell organelles. The long polypeptide chain can attain a unique 3D structure and the adopted configuration is often related to its function. Proteins those provide structural basis also attain suitable structure. Thus the structural stability and dynamics of proteins are two very important characteristics that keep the cell alive and healthy. However, under some physiological stress condition these proteins may assembled and produce different type of protein aggregates (Fig. **4** and Fig. **5**) [103 - 105]. The deposition of elongated fibrillar protein aggregates are linked to amyloid diseases and much other human pathology [39, 104 - 110]. The post mortem microscopic examination of organs and tissues of patients diagnosed with these severe maladies reveals amyloid plaques that contain long, unbranched, rod-like protein aggregates, known as amyloid fibrils [111, 112]. Polymorphisms of major amyloid aggregates maintained a core of cross-β pleated sheet structure of self-assembled proteins along the fibrillar axis of amyloid aggregates.

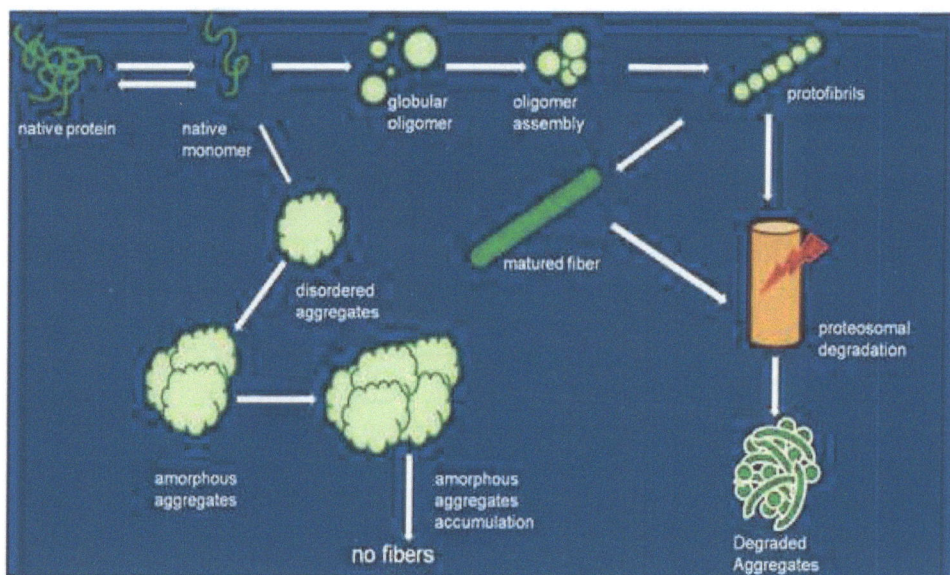

Fig. (4). Schematic representation of self-aggregated protein assembly structure.

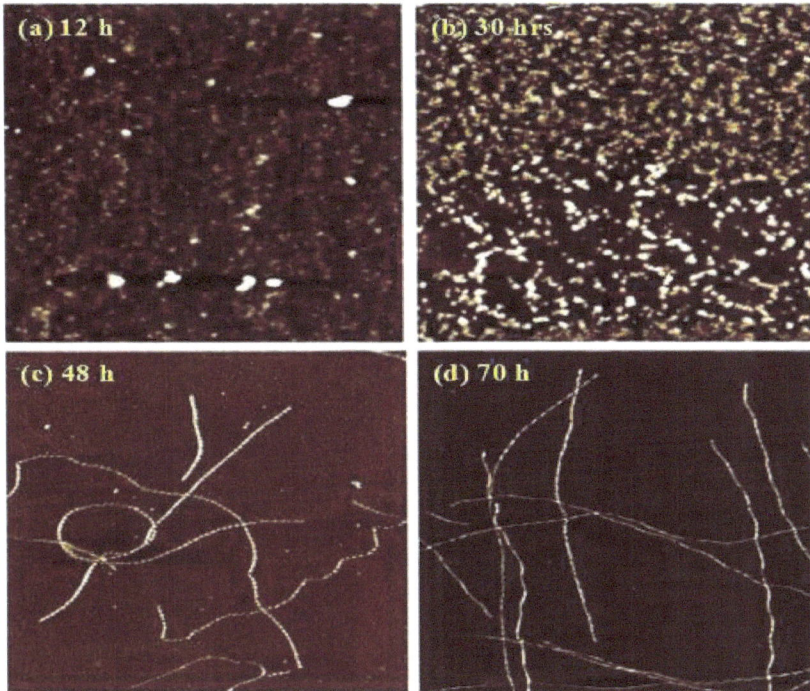

Fig. (5). Atomic force microscopy images of lysozyme in different forms of aggregate species: (a) spheroidal oligomers obtained after 12 hours of incubation; (b) spheroidal oligomers and pre-protofilaments after 30 hours of incubation; (c) premature fibril after 48 hours of incubation, and (d) mature fibril after 70 hours of incubation.

Amyloid deposits in several tissues interfered in complicated ways and disrupted the common cellular survival signalling pathways and finally initiate several types of health disorder. There are multiple diseases found to be linked with the aggregation of amyloids [39, 104, 107, 109, 110]. Neurodegenerative diseases, such as Parkinson's disease and Alzheimer's diseases are linked to aggregation of protein α-synuclein and amyloid β--peptides, respectively (Fig. **5**) [113 - 116]. Misfolded α-synuclein is the part of the abnormal protein aggregate found in Lewy bodies. Many globular proteins which have compacted fold with defined secondary and tertiary structure also linked to amyloid diseases. Example of such proteins includes β2-microglobulin, lysozyme, insulin, superoxide dismutase and many other proteins. Systemic amyloidoses, such as light chain and lysozyme amyloidoses also caused due to protein aggregation. Other human diseases such as familial amyloid polyneuropathy and dialysis-related amyloidosis, type II diabetes also linked to formation amyloid aggregates. The proteins or peptides having this typical property to form spontaneous amyloid aggregates are designated as amyloidogenic proteins or peptide [39, 104 - 110].

Protein aggregation and amyloid formation is generally sequence-sensitive. Both the physicochemical properties and the sequence arrangements play a significant role in protein aggregation and formation of fibrillar structure. Mutations in protein sequences that destabilize the secondary structure, in some cases promote protein aggregation. Mutations in the amyloid precursor protein and presenilin genes are associated with increased cellular production of β-amyloid(1-42), which is toxic to neurons. Again, modifications of the amino acid residues in some proteins retard aggregation especially if the mutation is a β-sheet breaker, such as proline. Chiti *et al.* reported that more than 40 human diseases linked to aggregation of proteins or protein fragments [106, 117, 118].

It is suggested that multivesicular bodies are the first site where protein aggregation occurs first. Uncontrollable growth of these aggregates leads to the destruction of the cell integrity and a release of fibril species into the extracellular space. The fibrilar species then propagate into amyloid fibrils. The amyloid fibrils themselves, however, are not necessarily the major toxic species. Growing line of evidences indicate that the oligomeric intermediates formed during protein fibrillization are more toxic and responsible for neurological damage in some neurodegenerative diseases. Therefore, the fibril formation is a defence mechanism that is aimed to isolate highly toxic misfolded proteins and their oligomers from the cell media, by 'packing' them into the much less toxic fibrillar form. Many natively unfolded proteins and folded proteins which are not linked to any diseases formation also are capable to form amyloid fiber under suitable solution condition, such as low pH, high ionic strength and in the presence of different co-solvents. There is therefore great interest in elucidating the mechanistic details of protein aggregation and the molecular structures formed along the aggregation pathway, for a more comprehensive understanding of the mechanisms of protein oligomerization and fibrillization and development of pharmacological means to ameliorate amyloid toxicity.

It is found that the amyloids are largely consisting of protein aggregates composed of insoluble protein fibers. The monomeric strands packed in a cross-β pattern that are stabilized by inter strand interactions such as hydrogen bonding, electrostatic interactions, aromatic interactions (π–π stacking), and hydrophobic interactions. The results of a variety of recent studies have indicated that amyloid fibril formation from globular native proteins occurs *via* partially unfolded intermediates that subsequently associate to form well-ordered mature fibrils which are a kind of irreversible conformation from its native form (Fig. **4** and Fig. **5**). The steps of fibrillation initiates with monomeric forms which interact among themselves to form different types of oligomeric species (Fig. **4**).

These will aggregate in due course to produce protofibrils which are found to be short, flexible, irregular in nature. These will ultimately mature and elongate into insoluble fibrils comprised of β-strands which are repeated in number and oriented perpendicularly to fiber axis. From many instances, similar structural features have been observed in amyloid fibrils which were obtained by *in vitro* studies from a large number of proteins. Although, many of them are not related to any known disease. This indicates that the capability to form amyloid fibrils is a generic property of any polypeptide chain. Wild type lysozyme can form amyloid fibril which is very similar to fibrils obtained from pathological deposits (Fig. **5**).

4.2.2. Raman Signature of Amyloid Aggregates of Human Proteins

A deeper understanding of molecular mechanisms which leads to aggregate formation is necessary to improve the therapeutic aspects of diseases as well as augment purify, store, and deliver protein-based drugs. The Raman methods are well suited for determining the structural features of proteins both in solution and as insoluble aggregates [10, 14, 100, 119 - 132]. The feasibility of the method is not hindered much either by the size of the assembly or by light-scattering artefacts that can severely restrict the use of other spectroscopic probes. Other advantages are that the spectral variations, including peak intensities, peak position or shape, can be directly related to changes in protein secondary and tertiary structure. This unique sensitivity to structural changes in unfolded or aggregated proteins makes Raman Spectroscopy a useful method for the prediction and monitoring of aggregation at all stages of drug production and amyloid formation.

It was earlier exploited to probe the assembly structure of viral capsids [29]. The high sensitivity of Raman spectroscopy to the secondary structure of proteins makes them particularly valuable for studying the conformational status in protein self-assembly and amyloid formation. The position of the amide I band depends on the protein conformation; the amide I band for proteins rich with α-helical secondary structure appears at $(1640-1654)$ cm^{-1}. The band locates in the range of $(1665-1680)$ cm^{-1} to a β-sheet rich protein structure. The band in the $(1654-1665)$ cm^{-1} range is assigned to unordered or disordered protein secondary structures. Attempts are made to distinguish parallel and antiparallel β-sheet conformations in protein aggregates based on amide I band position. The amide I bands of anti-parallel β-sheets may produce higher energy Raman values than the amide I band of parallel β-sheets due to weaker H-bonding stability of the anti-parallel β- sheet structure. Simulation studies also suggested that the peak position may also be changed significantly on the number of strands present in the sheet and it was prominent for the parallel β-sheet structure.

Using Raman spectroscopy as an analytical method Mungikar *et al.* reported the aggregation properties of therapeutic protein and developed a calibration model for thermal aggregation of the model protein [133]. Paul Carey's lab has performed pioneer investigation by Raman microscopy to explore transition structure that occurred during crystallization of native insulin into β-sheet conformation [26, 92, 134]. The study also showed predominant Amide I bands at 1657 cm^{-1} a signature of α-helical band shifted to 1669 cm$^{-1,}$ which is the characteristic β-sheet band. In the Amide III region, the β-sheet marker appeared at 1236 cm^{-1} and the distinctive α-helical band at 946 cm^{-1} disappeared. The Raman band positions thus establishing the fact that secondary conformation of insulin in the crystal changed to β-sheet structure from its α-helical form. Upon reduction of the disulfide bridge of insulin in crystals, a dramatic change was noticed in disulfide bonds in the region between 490 and 570 cm^{-1}. These bands disappear in the reduced insulin and a new band the 2573 cm^{-1} indicated the production of the free S-H group. Same group also utilized Raman method to find out the mechanistic details behind insulin fibrillation using pro-insulin. Generally, it is a good prototype used for understanding the fibrillation of globular protein [135]. Several recent articles also highlight additional details of insulin aggregates by Raman spectroscopic methods [17, 34].

Asher group showed that the amyloid fibrils grown from apo-α-lactalbumin were quite different than the mutant analogue [10, 136]. The mutant protein was the only one out of four disulfide linkages, -S-S-lactalbumin preserved no hydrophobic core in the fibrillar structure; however, the native protein preserved some of its hydrophobic core in the fibrillar form and showed both the amide II and amide II' (due to deuterated amide bond)Raman bands in the H/D solvent exchanged experiments. They also investigated the amide III band profile of protein samples obtained by deep UV laser excitation. They were able to classify the band into three sub-bands: amide III$_1$, amide III$_2$ and amide III$_3$ [127]. Using empirical relationships and the III$_3$ band position, they can determine the average psi (Ψ) dihedral angle of the protein backbone. They also showed that the band position is also influenced by the phi (Φ) dihedral angle. Their analysis further revealed that the amide III$_3$ band appears at ~1270 cm^{-1} and ~1246 cm^{-1}, respectively, for the 2.5$_1$ helix and PPII structure of the protein [22]. Benevides *et al.* noted a definitive tertiary signature of Raman band at 933 cm^{-1} denoting helix for HK97 capsid assembly [25, 137, 138]. The 800-2000 cm^{-1} region of the Raman band of Aβ40 oligomers was overcrowded with signatures at 929 cm^{-1} [36]. These bands denoted the occurrence of backbone tertiary helical conformation and the structural heterogeneity in the assembly of oligomeric forms. Earlier studies with helical viral proteins also observed a non-amide marker signature for the helix forms and the associated tertiary fold was assigned at ~1340 cm^{-1} [139]. This band emerged due to C-Cα-H valence angle bending. Isotope labelled study further affirmed that O=C-Cα-H stretching motion contribute significantly to this vibration band. The intensity of the signature band found to correlated with the content of α helical form and lower intensity signifies the lesser amount of α-helical fold.

The presence of Raman bands associated with side chain aromatic amino acid residues provides valuable information on the microenvironment; this amino acid residues are in the core of protein aggregates [37, 126, 140]. The Raman intensity of the Phe band at ~1002 cm^{-1} often used to probe the local environment of this amino acid in the aggregated protein. The intensity of the band obtained by DUVRR spectroscopic studies is comparatively low in the fibrillar structure of the protein and suggests that the phenyl ring gets exposed to water upon fibrillation. It was suggested that the Phe ring stretching mode that appears at 1002 cm^{-1} is very sensitive to solvent surrounding. The decrease in the Raman intensity of Tyr residue was also observed during insulin and amyloid β-aggregation. Similar observation was obtained for antibody aggregation. Gómez *et al.* demonstrated by comparing Raman spectral data sets of each native and insoluble aggregated IgG4 variant that the Raman regions of ~(760-770)cm^{-1}and (875-880) cm^{-1} assigned to tryptophan, displayed the most significant changes indicating the extent of solvent exposure [141]. SERS enabled tip enhanced Raman spectroscopic investigation further suggested that the aromatic amino acid residues distributed over the β-sheet and α-helix/unordered regions on the protein surface [78, 142].

(Fig. **5**) shows Raman spectra of Aβ40 fiber (20 mM phosphate buffer, pH 7.2) in the Raman region (1150-1800) cm^{-1}. Some of the characteristic Raman band assignments in three different states (monomer, oligomer and fibril) of the amyloid β- peptide (Aβ40) are provided in Table **2** and discussed in the recent article by Roy *et al.* [36]. The peptide Raman signature in the monomeric form was dominated by a broad amide I band at 1675 cm^{-1}. The band width at half height (BWHH) was ~35cm^{-1}. The band became much sharper compared to the same band in the oligomeric state of the peptide. Wälti *et al.* postulated that in Aβ42 fibril related disease, each layer of the fibril is comprised of two molecules and the residues spreading between 15-42 takes the shape of horseshoe like β-sheet structure [143].

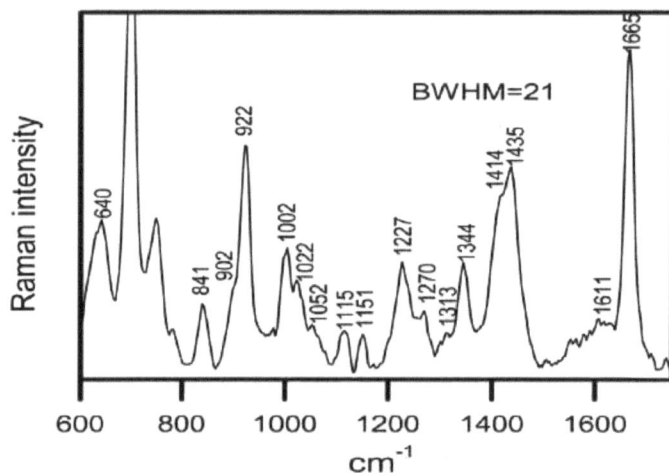

Fig. (6). 532 nm excited Raman spectra (600-1800 cm^{-1}) of Aβ40 fiber. Aβ40 fiber was prepared in 20 mM phosphate buffer of pH 7.2 incubation for 2 days at 25 °C, 10-30 μL solution was dropped onto a glass cover slip. Laser power at source 20 mW, ~2 mW at the sample, recording time 10 ×15 sec. The represented figures were obtained after baseline correction was done. Phosphate buffer/impurities contribute some signals at ~ 720 and ~ 920 cm^{-1}

Table 2. Raman bands (cm^{-1}) of Aβ40 as monomer, oligomer and late assembly (fibril) states.

Monomer	Early assembly (oligomer)	Late assembly (fibril)	Modes of Raman vibration
-	619	620	Phe
-	639	639	Tyr
-	723	-	unknown
-	-	768	Ala
819	834	818	Tyr
843	849	840	Tyr
873	-	-	unknown
-	903	904	υCC
-	915	915	unknown
946	929,944	929,947	CH$_2$ symmetric rock+ Cα-C stretching
-	955,968	962	CH$_2$ symmetric rock+ Cα-C stretching
1006	999	1000	Phe
-	1028	1027	Phe
1072	1080, 1094	1062	υCC, υCN, υCO
1136	1119	1120	υCC, Val,Ile
-	1150	1151	Ile, υCC

Monomer	Early assembly (oligomer)	Late assembly (fibril)	Modes of Raman vibration
1168	1174	1165	Try, Phe
1216	1206, 1232	1221, 1232	Amide III, -turn, β-sheet
1245	-	-	Amide III, polyproline II
1275, 1297	1258,1277	1272	Amide III, helix
-	-	1308	CH_2wtist/wag
1325	1319	1321	Amide III, CH_2wtist/wag
-	1336	1339	CH_2wtist or wag+Cα-H bending+ Cα-C streching
1362	-	-	unknown
-	1403	1406	Symmetric, υCO_2^-
1449	1434	1434, 1456	CH_2,CH_3 deformation,CH_2 scissoring
1573	1560	1547	Phe
-	1604	1604	Phe
1614	-	-	Tyr
-	1667	1665	Amide I, β-sheet
1677	-	-	Amide I, polyproline II, random β-space

Our Raman signature analysis observed minor amide III signals at 1275 cm^{-1}, 1277 cm^{-1} and 1272 cm^{-1} confirming monomer, oligomer and fiber, respectively, and attributed to collagen triple helix type/extended 2.5$_1$-helix type of structure. Asher *et al*.l. depicted a stable 2.5$_1$-helix conformation in a model peptide (PGA) and noted a signature band at 1271 cm^{-1} [22]. The structure possesses similarity with PPII like structure, although bigger separation distances occurred among the charge groups in the 2.5$_1$-helical form. This region may be present within all the three conformation (monomer, oligomer and fiber) of the peptide. This is PPII like conformation is highly extended and confined itself in the N-terminal residues.

4.2.3. Raman Features of Protein Aggregates Produced by Lysozyme

Hen egg-white lysozyme (HEWL) is a small (14.3 kD) globular protein and it is widely used to study the mechanisms of protein folding and amyloid aggregation [140, 144 - 152]. human ortholog is found abundantly in tears, saliva, and mucus and is a part of the human body's defence against some bacteria [153]. The mutations in human lysozyme which are familial, are found to be linked with systemic non-neuropathic amyloidosis deposits in the kidney, liver and gastrointestinal tract as an amyloid fibril [108, 154, 155]. The native form of HEWL is comprised of two domains, α, and β and linked by four disulfide linkages [147, 156]. Many studies have successfully demonstrated that under heat and acidic conditions, this protein easily transformed into the amyloid-like fibrillar conformation [157]. Wild type human lysozyme can form amyloid fibril which is very similar to fibrils obtained from pathological deposits. A recent report has shown that mature fiber of HEWL developed by fragments of sequences 49-101 and 53-101 position ascribed to acid hydrolysis under acidic conditions and high temperature [158, 159]. Oligomeric amyloid is an intermediate step in the formation of mature fibrils. Amyloid fibrils aggregates are mostly found in the extracellular space in the tissue.

Engaging Raman microscopy-based drop-coating deposition Raman (DCDR) method Dolui *et al*. analyzed the secondary structure and tertiary alignment of HEWL that assembled to form oligomers and compared that information with the protein aggregates [42]. Incubation of freshly prepared hen egg-white lysozyme(HEWL) protein solution in acidic buffer at 60°C for several hours produced oligomeric structure in its early phase of incubation. AFM images show the heterogeneous mixture of small and large spherical oligomers. These oligomers were in the range of 5-10 nm diameters (Fig. **5**). Some of them were 15-25 nm upon further longer incubation, at a similar temperature for a period of 30 h, it produced pre-protofiber species with a diameter of 7-12 nm. The premature fibril species appeared as a thread-like structure after 48 h of incubation. These organized fiber species were of few nanometres thickness and micrometres in their length. After 70 h of incubation long mature amyloid-likefibers were formed. Raman spectra of these are recorded under Raman microscope and compared with the Raman spectra of lysozyme single-crystal gown in two different pHs, pH 7.4 and pH 4.8 (Fig. **3**). Table **1** highlights the band positions with assignments of vibration characteristics [37].

The protein in its highly monomeric state showed a distinct amide I band at 1656 cm^{-1}, similar to the monomeric protein in crystals. The band positions denoted the existence of major α-helical conformation of the protein in all three solution conditions [34, 37]. X-ray crystal structure reveals the presence of ~45% of α-helical structures in lysozyme and distributed in four α-helixes in the α-domain of lysozyme. The band position of amide I of the two crystals were quite similar except slight broadening of the tryptophan vibration bands those appeared at 755, 1340 and 1550 cm-1 (Fig. **3**).

Lysozyme monomer (0 h incubation) and oligomer both produced similar overlapping bands in amide I (1620-1800) cm^{-1} and amide III (1200-340) cm^{-1} regions (Fig. **7** and Table **1**) [42]. The protein oligomers as formed upon incubation (Fig. **5**) for 12 h showed, the amide I vibration band maxima shifted slightly to a higher frequency and it appeared at 1659 cm^{-1}; the bandwidth at half maxima (BWHM) was ~40 cm^{-1}. A Raman signature near 1250/1245 cm^{-1} corresponds to polyproline II (PPII) protein secondary conformation [7, 24]. Lednev *et al.* showed, however, the signature position at ~1272 cm^{-1} as the indication of 2.5$_1$ hellix structure [95]. The 0 h sample and 12 h samples showed the broadband position at 1250 cm^{-1} and 1245 cm^{-1}, respectively. It depicted that during initial phase of aggregation, the PPII conformation was predominant similar to poly L glutamic acid [10, 31]. Protofibrils emerged after a longer period of incubation (Fig. **5**). The shifting of amide I band position at a higher frequency (1665 cm^{-1}) suggested a drop in the α-helical component of the protein.

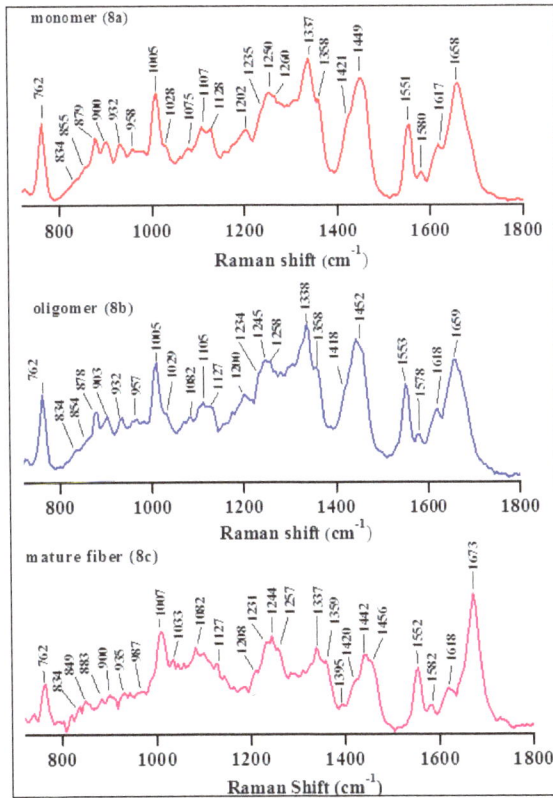

Fig. (7). 532 nm excited Raman spectra (700-1800 cm^{-1}) of lysozyme monomer, oligomer and fiber: (a) monomeric lysozyme solution was prepared in pH 1.65 (25 mM HCl) and incubated for 0 hrs. (b) oligomerized lysozyme solution was prepared in pH 1.65 (25 mM HCl) and incubated for 12 hours at 60 °C.(c) fiber was prepared in similar buffer condition, incubation time 70 hrs at 60 °C. ~20 µL solution was dropped onto a glass coverslip. Laser power at source 20 mW, ~1 mW at the sample, recording time of 100 sec. The displayed spectra were an average of multiple scans and a suitable baseline correction was made. A similar figure is published by Dolui *et al.* [42].

With a longer incubation period (70 h) there is an enrichment of fibrillar morphology and distinct change in the intensity and sharpness of amide I band found, the α-helical 1657 cm^{-1} band moved and was replaced by band at 1673 cm^{-1}, suggesting presence of β-sheet. The narrow and sharp amide I signature at 1673 cm^{-1} (FWHM, 27 cm^{-1}) corresponded to cross β-sheet conformation. The bands found at 1231 cm^{-1} became predominant, suggesting a more compact β-sheet structure in the fibrillar form [8, 9, 160]. This signature was not prominent in incubated samples of 0 h stage and in the monomeric condition. However, the lower shoulder band at ~1260 cm^{-1}was found in all the scenarios, and this signature may be attached to an unfolded component of the crystal. The α-helical signature at 932 cm^{-1} also became lower in the fibrillar conformation. Two very sensitive bands at ~900 and ~932 cm^{-1}are often attributed to the N-Cα-C stretching vibrations of the protein skeleton and the band position at ~932 cm^{-1}is associated with α-helical structures [33, 34, 36, 160]. These two bands were also equally present in both the monomer and the oligomeric state. The band intensities decrease significantly in fibrillar assembly conformations. These are nicely discussed in the recent research article by Dolui *et al.* [42].

The Cα-H bending band often appears at ~1400 cm^{-1}; the band intensity is significantly enhanced in ultraviolet resonance Raman spectra. The band is also very much sensitive to protein backbone conformation. The β sheet conformation gives the band at ~1395 cm^{-1}. However, in the α-helical conformation, the Cα-H bending gets to mix with N-H vibration and is coupled; consequently, the band intensity becomes weaker. The Cα-H bending band intensity could be seen in the fibrillar state at1395 cm^{-1}and it was due to the enhancement of the β-sheet structure. It was neither prominent in the oligomeric or in the monomeric form of the protein nor suggested that the cross-β sheet structure was not dominated in its oligomeric state. Raman spectra of lysozyme also contain vibrational bands originated from aromatic phenylalanine, tyrosine and tryptophan residues (Fig. 7) [42].

Comparatively a prominent peak at 762 cm^{-1} was due to in-phase symmetric breathings vibration of benzene and pyrrole (indole ring) of the TRP residue. The bandwidth at half maxima (BWHM) for the tryptophan marker band assigned at 762 cm^{-1}was mostly remained unchanged in the oligomeric stage and denoted that orientation of the group was similar and the intensity was not altered to a significant level. The overall microenvironment and surroundings around the phenyl/indole rings are also responsible for the intensity and position of the band. Lysozyme contains six Trp as residues 28, 62, 63, 108, 111, and 123. Four of them are located in the α-domain and the remaining two occupies in the β-domain facing the cleft that comprises the active site and divides the protein into two (α and β) domains. These residues preferably remain buried inside the hydrophobic microenvironment as both the phenyl and indole groups are hydrophobic in nature. Three phenylalanine residues also remain buried inside hydrophobic pocket due to the same reason. Thus, changes in the microenvironment of these residues affect the Raman signature of the characteristic bands and often act as the guide of the tertiary structural rearrangements.

We observed that the tryptophan doublet bands appeared at 1337 and 1358 cm^{-1} for both the monomer and oligomer and the intensity ratio also remained similar. We also observed tryptophan W17 mode of vibration band at ~879 cm^{-1} and it was similar both in the oligomeric intermediate state and suggested that no major changes in the position of the Trp ring position. The hydrogen bond found inside the indole ring of nitrogen is also sensitive to the band intensity and position. The band position shifts toward a higher wave number (~883 cm^{-1}) for non H-bonded tryptophan and shifts to a lower wave number (~871 cm^{-1}) as the H-bonding strength increases [40]. Using ultraviolet resonance Raman spectroscopy (UVVR), Zu *et al*. derived protein backbone structure and assessed the local microenvironment of some side-chain of aromatic amino acid residues in the fibrilization of lysozyme [52]. One of the important observations was that the phenylalanine became more solvent-exposed in fibrils compared to native protein in the solution state.

4.2.4. An Estimation of Protein Secondary Structure by Amide I Band Analysis

As mentioned, the Raman method has a great advantage to measure the high-quality Raman spectrum of protein samples under Raman microscope. It has the potential to provide structural information of native proteins and intermediates formed in the fibrillization pathway such as oligomers, protofibrils and fibrils. A band fitting analysis has been proposed based on the Amide I band fitting for quantitative analysis of the content of secondary structure components. The fitting was conducted using of Levenberg–Marquardt non-linear least-squares method as applied and incorporated in Curve Fit. Ab routine of GRAMS/AI 9.02 software. The quantitative estimation of secondary structure distribution in the monomeric, oligomeric and fibrillar state of aggregation was done for different proteins.

Lysozyme (HEWL) under acidic condition produce fibrillar aggregates and the aggregation processes involve several intermediates [42]. (Fig. **8**) shows multi-component amide I Raman band fitting analysis to derive the content of secondary structures present in different assembly state of the protein lysozyme. The band assigned to helical conformation was centered at 1657 cm^{-1} and in the monomeric (0 h of incubation) state, it was comprised 45% of the total amide I band intensity. The component band at 1672 cm^{-1} was for the contribution from the β-sheet secondary structure and it was ~17% of the total band intensity, ~20% of the band intensity was due to extended like PPII structure and it peaked at 1689 cm⁻¹. The remaining 18% of the total amide I band intensity was related with the band at 1639 cm⁻¹ and this might be a solvent-exposed extended structure [81, 103]. The presence of α-helical component was similar to the reported results of lysozyme crystals. In the oligomeric conformation, the helical content was ~40% with the component band at 1657 cm⁻¹ and it was quiet similar with monomeric stage of the protein. The β-sheet amount was 20% (component band at 1672 cm⁻¹) and PPII like structure or disordered amount was 21% with marker band at 1685 cm⁻¹. Thus, the secondary conformation in the oligomeric state of the protein was similar to the monomer (0 h incubated sample) with a little increment of β-sheet content. In the fibrillar assembly state, β-sheet content was increased to 45% (component band at 1673 cm⁻¹) and the α-helical component at 1658 cm⁻¹ fell down to 26%. The PPII and extended strand structure also decreased to 16% with a distinct marker band at 1687 cm⁻¹. Thus the changes in conformation were continuous processes (Fig. 9).

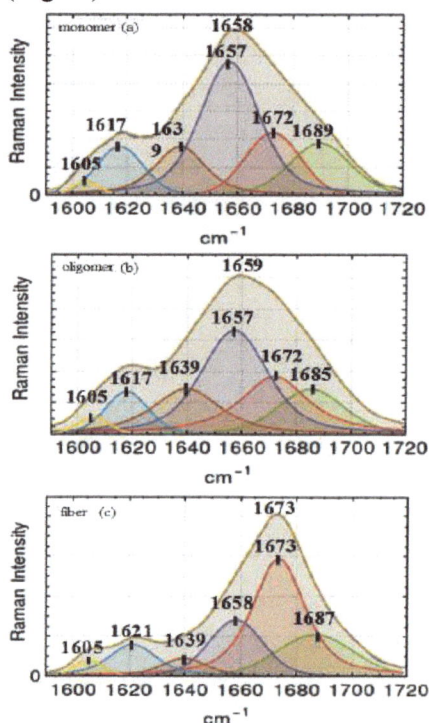

Fig. (8). Curve fitting analysis of the Raman amide I (1590-1720 cm⁻¹) of separated lysozyme aggregates. (a) monomer, (b) oligomer, and (c) fiber on the glass coverslip. The yellow and deep blue lines are the experimental and fitted spectra. Four component bands that represent amide I regions are shown in the green, orange, violet, and brown. The fitted peak positions are also assigned aromatic residues are in cyano and light yellow. This figure is for representation purposes and may contain some error. Please see the recent article published by Dolui *et al.* [42].

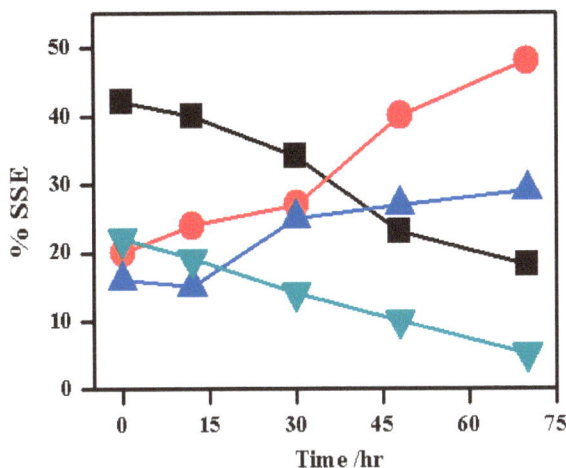

Fig. (9). Time-wise changes of the content of different secondary structure elements present in incubated lysozyme over time as derived from the curve fitting analysis of Raman amide I band. Figure shows the secondary structural component in percentage (%) of the total amide I band intensity against incubation time: helix (black), β-sheet (red), loose β-strand, PPII and disordered (blue) and vibronic coupling component band intensity (green). For detailed component band analysis, see the recently published articles [42, 63, 123].

4.2.5. Amyloid Fibril Formation Mechanism by Raman Analysis

The mechanism of aggregation of different amyloidogenic proteins into the amyloid fibers is yet not very clear. The protein fibril formation induced by different biophysical conditions or environmental modification most likely initiates with a partial unfolding state of the proteins. Partially unfolded proteins are changed into intermediates oligomers, which are further turned into protofiber and finally into the mature fibril. Recently several studies highlighted this model of aggregation (Fig. **4**). Using dynamic light scattering and atomic force microscopy (AFM) experiments, Jain and Udgaonkar provided conclusive evidence that the fibrillation of lysozyme follow a multistep mechanism [161, 162]. On-pathway aggregation that leads to the production of amyloid like fibril formation of lysozyme started from soluble monomers and involved intermediates such as oligomers, protofibrils. Mangialardo *et al.* and many others inspected the secondary structure of hen white egg lysozyme protein in its native and fibrillar state by off-resonance Raman spectroscopy. Several other studies defined structural integrity of intermediates in fibril formation pathways by Raman spectroscopy. However, the nucleation step and formation of oligomeric intermediates are poorly understood. Shasilov *et al.* used 2D-correlation deep UV resonance Raman to directly probe the initial steps of lysozyme fibrillation [120]. They analyzed amide I, II, and III bands in Raman spectra obtained by deep UV laser excitation and established that the fibrilization processes were highly correlated events. Initially, the protein molecule attained a disordered coil structure and it was not well defined state of the protein. Subsequently, it produced oligomers with a slight increment in β-sheet structure. Several such oligomers assembled further and created nucleus and rapidly transformed into amyloid like fibrillar structure. Xing *et al.* highlighted some of the early events and kinetics of protein fibrillation by off-resonance Raman method [148]. They analyzed the Raman marker changes of the skeleton and side chain vibrations over incubation time and provided strong support in favor of a multistep kinetic mechanism of fibrillation of lysozyme under heat and acidic conditions. They utilized unique N-Cα-C stretching (~900 cm^{-1} and ~940 cm^{-1}) and amide I band (1640-1680) cm^{-1}vibration bands to examine the conformational transformations of the protein. Because the N-Cα-C stretching intensity at ~940 cm^{-1} is proportional to the population of α-helical secondary structures, they utilized this to monitor the α-helical structure transformation into β-sheet rich amyloid formation. They also utilized the vibrational bands associated with Trp and Phe amino acid residues to probe the local micro-environment of some of these residues and it leads them to find a good coorelation with the tertiary conformation of the protein in the initial and late stage of aggregation. Further investigation concluded that the kinetics of lysozyme fibrillation under heat and acidic conditions follow a four stagestep by step transformation mechanism.

In a recent investigation, Maiti *et al.*, found that the protein loses its tertiary structure keeping major secondary folds intact in the oligomeric state [42]. It is proposed in the presence of strong acidic pH and high temperature, the tertiary and compact globular form of the protein loosens up and as a consequence, a certain increase in PPII like secondary structure, a killer conformation occurred [112]. In this configuration, cleavage of some of the disulfide linkages started and Trp, Tyr residues were exposed to heterogeneous microenvironment. This allowed most fibrillogenic residues to expose and oligomerization; subsequently, it quickly transformed into amyloid fibril. Using UV-CD and intrinsic tryptophan fluorescence analysis Mishra *et al.* showed that thermal unfolding of the protein and the Trp and Tyr residues assumed a very asymmetric solvent environment at high temperature [158]. It leads them to conclude that the tertiary structure of the protein is broken at a slightly lower temperature before the collapse of the secondary structure [158].

Two small but prominent bands emerged at ~900 and ~930 cm^{-1} in the Raman spectra of the lysozyme crystals in hanging drop condition (Fig. **3**). These two bands are assigned to N-Cα-C stretching vibrations of the protein skeleton in its α-helical conformations and the band at ~930 cm^{-1}, in particular, is used for determining the existence of α-helix in protein. Kocherbitov *et al.* also noticed common pattern during their investigation on hydration status of crystals [37].

Raman spectra of lysozyme as discussed earlier, divulged that in the oligomeric stage ~40% residues adopted α-helical conformation (Fig. **8**). This α-helical conformation was also clear from CD spectra of the lysozyme. However, tertiary collapse during the formation of oligomeric assembly was denoted by most of the Raman band, including broadened disulfide marker band at ~ 505 cm^{-1} [42]. The band intensities lowered significantly during the formation of fibrillar assembly. The obtained results denoted that in the oligomeric state, lysozyme maintained most of its helical domains, however, maybe in not very compact form.

Several investigations also pointed that lysozyme becomes fragmented under acidic conditions and high temperature and, the fragments may be more prone to form amyloid fibril [163]. It was also observed that the fragmentation of the protein was not instantaneous and it depends on several factors such as incubation condition. Although, the process of fragmentation and nicking of the protein may get faster in the tertiary collapsed molten globule stage (oligomeric state) with loose helical folds and PPII configuration of the extended region. The disulfide marker Raman band intensity at ~505 cm^{-1} substantially decreased after oligomerization [42]. This suggests a concept of formation of fibrils with fragmented peptides; however, intact protein can form s fibrils depending upon the surroundings.

A significant loss in disulfide (-S-S-) linkages was noticed and the Raman band at 505 cm^{-1} decreased in the fibrillar state when it was compared with other states of the protein [42]. It was suggested that fibril formation was accelerated upon fragmentation of the peptide at low pH conditions. The higher energy component band of tryptophan Fermi doublet at 1360 cm^{-1} was enhanced in the fibrillar state compared to its oligomeric condition and indicated that tryptophan microenvironment became more hydrophobic in the fibrillar state. Raman bands at ~875 cm^{-1} of the residue shifted to high energy side and became sharper in the fibrillar state. It also suggested that tryptophan surroundings became more hydrophobic. This thing upholds the theory that hydrophobic zipping occurred as a thermodynamically most stable fibrillar structure [164, 165].

It was found that the manifestation of β conformation initiated only after 30 hrs of incubation of the protein (Fig. **7**). After the lagging stage, the faster elevation of β-sheet structure formed. This period was marked as a transition state from the oligomeric stage to pre-protofiber stage. This structural elevation reaches to a static phase after 70 hrs of incubation. It was found that the mature fibril of the lysozyme was highly enriched with β-sheet conformation. Although the oligomers that formed after 12 hrs of incubation did not have any exposed β-sheet conformation, it increased rapidly after the protofibril stage [42]. The protofiber amide I band showed a transition from the initial α-helix marked by a shift in the peak position. The amide I band position emerged at a higher frequency (1664 cm^{-1}), depicting a drop in the α-helical conformation. The narrow and sharp amide I band at 1673 cm^{-1} in the spectrum of fibrillar lysozyme was assigned to the cross β-sheet structure.

Raman spectra of samples incubated for 0 h time and 12 h looked alike [42]. However, the Raman difference spectra (0 h-12h) exhibited some important changes both in position and intensities in the conformation linked amide I and amide III region of the spectra [42]. Xu *et al.* also observed a decrease of this band intensity along with the protein transformation into aggregates of different morphology; however, they used deep UV excitation to record the Raman spectra [140]. The width of the amide I band increased and depicted heterogeneity in structural state. The separate spectrum of 0 h and 12 h incubated samples produce a positive feature at 1654 cm^{-1} and a very small negative pattern at 1671 cm^{-1}. It suggested an increase in conformational heterogeneity and a slight rise of β-sheet conformation [42].

Amide III region also exhibits Raman intensity differences and signature of band broadening. However, these changes were very much distinct in the subtracted spectrum received from an incubated sample of 0h and fibrils obtained at 70h of incubation [42]. The amide I region exhibits a strong positive peak at 1652 cm^{-1} and a negative band at 1674 cm^{-1} and confirming that major structural changes occurred upon fibrilization. Also, a negative feature in the difference spectra at 1225 cm^{-1} suggested the existence of β-sheet structure.

A multi-component amide I Raman band fitting analysis [42] suggested that in monomeric (0 h of incubation) stage, the helical component was ~42% of the total amide I band intensity. The β-sheet secondary structure component band at 1672 cm^{-1} was comprised of ~20% of the total band intensity and ~16% of the band intensity was allocated to extended and PPII like structure (1689 cm^{-1}). The remaining 22% of the total amide I band intensity was assigned with the band at 1639 cm^{-1} and this might be a solvent-exposed extended structure [16, 166]. In the oligomeric stage, the helical content was ~40% and it was similar to the monomeric stage of the protein. The β-sheet amount was 24% and PPII like structure or disordered amount was 15% with the signature band at 1685 cm^{-1}. Thus, the secondary structural feature in the oligomeric stage of the protein was very much similar to the monomer (0 h incubated sample) with a slight increment of β-sheet amount. In the protofibril stage, the helical amount was ~ 34% with the component band at position 1656 cm^{-1} while the β-sheet content was 25% (component band at 1671 cm^{-1}) and PPII like structure or unordered amount was 18% with the signature spectra at 1688 cm^{-1}.

In the fibrillar assembly stage, β-sheet content was elevated to 48% (component band at 1673 cm^{-1}) and the α-helical amount at 1658 cm^{-1} was decreased to 26%. The PPII and extended strand structure were also decreased to 16% with a signature band at 1687 cm^{-1}. Time dependent alteration of various secondary structural features of incubated lysozyme over time as received from the curve fitting analysis of Raman amide I band is depicted in Fig. (**9**).

Thus Raman signature suggested the faster conversion of monomer to early assembly formation of oligomeric stage could be the reconstruction of the highly energetic relatively stable molten globular like structures such as helical fold and β-sheet/strand conformations. Partial compact conformations were also noticed for other amyloidogenic proteins. In the recent past, IR studies revealed that Josephin domain of ataxin-3 transformed into amyloid fiber *via* creation of native oligomers which has both the random coil and α-helix conformation [167]. These features suggested that, like many other proteins, lysozyme may also attain a pre-molten globule stage prior to formaion of oligomers and some of these structural features may be transmitted in the initial stage of soluble oligomer which eventually transformed into insoluble amyloid fibril. Hydrophobic zipping allowed the formation of thermodynamically most stable fibrillar conformation (Fig. **10**).

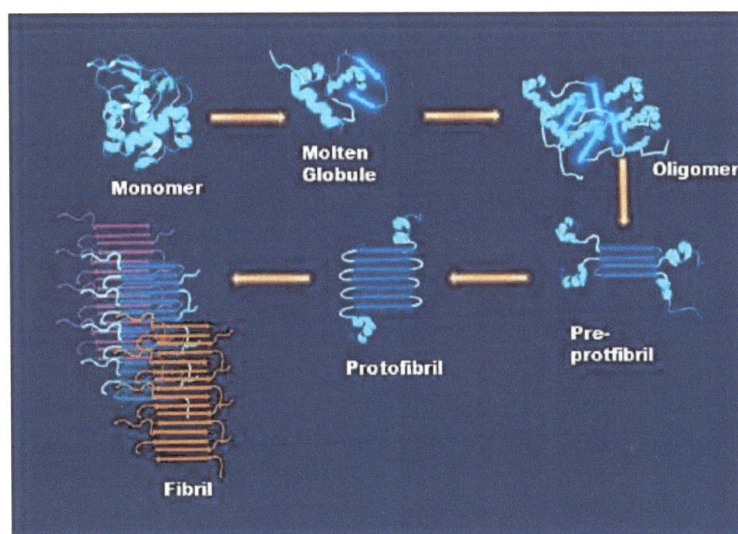

Fig. (10). Schematic representation of heat induced lysozyme amyloid fibril formation mechanism.

CONCLUSION AND PERSPECTIVES

Amyloid deposits in several tissues interfered in complicated ways and disrupted the common cellular survival signalling pathways. It thus causes many human disorders and diseases, including cancers, type II diabetes, and neurodegenerative disorders. Inter-conversion of soluble well ordered structured proteins like lysozyme, insulin, TTR, Huntington and β-2 microglobular peptides like Aβ, tau-fragments, and amylin, etc all belong to the amyloidogenic group of proteins. Therefore, it is very important to address the conformation and mechanism of amyloid formation to realize the cause of several diseases linked to amyloid formation. In this article, we discussed the Raman spectroscopic analysis of protein secondary conformation present in different forms of the assembly structure produced by different amylodogenic proteins such as lysozyme. Also, the review highlighted the aggregation mechanism of some points by Raman spectroscopic data. One important result is that Raman analysis strongly suggests a notable amount of α-helical folds are present during the oligomeric assembly stage of globular proteins such as HEWL, although the α-helical folds may not be as compact as in the native state. This observation is important as the oligomer structure of several proteins is thought to be more toxic than the β-sheet rich amyloid fibrils. Oligomers may have a suitable structure to interact effectively with cell membranes. The Raman analysis exhibited that the globular fold of the protein remained not fully unscathed rather, the individual helical folds are retained to a large extent. The amide I band fitting analysis showed that ~40% residues remained in α-helical secondary conformation in the oligomeric stage of lysozyme. The presence of α-helical conformation was also apparent in the CD spectra of the protein observed at different time points of incubation. In the wet crystals these helical folds are held tightly. Thus the opened up helical fold may make potent interaction with several membrane-bound cell organelles and consequently become more potent/toxic in a biological environment. So, like the fusion of viral protein with cell membrane, the helical segments which are present in the oligomeric conformation may cause the transfusion of virulent oligomer from one cell to another. Therefore, to design the potential therapeutic selection of the protein regions, predominantly proteins showing helical propensities and originating structural evaluation in the processes of fibrillation may be innovative and new approaches to control amyloid formation concerned with numerous human diseases.

CONSENT FOR PUBLICATION

Not applicable.

CONFLICT OF INTEREST

The author declares no conflict of interest, financial or otherwise.

ACKNOWLEDGEMENTS

Declared none.

REFERENCES

[1]   Raman CV, Krishnan KS. A New Type of Secondary Radiation. Nature 1928; 121: 501-2.
        [http://dx.doi.org/10.1038/121501c0]

[2]   Hendra PJ, Chalmers JP, Griffiths PR. Handbook of vibrational spectroscopy. 2002.

[3]   Kitagawa T, Hirota S. Handbook of Vibrational Spectroscopy. 2006.

[4]   Asher SA. Biochemical Applications of Raman and Resonance Raman Spectroscopies. P. R. Carey,
        Bernard Horecker Raman Spectroscopy in Biology: Principles and Applications. Anthony T. Tu. Q
        Rev Biol 1984; 59: 314.
        [http://dx.doi.org/10.1086/413923]

[5]   Volkmer A, Book L, Xie X. Time-resolved coherent anti-Stokes Raman scattering microscopy:
        Imaging based on Raman free induction decay. Appl Phys Lett 2002; 80: 1505-7.
        [http://dx.doi.org/10.1063/1.1456262]

[6]   Thomas GJ Jr. Raman spectroscopy of protein and nucleic acid assemblies. Annu Rev Biophys Biomol
        Struct 1999; 28: 1-27.
        [http://dx.doi.org/10.1146/annurev.biophys.28.1.1] [PMID: 10410793]

[7]   Tuma R. Raman spectroscopy of proteins: from peptides to large assemblies. J Raman Spectrosc 2005;
        36: 307-19.
        [http://dx.doi.org/10.1002/jrs.1323]

[8]   Yu N-T, Jo BH. Comparison of protein structure in crystals and in solution by laser Raman scattering.
        II. Ribonuclease A and carboxypeptidase A. J Am Chem Soc 1973; 95(15): 5033-7.
        [http://dx.doi.org/10.1021/ja00796a041] [PMID: 4741287]

[9]   Chen MC, Lord RC, Mendelsohn R. Laser-excited raman spectroscopy of biomolecules: IV Thermal
        denaturation of aqueous lysozyme Biochim Biophys Acta BBA - Protein Struct 1973; 328: 252-60.

[10]   Oladepo SA, Xiong K, Hong Z, Asher SA, Handen J, Lednev IK. UV resonance Raman investigations
        of peptide and protein structure and dynamics. Chem Rev 2012; 112(5): 2604-28.
        [http://dx.doi.org/10.1021/cr200198a] [PMID: 22335827]

[11]   Carey PR. Raman crystallography and other biochemical applications of Raman microscopy. Annu
        Rev Phys Chem 2006; 57: 527-54.
        [http://dx.doi.org/10.1146/annurev.physchem.57.032905.104521] [PMID: 16599820]

[12]   Sereda V, Lednev IK. Polarized Raman spectroscopy of aligned insulin fibrils. J Raman Spectrosc
        2014; 45(8): 665-71.
        [http://dx.doi.org/10.1002/jrs.4523] [PMID: 25316956]

[13]   Lednev IK, Shashilov V, Xu M. Ultraviolet Raman spectroscopy is uniquely suitable for studying
        amyloid diseases. Curr Sci 2009; 97: 180-5.

[14]   Spiro TG, Gaber BP. Laser Raman scattering as a probe of protein structure. Annu Rev Biochem
        1977; 46: 553-72.
        [http://dx.doi.org/10.1146/annurev.bi.46.070177.003005] [PMID: 332065]

[15]   Kitagawa T, Hirota S. Raman Spectroscopy of Proteins. In: Handbook of Vibrational Spectroscopy

American Cancer Society. Epub ahead of print 2006.
[http://dx.doi.org/10.1002/0470027320.s8202]

[16]   Chang SG, Choi KD, Jang S-H, Shin HC. Role of disulfide bonds in the structure and activity of human insulin. Mol Cells 2003; 16(3): 323-30.
[PMID: 14744022]

[17]   Ratha BN, Kar RK, Bednarikova Z, *et al.* Molecular Details of a Salt Bridge and Its Role in Insulin Fibrillation by NMR and Raman Spectroscopic Analysis. J Phys Chem B 2020; 124(7): 1125-36.
[http://dx.doi.org/10.1021/acs.jpcb.9b10349] [PMID: 31958230]

[18]   Duguid J, Bloomfield V, Benevides J, *et al.* Raman Spectral Studies Of Nucleic-Acids. 44. Raman-Spectroscopy Of Dna-Metal Complexes. 1. Interactions And Conformational Effects Of The Divalent-Cations - Mg, Ca, Sr, Ba, Mn, Co, Ni, Cu, Pd, and Cd. Biophys J 1993; 65: 1916-28.
[http://dx.doi.org/10.1016/S0006-3495(93)81263-3] [PMID: 8298021]

[19]   Shashilov V, Lednev IK. Biospectroscopy - Deep-UV Raman spectroscopy directly probes a fibrillation nucleus. Laser Focus World 2007; 43: 87-90.

[20]   Downes A, Elfick A. Raman spectroscopy and related techniques in biomedicine. Sensors (Basel) 2010; 10(3): 1871-89.
[http://dx.doi.org/10.3390/s100301871] [PMID: 21151763]

[21]   Shipp DW, Sinjab F, Notingher I. Raman spectroscopy: techniques and applications in the life sciences. Adv Opt Photonics 2017; 9: 315.
[http://dx.doi.org/10.1364/AOP.9.000315]

[22]   Mikhonin AV, Myshakina NS, Bykov SV, Asher SA. UV resonance Raman determination of polyproline II, extended 2.5(1)-helix, and β-sheet Ψ angle energy landscape in poly-L-lysine and poly-L-glutamic acid. J Am Chem Soc 2005; 127(21): 7712-20.
[http://dx.doi.org/10.1021/ja044636s] [PMID: 15913361]

[23]   Wang M, Jiji RD. Resolution of localized small molecule-Aβ interactions by deep-ultraviolet resonance Raman spectroscopy. Biophys Chem 2011; 158(2-3): 96-103.
[http://dx.doi.org/10.1016/j.bpc.2011.05.017] [PMID: 21652140]

[24]   Hahn MB, Solomun T, Wellhausen R, *et al.* Influence of the Compatible Solute Ectoine on the Local Water Structure: Implications for the Binding of the Protein G5P to DNA. J Phys Chem B 2015; 119(49): 15212-20.
[http://dx.doi.org/10.1021/acs.jpcb.5b09506] [PMID: 26555929]

[25]   Benevides JM, Overman SA, Thomas GJ Jr. Raman spectroscopy of proteins. Curr Protoc Protein Sci 2004; Chapter 17: 8.
[PMID: 18429253]

[26]   Zoete V, Meuwly M, Karplus M. A comparison of the dynamic behavior of monomeric and dimeric insulin shows structural rearrangements in the active monomer. J Mol Biol 2004; 342(3): 913-29.
[http://dx.doi.org/10.1016/j.jmb.2004.07.033] [PMID: 15342246]

[27]   Bandekar J. Amide modes and protein conformation. Biochim Biophys Acta 1992; 1120(2): 123-43.
[http://dx.doi.org/10.1016/0167-4838(92)90261-B] [PMID: 1373323]

[28]   Bonifacio A, Sergo V. Effects of Sample Orientation in Raman Microspectroscopy of Collagen Fibers and Their Impact on the Interpretation of the Amide III Band. Vib Spectrosc 2010; 53: 314.
[http://dx.doi.org/10.1016/j.vibspec.2010.04.004]

[29]   Overman SA, Thomas GJ Jr. Amide modes of the α-helix: Raman spectroscopy of filamentous virus fd containing peptide 13C and 2H labels in coat protein subunits. Biochemistry 1998; 37(16): 5654-65.
[http://dx.doi.org/10.1021/bi972339c] [PMID: 9548951]

[30]   Kurouski D, Van Duyne RP, Lednev IK. Exploring the structure and formation mechanism of amyloid fibrils by Raman spectroscopy: a review. Analyst (Lond) 2015; 140(15): 4967-80.
[http://dx.doi.org/10.1039/C5AN00342C] [PMID: 26042229]

[31] Mikhonin AV, Ahmed Z, Ianoul A, *et al.* Assignments and Conformational Dependencies of the Amide III Peptide Backbone UV Resonance Raman Bands. J Phys Chem B 2004; 108: 19020-8.
[http://dx.doi.org/10.1021/jp045959d]

[32] Asher SA, Ianoul A, Mix G, *et al.* Dihedral psi angle dependence of the amide III vibration: a uniquely sensitive UV resonance Raman secondary structural probe. J Am Chem Soc 2001; 123(47): 11775-81.
[http://dx.doi.org/10.1021/ja0039738] [PMID: 11716734]

[33] Maiti NC, Apetri MM, Zagorski MG, Carey PR, Anderson VE. Raman spectroscopic characterization of secondary structure in natively unfolded proteins: α-synuclein. J Am Chem Soc 2004; 126(8): 2399-408.
[http://dx.doi.org/10.1021/ja0356176] [PMID: 14982446]

[34] Dolui S, Roy A, Pal U, Saha A, Maiti NC. Structural Insight of Amyloidogenic Intermediates of Human Insulin. ACS Omega 2018; 3(2): 2452-62.
[http://dx.doi.org/10.1021/acsomega.7b01776] [PMID: 30023834]

[35] Apetri MM, Maiti NC, Zagorski MG, Carey PR, Anderson VE. Secondary structure of α-synuclein oligomers: characterization by raman and atomic force microscopy. J Mol Biol 2006; 355(1): 63-71.
[http://dx.doi.org/10.1016/j.jmb.2005.10.071] [PMID: 16303137]

[36] Roy A, Chandra K, Dolui S, Maiti NC. Envisaging the Structural Elevation in the Early Event of Oligomerization of Disordered Amyloid β Peptide. ACS Omega 2017; 2(8): 4316-27.
[http://dx.doi.org/10.1021/acsomega.7b00522] [PMID: 31457723]

[37] Kocherbitov V, Latynis J, Misiūnas A, Barauskas J, Niaura G. Hydration of lysozyme studied by Raman spectroscopy. J Phys Chem B 2013; 117(17): 4981-92.
[http://dx.doi.org/10.1021/jp4017954] [PMID: 23557185]

[38] Tsuboi M, Suzuki M, Overman SA, Thomas GJ Jr. Intensity of the polarized Raman band at 1340-1345 cm-1 as an indicator of protein α-helix orientation: application to Pf1 filamentous virus. Biochemistry 2000; 39(10): 2677-84.
[http://dx.doi.org/10.1021/bi9918846] [PMID: 10704218]

[39] De Felice FG, Vieira MN, Meirelles MN, Morozova-Roche LA, Dobson CM, Ferreira ST. Formation of amyloid aggregates from human lysozyme and its disease-associated variants using hydrostatic pressure. FASEB J 2004; 18(10): 1099-101.
[http://dx.doi.org/10.1096/fj.03-1072fje] [PMID: 15155566]

[40] Miura T, Takeuchi H, Harada I. Raman spectroscopic characterization of tryptophan side chains in lysozyme bound to inhibitors: role of the hydrophobic box in the enzymatic function. Biochemistry 1991; 30(24): 6074-80.
[http://dx.doi.org/10.1021/bi00238a035] [PMID: 1646007]

[41] Miura T, Takeuchi H, Harada I. Raman spectroscopic characterization of tryptophan side chains in lysozyme bound to inhibitors: role of the hydrophobic box in the enzymatic function. Biochemistry 1991; 30(24): 6074-80.
[http://dx.doi.org/10.1021/bi00238a035] [PMID: 1646007]

[42] Dolui S, Mondal A, Roy A, *et al.* Order, Disorder, and Reorder State of Lysozyme: Aggregation Mechanism by Raman Spectroscopy. J Phys Chem B 2020; 124(1): 50-60.
[http://dx.doi.org/10.1021/acs.jpcb.9b09139] [PMID: 31820990]

[43] Harada I, Miura T, Takeuchi H. Origin of the doublet at 1360 and 1340 cm−1 in the Raman spectra of tryptophan and related compounds. Spectrochim Acta Part Mol Spectrosc 1986; 42: 307-12.
[http://dx.doi.org/10.1016/0584-8539(86)80193-3]

[44] Siamwiza MN, Lord RC, Chen MC, *et al.* Interpretation of the doublet at 850 and 830 cm-1 in the Raman spectra of tyrosyl residues in proteins and certain model compounds. Biochemistry 1975; 14(22): 4870-6.
[http://dx.doi.org/10.1021/bi00693a014] [PMID: 241390]

[45] Rodriguez-Mendieta IR, Spence GR, Gell C, Radford SE, Smith DA. Ultraviolet resonance Raman studies reveal the environment of tryptophan and tyrosine residues in the native and partially folded states of the E colicin-binding immunity protein Im7. Biochemistry 2005; 44(9): 3306-15.
[http://dx.doi.org/10.1021/bi047746k] [PMID: 15736941]

[46] Hernandez B, Coic Y-M, Pflueger F, *et al.* All characteristic Raman markers of tyrosine and tyrosinate originate from phenol ring fundamental vibrations. J Raman Spectrosc 2016; 47: 210-20.
[http://dx.doi.org/10.1002/jrs.4776]

[47] Arp Z, Autrey D, Laane J, Overman SA, Thomas GJ Jr. Tyrosine Raman signatures of the filamentous virus Ff are diagnostic of non-hydrogen-bonded phenoxyls: demonstration by Raman and infrared spectroscopy of p-cresol vapor. Biochemistry 2001; 40(8): 2522-9.
[http://dx.doi.org/10.1021/bi0023753] [PMID: 11327874]

[48] Kurouski D, Washington J, Ozbil M, Prabhakar R, Shekhtman A, Lednev IK. Disulfide bridges remain intact while native insulin converts into amyloid fibrils. PLoS One 2012; 7(6): e36989. Epub ahead of print.
[http://dx.doi.org/10.1371/journal.pone.0036989] [PMID: 22675475]

[49] Nakamura K, Era S, Ozaki Y, Sogami M, Hayashi T, Murakami M. Conformational changes in seventeen cystine disulfide bridges of bovine serum albumin proved by Raman spectroscopy. FEBS Lett 1997; 417(3): 375-8.
[http://dx.doi.org/10.1016/S0014-5793(97)01326-4] [PMID: 9409755]

[50] Sugeta H, Go A, Miyazawa T. Vibrational Spectra and Molecular Conformations of Dialkyl Disulfides. Bull Chem Soc Jpn 1973; 46: 3407.
[http://dx.doi.org/10.1246/bcsj.46.3407]

[51] Zheng R, Zheng X, Dong J, Carey PR. Proteins can convert to β-sheet in single crystals. Protein Sci 2004; 13(5): 1288-94.
[http://dx.doi.org/10.1110/ps.03550404] [PMID: 15096634]

[52] Xu M, Ermolenkov VV, He W, Uversky VN, Fredriksen L, Lednev IK. Lysozyme fibrillation: deep UV Raman spectroscopic characterization of protein structural transformation. Biopolymers 2005; 79(1): 58-61.
[http://dx.doi.org/10.1002/bip.20330] [PMID: 15962278]

[53] Desroches J, Jermyn M, Pinto M, *et al. et al.*A new method using Raman spectroscopy for *in vivo* targeted brain cancer tissue biopsy Sci Rep; 8 EPUB ahead of print 29 January 2018.
[http://dx.doi.org/10.1038/s41598-018-20233-3]

[54] Fu Y, Kuppe C, Valev VK, Fu H, Zhang L, Chen J. Surface-Enhanced Raman Spectroscopy: A Facile and Rapid Method for the Chemical Component Study of Individual Atmospheric Aerosol. Environ Sci Technol 2017; 51(11): 6260-7.
[http://dx.doi.org/10.1021/acs.est.6b05910] [PMID: 28498657]

[55] Das AK, Rawat A, Bhowmik D, Pandit R, Huster D, Maiti S. An early folding contact between Phe19 and Leu34 is critical for amyloid-β oligomer toxicity. ACS Chem Neurosci 2015; 6(8): 1290-5.
[http://dx.doi.org/10.1021/acschemneuro.5b00074] [PMID: 25951510]

[56] Deckert-Gaudig T, Kämmer E, Deckert V. Tracking of nanoscale structural variations on a single amyloid fibril with tip-enhanced Raman scattering. J Biophotonics 2012; 5(3): 215-9.
[http://dx.doi.org/10.1002/jbio.201100142] [PMID: 22271749]

[57] Sathyavathi R, Dingari NC, Barman I, *et al.* Raman spectroscopy provides a powerful, rapid diagnostic tool for the detection of tuberculous meningitis in *ex vivo* cerebrospinal fluid samples. J Biophotonics 2013; 6(8): 567-72.
[http://dx.doi.org/10.1002/jbio.201200110] [PMID: 22887773]

[58] van Howe J, Xu C. Ultrafast optical signal processing based upon space-time dualities. J Lit Technol 2006; 24: 2649-62.

[http://dx.doi.org/10.1109/JLT.2006.875229]

[59]   Ryzhikova E, Kazakov O, Halamkova L, *et al.* Raman spectroscopy of blood serum for Alzheimer's disease diagnostics: specificity relative to other types of dementia. J Biophotonics 2015; 8(7): 584-96.
[http://dx.doi.org/10.1002/jbio.201400060] [PMID: 25256347]

[60]   Evans CL, Xie XS. Coherent anti-stokes Raman scattering microscopy: chemical imaging for biology and medicine. Annu Rev Anal Chem (Palo Alto, Calif) 2008; 1: 883-909.
[http://dx.doi.org/10.1146/annurev.anchem.1.031207.112754] [PMID: 20636101]

[61]   Pozzi EA, Sonntag MD, Jiang N, Klingsporn JM, Hersam MC, Van Duyne RP. Tip-enhanced Raman imaging: an emergent tool for probing biology at the nanoscale. ACS Nano 2013; 7(2): 885-8.
[http://dx.doi.org/10.1021/nn400560t] [PMID: 23441673]

[62]   Kumar S, Matange N, Umapathy S, Visweswariah SS. Linking carbon metabolism to carotenoid production in mycobacteria using Raman spectroscopy. FEMS Microbiol Lett 2015; 362(3): 1-6. Epub ahead of print.
[http://dx.doi.org/10.1093/femsle/fnu048] [PMID: 25673658]

[63]   Apetri MM, Maiti NC, Zagorski MG, Carey PR, Anderson VE. Secondary structure of alpha-synuclein oligomers: characterization by raman and atomic force microscopy. J Mol Biol 2006; 355(1): 63-71.
[http://dx.doi.org/10.1016/j.jmb.2005.10.071] [PMID: 16303137]

[64]   Djaker N, Lenne P, Rigneault H. Vibrational imaging coherent anti-Stokes Raman scattering (CARS) microscopy. In: Avrillier S, Tualle JM, Eds. FEMTOSECOND LASER APPLICATIONS IN BIOLOGY. 2004; pp. 133-9.
[http://dx.doi.org/10.1117/12.544999]

[65]   Kuhar N, Sil S, Verma T, *et al.* Challenges in application of Raman spectroscopy to biology and materials. RSC Advances 2018; 8: 25888-908.
[http://dx.doi.org/10.1039/C8RA04491K]

[66]   Devitt G, Howard K, Mudher A, Mahajan S. Raman Spectroscopy: An Emerging Tool in Neurodegenerative Disease Research and Diagnosis. ACS Chem Neurosci 2018; 9(3): 404-20.
[http://dx.doi.org/10.1021/acschemneuro.7b00413] [PMID: 29308873]

[67]   Hoyer W, Cherny D, Subramaniam V, Jovin TM. Rapid self-assembly of α-synuclein observed by *in situ* atomic force microscopy. J Mol Biol 2004; 340(1): 127-39.
[http://dx.doi.org/10.1016/j.jmb.2004.04.051] [PMID: 15184027]

[68]   Sanders A, Zhang L, Bowman RW, Herrmann LO, Baumberg JJ. Facile Fabrication of Spherical Nanoparticle-Tipped AFM Probes for Plasmonic Applications. Part Part Syst Charact 2015; 32(2): 182-7.
[http://dx.doi.org/10.1002/ppsc.201400104] [PMID: 26213449]

[69]   Patra PP, Chikkaraddy R, Tripathi RPN, Dasgupta A, Kumar GV. Plasmofluidic single-molecule surface-enhanced Raman scattering from dynamic assembly of plasmonic nanoparticles. Nat Commun 2014; 5: 4357. Epub ahead of print.
[http://dx.doi.org/10.1038/ncomms5357] [PMID: 25000476]

[70]   Bailo E, Deckert V. Tip-enhanced Raman spectroscopy of single RNA strands: towards a novel direct-sequencing method. Angew Chem Int Ed Engl 2008; 47(9): 1658-61.
[http://dx.doi.org/10.1002/anie.200704054] [PMID: 18188855]

[71]   Bailo E, Deckert V. Tip-enhanced Raman scattering. Chem Soc Rev 2008; 37(5): 921-30.
[http://dx.doi.org/10.1039/b705967c] [PMID: 18443677]

[72]   Blum C, Schmid T, Opilik L, *et al.* Missing Amide I Mode in Gap-Mode Tip-Enhanced Raman Spectra of Proteins. J Phys Chem C 2012; 116: 23061-6.
[http://dx.doi.org/10.1021/jp306831p]

[73]   Hayazawa N, Inouye Y, Sekkat Z, *et al.* Near-field Raman scattering enhanced by a metallized tip. Chem Phys Lett 2001; 335: 369-74.

[http://dx.doi.org/10.1016/S0009-2614(01)00065-3]

[74]   Barbillat J, Dhamelincourt P, Delhaye M, *et al.* Raman Confocal Microprobing, Imaging and Fiberoptic Remote-Sensing - A Further Step in Molecular Analysis. J Raman Spectrosc 1994; 25: 3-11.
[http://dx.doi.org/10.1002/jrs.1250250103]

[75]   Caspers PJ, Lucassen GW, Carter EA, Bruining HA, Puppels GJ. *In vivo* confocal Raman microspectroscopy of the skin: noninvasive determination of molecular concentration profiles. J Invest Dermatol 2001; 116(3): 434-42.
[http://dx.doi.org/10.1046/j.1523-1747.2001.01258.x] [PMID: 11231318]

[76]   Pudney PDA, Mélot M, Caspers PJ, Van Der Pol A, Puppels GJ. An *in vivo* confocal Raman study of the delivery of trans retinol to the skin. Appl Spectrosc 2007; 61(8): 804-11.
[http://dx.doi.org/10.1366/000370207781540042] [PMID: 17716398]

[77]   Schuster KC, Reese I, Urlaub E, Gapes JR, Lendl B. Multidimensional information on the chemical composition of single bacterial cells by confocal Raman microspectroscopy. Anal Chem 2000; 72(22): 5529-34.
[http://dx.doi.org/10.1021/ac000718x] [PMID: 11101227]

[78]   Kurouski D, Postiglione T, Deckert-Gaudig T, Deckert V, Lednev IK. Amide I vibrational mode suppression in surface (SERS) and tip (TERS) enhanced Raman spectra of protein specimens. Analyst (Lond) 2013; 138(6): 1665-73.
[http://dx.doi.org/10.1039/c2an36478f] [PMID: 23330149]

[79]   Kurouski D, Deckert-Gaudig T, Deckert V, Lednev IK. Surface characterization of insulin protofilaments and fibril polymorphs using tip-enhanced Raman spectroscopy (TERS). Biophys J 2014; 106(1): 263-71.
[http://dx.doi.org/10.1016/j.bpj.2013.10.040] [PMID: 24411258]

[80]   Stiles PL, Dieringer JA, Shah NC, Van Duyne RP. Surface-enhanced Raman spectroscopy. Annu Rev Anal Chem (Palo Alto, Calif) 2008; 1: 601-26.
[http://dx.doi.org/10.1146/annurev.anchem.1.031207.112814] [PMID: 20636091]

[81]   McFarland AD, Young MA, Dieringer JA, Van Duyne RP. Wavelength-scanned surface-enhanced Raman excitation spectroscopy. J Phys Chem B 2005; 109(22): 11279-85.
[http://dx.doi.org/10.1021/jp050508u] [PMID: 16852377]

[82]   Chattopadhyay S, Sabharwal PK, Jain S, Kaur A, Singh H. Functionalized polymeric magnetic nanoparticle assisted SERS immunosensor for the sensitive detection of S. typhimurium. Anal Chim Acta 2019; 1067: 98-106.
[http://dx.doi.org/10.1016/j.aca.2019.03.050] [PMID: 31047154]

[83]   Böhme R, Mkandawire M, Krause-Buchholz U, *et al.* Characterizing cytochrome c states--TERS studies of whole mitochondria. Chem Commun (Camb) 2011; 47(41): 11453-5.
[http://dx.doi.org/10.1039/c1cc15246g] [PMID: 21947234]

[84]   Kočišová E, Procházka M, Vaculčiaková L. Drop-Coating Deposition Raman (DCDR) Spectroscopy as a Tool for Membrane Interaction Studies: Liposome-Porphyrin Complex. Appl Spectrosc 2015; 69(8): 939-45.
[http://dx.doi.org/10.1366/14-07836] [PMID: 26163374]

[85]   Filik J, Stone N. Drop coating deposition Raman spectroscopy of protein mixtures. Analyst (Lond) 2007; 132(6): 544-50.
[http://dx.doi.org/10.1039/b701541k] [PMID: 17525811]

[86]   Ortiz C, Zhang D, Xie Y, Ribbe AE, Ben-Amotz D. Validation of the drop coating deposition Raman method for protein analysis. Anal Biochem 2006; 353(2): 157-66.
[http://dx.doi.org/10.1016/j.ab.2006.03.025] [PMID: 16674909]

[87]   Lednev IK, Ermolenkov VV, He W, Xu M. Deep-UV Raman spectrometer tunable between 193 and

205 nm for structural characterization of proteins. Anal Bioanal Chem 2005; 381(2): 431-7.
[http://dx.doi.org/10.1007/s00216-004-2991-5] [PMID: 15625596]

[88]   Ahmed Z, Beta IA, Mikhonin AV, Asher SA. UV-resonance raman thermal unfolding study of Trp-cage shows that it is not a simple two-state miniprotein. J Am Chem Soc 2005; 127(31): 10943-50.
[http://dx.doi.org/10.1021/ja050664e] [PMID: 16076200]

[89]   Shashilov VA, Sikirzhytski V, Popova LA, Lednev IK. Quantitative methods for structural characterization of proteins based on deep UV resonance Raman spectroscopy. Methods 2010; 52(1): 23-37.
[http://dx.doi.org/10.1016/j.ymeth.2010.05.004] [PMID: 20580825]

[90]   Chi Z, Asher SA. UV resonance Raman determination of protein acid denaturation: selective unfolding of helical segments of horse myoglobin. Biochemistry 1998; 37(9): 2865-72.
[http://dx.doi.org/10.1021/bi971161r] [PMID: 9485437]

[91]   Mikhonin AV, Asher SA, Bykov SV, Murza A. UV Raman spatially resolved melting dynamics of isotopically labeled polyalanyl peptide: slow alpha-helix melting follows 3(10)-helices and pi-bulges premelting. J Phys Chem B 2007; 111(12): 3280-92.
[http://dx.doi.org/10.1021/jp0654009] [PMID: 17388440]

[92]   Dong J, Wan ZL, Chu YC, *et al.* Isotope-Edited Raman Spectroscopy of Proteins: A General Strategy to Probe Individual Peptide Bonds with Application to Insulin. 2001; 123: p. 7919.

[93]   Wokaun AB. Schrader: Infrared and Raman Spectroscopy - Methods and Applications. VCH, Weinheim, 1995, DM 298,-, ISBN 3-527-26446-9. Ber Bunsenges Phys Chem 1996; 100: 1268-8.
[http://dx.doi.org/10.1002/bbpc.19961000733]

[94]   Lykina A, Artemyev D, Kukushkin V, *et al.* Multivariate analysis of tissues Raman spectra using regression methods. J Phys Conf Ser 2019; 1368: 022042.
[http://dx.doi.org/10.1088/1742-6596/1368/2/022042]

[95]   Lee HJ, Cheng J-X. Imaging chemistry inside living cells by stimulated Raman scattering microscopy. Methods 2017; 128: 119-28.
[http://dx.doi.org/10.1016/j.ymeth.2017.07.020] [PMID: 28746829]

[96]   Djaker N, Lenne P-F, Marguet D, *et al.* Coherent anti-Stokes Raman scattering microscopy (CARS): Instrumentation and applications Nucl Instrum METHODS Phys Res Sect -Accel Spectrometers Detect Assoc Equip 2007; 571: 177-81.

[97]   Alexander KD, Skinner K, Zhang S, Wei H, Lopez R. Tunable SERS in gold nanorod dimers through strain control on an elastomeric substrate. Nano Lett 2010; 10(11): 4488-93.
[http://dx.doi.org/10.1021/nl1023172] [PMID: 20923232]

[98]   Cai S, Singh BR. A distinct utility of the amide III infrared band for secondary structure estimation of aqueous protein solutions using partial least squares methods. Biochemistry 2004; 43(9): 2541-9.
[http://dx.doi.org/10.1021/bi030149y] [PMID: 14992591]

[99]   Guimarães LL, Moreira LP, Lourenço BF, *et al.* Multivariate Method Based on Raman Spectroscopy for Quantification of Dipyrone in Oral Solutions. J Spectrosc 2018; 2018: 3538171.
[http://dx.doi.org/10.1155/2018/3538171]

[100]   Oshokoya OO, Roach CA, Jiji RD. Quantification of protein secondary structure content by multivariate analysis of deep-ultraviolet resonance Raman and circular dichroism spectroscopies. Anal Methods 2014; 6: 1691-9.
[http://dx.doi.org/10.1039/C3AY42032A]

[101]   Pichardo-Molina JL, Frausto-Reyes C, Barbosa-García O, *et al.* Raman spectroscopy and multivariate analysis of serum samples from breast cancer patients. Lasers Med Sci 2007; 22(4): 229-36.
[http://dx.doi.org/10.1007/s10103-006-0432-8] [PMID: 17297595]

[102]   Morris AM, Watzky MA, Finke RG. Protein aggregation kinetics, mechanism, and curve-fitting: a review of the literature. Biochim Biophys Acta 2009; 1794(3): 375-97.

[http://dx.doi.org/10.1016/j.bbapap.2008.10.016] [PMID: 19071235]

[103] Das S, Pal U, Das S, *et al.* Sequence complexity of amyloidogenic regions in intrinsically disordered human proteins. PLoS One 2014; 9(3): e89781. Epub ahead of print.
[http://dx.doi.org/10.1371/journal.pone.0089781] [PMID: 24594841]

[104] Obici L, Perfetti V, Palladini G, Moratti R, Merlini G. Clinical aspects of systemic amyloid diseases. Biochim Biophys Acta 2005; 1753(1): 11-22.
[http://dx.doi.org/10.1016/j.bbapap.2005.08.014] [PMID: 16198646]

[105] Baskakov IV, Sipe JD, Bucciantini M, Giannoni E, Chiti F. Amyloid Proteins: The Beta Sheet Conformation and Disease. 2005. *et al.* Inherent toxicity of aggregates implies a common mechanism for protein misfolding diseases NATURE 2005; 416: 507-11.

[106] Bucciantini M, Giannoni E, Chiti F, *et al.* Inherent toxicity of aggregates implies a common mechanism for protein misfolding diseases. Nature 2002; 416(6880): 507-11.
[http://dx.doi.org/10.1038/416507a] [PMID: 11932737]

[107] Iram A, Naeem A. Protein folding, misfolding, aggregation and their implications in human diseases: discovering therapeutic ways to amyloid-associated diseases. Cell Biochem Biophys 2014; 70(1): 51-61.
[http://dx.doi.org/10.1007/s12013-014-9904-9] [PMID: 24639112]

[108] Jean E, Ebbo M, Valleix S, *et al.* A new family with hereditary lysozyme amyloidosis with gastritis and inflammatory bowel disease as prevailing symptoms. BMC Gastroenterol 2014; 14: 159.
[http://dx.doi.org/10.1186/1471-230X-14-159] [PMID: 25217048]

[109] Eisenberg D, Jucker M. The amyloid state of proteins in human diseases. Cell 2012; 148(6): 1188-203.
[http://dx.doi.org/10.1016/j.cell.2012.02.022] [PMID: 22424229]

[110] Ke PC, Sani M-A, Ding F, *et al.* Implications of peptide assemblies in amyloid diseases. Chem Soc Rev 2017; 46(21): 6492-531.
[http://dx.doi.org/10.1039/C7CS00372B] [PMID: 28702523]

[111] Cui M. Past and recent progress of molecular imaging probes for β-amyloid plaques in the brain. Curr Med Chem 2014; 21(1): 82-112.
[http://dx.doi.org/10.2174/09298673113209990216] [PMID: 23992340]

[112] Koffie RM, Meyer-Luehmann M, Hashimoto T, *et al.* Oligomeric amyloid β associates with postsynaptic densities and correlates with excitatory synapse loss near senile plaques. Proc Natl Acad Sci USA 2009; 106(10): 4012-7.
[http://dx.doi.org/10.1073/pnas.0811698106] [PMID: 19228947]

[113] Dauer W, Przedborski S. Parkinson's disease: mechanisms and models. Neuron 2003; 39(6): 889-909.
[http://dx.doi.org/10.1016/S0896-6273(03)00568-3] [PMID: 12971891]

[114] Kalia LV, Kalia SK, McLean PJ, Lozano AM, Lang AE. α-Synuclein oligomers and clinical implications for Parkinson disease. Ann Neurol 2013; 73(2): 155-69.
[http://dx.doi.org/10.1002/ana.23746] [PMID: 23225525]

[115] Bernstein SL, Dupuis NF, Lazo ND, *et al.* Amyloid-β protein oligomerization and the importance of tetramers and dodecamers in the aetiology of Alzheimer's disease. Nat Chem 2009; 1(4): 326-31.
[http://dx.doi.org/10.1038/nchem.247] [PMID: 20703363]

[116] Hardy JA, Higgins GA. Alzheimer's disease: the amyloid cascade hypothesis. Science 1992; 256(5054): 184-5.
[http://dx.doi.org/10.1126/science.1566067] [PMID: 1566067]

[117] Chiti F, Dobson CM. Protein misfolding, functional amyloid, and human disease. Annu Rev Biochem 2006; 75: 333-66.
[http://dx.doi.org/10.1146/annurev.biochem.75.101304.123901] [PMID: 16756495]

[118] Chiti F, Dobson CM. Protein Misfolding, Amyloid Formation, and Human Disease: A Summary of

Progress Over the Last Decade. Annu Rev Biochem 2017; 86: 27-68.
[http://dx.doi.org/10.1146/annurev-biochem-061516-045115] [PMID: 28498720]

[119] Yu N-T, Jo BH. Comparison of protein structure in crystals and in solution by laser raman scattering. I. Lysozyme. Arch Biochem Biophys 1973; 156(2): 469-74.
[http://dx.doi.org/10.1016/0003-9861(73)90296-8] [PMID: 4718781]

[120] Shashilov VA, Lednev IK. Two-dimensional correlation Raman spectroscopy for characterizing protein structure and dynamics. J Raman Spectrosc 2009; 40: 1749-58.
[http://dx.doi.org/10.1002/jrs.2544]

[121] Sane SU, Cramer SM, Przybycien TM. A holistic approach to protein secondary structure characterization using amide I band Raman spectroscopy. Anal Biochem 1999; 269(2): 255-72.
[http://dx.doi.org/10.1006/abio.1999.4034] [PMID: 10221997]

[122] Handen JD, Lednev IK, Deep UV. Deep UV Resonance Raman Spectroscopy for Characterizing Amyloid Aggregation. Methods Mol Biol 2016; 1345: 89-100.
[http://dx.doi.org/10.1007/978-1-4939-2978-8_6] [PMID: 26453207]

[123] Syme CD, Blanch EW, Holt C, *et al.* A Raman optical activity study of rheomorphism in caseins, synucleins and tau. New insight into the structure and behaviour of natively unfolded proteins. Eur J Biochem 2002; 269(1): 148-56.
[http://dx.doi.org/10.1046/j.0014-2956.2001.02633.x] [PMID: 11784308]

[124] Lippert JL, Tyminski D, Desmeules PJ. Determination of the secondary structure of proteins by laser Raman spectroscopy. J Am Chem Soc 1976; 98(22): 7075-80.
[http://dx.doi.org/10.1021/ja00438a057] [PMID: 965667]

[125] Kinalwa MN, Blanch EW, Doig AJ. Accurate determination of protein secondary structure content from Raman and Raman optical activity spectra. Anal Chem 2010; 82(15): 6347-9.
[http://dx.doi.org/10.1021/ac101334h] [PMID: 20669990]

[126] Jakubek RS, Handen J, White SE, Asher SA, Lednev IK. Ultraviolet Resonance Raman Spectroscopic Markers for Protein Structure and Dynamics. Trends Analyt Chem 2018; 103: 223-9.
[http://dx.doi.org/10.1016/j.trac.2017.12.002] [PMID: 32029956]

[127] Bonnier F, Byrne HJ. Understanding the molecular information contained in principal component analysis of vibrational spectra of biological systems. Analyst (Lond) 2012; 137(2): 322-32.
[http://dx.doi.org/10.1039/C1AN15821J] [PMID: 22114757]

[128] Miao K, Wei L. Live-Cell Imaging and Quantification of PolyQ Aggregates by Stimulated Raman Scattering of Selective Deuterium Labeling. ACS Cent Sci 2020; 6(4): 478-86.
[http://dx.doi.org/10.1021/acscentsci.9b01196] [PMID: 32341997]

[129] Wen C-I, Hiramatsu H. The 532-nm-excited hyper-Raman spectroscopy of globular protein and aromatic amino acids. J Raman Spectrosc 2020; 51: 274-8.
[http://dx.doi.org/10.1002/jrs.5777]

[130] Kögler M, Itkonen J, Viitala T, Casteleijn MG. Assessment of recombinant protein production in E. coli with Time-Gated Surface Enhanced Raman Spectroscopy (TG-SERS). Sci Rep 2020; 10(1): 2472.
[http://dx.doi.org/10.1038/s41598-020-59091-3] [PMID: 32051493]

[131] Fang C, Tang L. Mapping Structural Dynamics of Proteins with Femtosecond Stimulated Raman Spectroscopy. Annu Rev Phys Chem 2020; 71: 239-65.
[http://dx.doi.org/10.1146/annurev-physchem-071119-040154] [PMID: 32075503]

[132] Buhrke D, Hildebrandt P. Probing Structure and Reaction Dynamics of Proteins Using Time-Resolved Resonance Raman Spectroscopy. Chem Rev 2020; 120(7): 3577-630.
[http://dx.doi.org/10.1021/acs.chemrev.9b00429] [PMID: 31814387]

[133] Wen ZQ. Raman spectroscopy of protein pharmaceuticals. J Pharm Sci 2007; 96(11): 2861-78.
[http://dx.doi.org/10.1002/jps.20895] [PMID: 17847076]

[134] Dong J, Wan Z, Popov M, Carey PR, Weiss MA. Insulin assembly damps conformational fluctuations: Raman analysis of amide I linewidths in native states and fibrils. J Mol Biol 2003; 330(2): 431-42.
[http://dx.doi.org/10.1016/S0022-2836(03)00536-9] [PMID: 12823980]

[135] Huang K, Maiti NC, Phillips NB, Carey PR, Weiss MA. Structure-specific effects of protein topology on cross-beta assembly: studies of insulin fibrillation. Biochemistry 2006; 45(34): 10278-93.
[http://dx.doi.org/10.1021/bi060879g] [PMID: 16922503]

[136] de Laureto PP, Frare E, Battaglia F, Mossuto MF, Uversky VN, Fontana A. Protein dissection enhances the amyloidogenic properties of alpha-lactalbumin. FEBS J 2005; 272(9): 2176-88.
[http://dx.doi.org/10.1111/j.1742-4658.2005.04638.x] [PMID: 15853802]

[137] Benevides JM, Bondre P, Duda RL, Hendrix RW, Thomas GJ Jr. Domain structures and roles in bacteriophage HK97 capsid assembly and maturation. Biochemistry 2004; 43(18): 5428-36.
[http://dx.doi.org/10.1021/bi0302494] [PMID: 15122908]

[138] Benevides JM, Weiss MA, Thomas GJ Jr. Design of the helix-turn-helix motif: nonlocal effects of quaternary structure in DNA recognition investigated by laser Raman spectroscopy. Biochemistry 1991; 30(18): 4381-8.
[http://dx.doi.org/10.1021/bi00232a003] [PMID: 2021630]

[139] Tsuboi M, Suzuki M, Overman SA, Thomas GJ Jr. Intensity of the polarized Raman band at 1340-1345 cm-1 as an indicator of protein alpha-helix orientation: application to Pf1 filamentous virus. Biochemistry 2000; 39(10): 2677-84.
[http://dx.doi.org/10.1021/bi9918846] [PMID: 10704218]

[140] Xu M, Shashilov VA, Ermolenkov VV, Fredriksen L, Zagorevski D, Lednev IK. The first step of hen egg white lysozyme fibrillation, irreversible partial unfolding, is a two-state transition. Protein Sci 2007; 16(5): 815-32.
[http://dx.doi.org/10.1110/ps.062639307] [PMID: 17400924]

[141] Ettah I, Ashton L. Engaging with Raman Spectroscopy to Investigate Antibody Aggregation. Antibodies (Basel) 2018; 7(3): E24. Epub ahead of print.
[http://dx.doi.org/10.3390/antib7030024] [PMID: 31544876]

[142] Schwenk N, Mizaikoff B, Cárdenas S, López-Lorente ÁI. Gold-nanostar-based SERS substrates for studying protein aggregation processes. Analyst (Lond) 2018; 143(21): 5103-11.
[http://dx.doi.org/10.1039/C8AN00804C] [PMID: 30178815]

[143] Wälti MA, Ravotti F, Arai H, *et al*. Atomic-resolution structure of a disease-relevant Aβ(1-42) amyloid fibril. Proc Natl Acad Sci USA 2016; 113(34): E4976-84.
[http://dx.doi.org/10.1073/pnas.1600749113] [PMID: 27469165]

[144] Jahn TR, Radford SE. Folding *versus* aggregation: polypeptide conformations on competing pathways. Arch Biochem Biophys 2008; 469(1): 100-17.
[http://dx.doi.org/10.1016/j.abb.2007.05.015] [PMID: 17588526]

[145] Amani S, Naeem A. Understanding protein folding from globular to amyloid state: Aggregation: Darker side of protein. Process Biochem 2013; 48: 1651-64.
[http://dx.doi.org/10.1016/j.procbio.2013.08.011]

[146] Chaturvedi SK, Siddiqi MK, Alam P, *et al*. Protein misfolding and aggregation: Mechanism, factors and detection. Process Biochem 2016; 51: 1183-92.
[http://dx.doi.org/10.1016/j.procbio.2016.05.015]

[147] Fink AL. Natively unfolded proteins. Curr Opin Struct Biol 2005; 15(1): 35-41.
[http://dx.doi.org/10.1016/j.sbi.2005.01.002] [PMID: 15718131]

[148] Xing L, Fan W, Chen N, *et al*. Amyloid formation kinetics of hen egg white lysozyme under heat and acidic conditions revealed by Raman spectroscopy. J Raman Spectrosc 2019; 50: 629-40.
[http://dx.doi.org/10.1002/jrs.5567]

[149] Bai P, Peng Z. Cooperative folding of the isolated alpha-helical domain of hen egg-white lysozyme. J Mol Biol 2001; 314(2): 321-9.
[http://dx.doi.org/10.1006/jmbi.2001.5122] [PMID: 11718563]

[150] Ma B, Zhang F, Wang X, Zhu X. Investigating the inhibitory effects of zinc ions on amyloid fibril formation of hen egg-white lysozyme. Int J Biol Macromol 2017; 98: 717-22.
[http://dx.doi.org/10.1016/j.ijbiomac.2017.01.128] [PMID: 28163126]

[151] Vernaglia BA, Huang J, Clark ED. Guanidine hydrochloride can induce amyloid fibril formation from hen egg-white lysozyme. Biomacromolecules 2004; 5(4): 1362-70.
[http://dx.doi.org/10.1021/bm0498979] [PMID: 15244452]

[152] Xu M, Ermolenkov VV, Uversky VN, Lednev IK. Hen egg white lysozyme fibrillation: a deep-UV resonance Raman spectroscopic study. J Biophotonics 2008; 1(3): 215-29.
[http://dx.doi.org/10.1002/jbio.200710013] [PMID: 19412971]

[153] Proctor VA, Cunningham FE. The chemistry of lysozyme and its use as a food preservative and a pharmaceutical. Crit Rev Food Sci Nutr 1988; 26(4): 359-95.
[http://dx.doi.org/10.1080/10408398809527473] [PMID: 3280250]

[154] Pepys MB, Hawkins PN, Booth DR, *et al.* Human lysozyme gene mutations cause hereditary systemic amyloidosis. Nature 1993; 362(6420): 553-7.
[http://dx.doi.org/10.1038/362553a0] [PMID: 8464497]

[155] Gillmore JD, Booth DR, Madhoo S, Pepys MB, Hawkins PN. Hereditary renal amyloidosis associated with variant lysozyme in a large English family. Nephrol Dial Transplant 1999; 14(11): 2639-44.
[http://dx.doi.org/10.1093/ndt/14.11.2639] [PMID: 10534505]

[156] Van Dael H, Haezebrouck P, Morozova L, Arico-Muendel C, Dobson CM. Partially folded states of equine lysozyme. Structural characterization and significance for protein folding. Biochemistry 1993; 32(44): 11886-94.
[http://dx.doi.org/10.1021/bi00095a018] [PMID: 8218261]

[157] Chiti F, Bucciantini M, Capanni C, Taddei N, Dobson CM, Stefani M. Solution conditions can promote formation of either amyloid protofilaments or mature fibrils from the HypF N-terminal domain. Protein Sci 2001; 10(12): 2541-7.
[http://dx.doi.org/10.1110/ps.ps.10201] [PMID: 11714922]

[158] Krebs MR, Wilkins DK, Chung EW, *et al.* Formation and seeding of amyloid fibrils from wild-type hen lysozyme and a peptide fragment from the beta-domain. J Mol Biol 2000; 300(3): 541-9.
[http://dx.doi.org/10.1006/jmbi.2000.3862] [PMID: 10884350]

[159] Frare E, Polverino De Laureto P, Zurdo J, Dobson CM, Fontana A. A highly amyloidogenic region of hen lysozyme. J Mol Biol 2004; 340(5): 1153-65.
[http://dx.doi.org/10.1016/j.jmb.2004.05.056] [PMID: 15236974]

[160] Ding F, LaRocque JJ, Dokholyan NV. Direct observation of protein folding, aggregation, and a prion-like conformational conversion. J Biol Chem 2005; 280(48): 40235-40.
[http://dx.doi.org/10.1074/jbc.M506372200] [PMID: 16204250]

[161] Jain S, Udgaonkar JB. Evidence for stepwise formation of amyloid fibrils by the mouse prion protein. J Mol Biol 2008; 382(5): 1228-41.
[http://dx.doi.org/10.1016/j.jmb.2008.07.052] [PMID: 18687339]

[162] Kumar S, Udgaonkar JB. Mechanisms of Amyloid Fibril Formation by Proteins. Curr Sci 2010; 98: 639.

[163] Mishra R, Sörgjerd K, Nyström S, Nordigården A, Yu YC, Hammarström P. Lysozyme amyloidogenesis is accelerated by specific nicking and fragmentation but decelerated by intact protein binding and conversion. J Mol Biol 2007; 366(3): 1029-44.
[http://dx.doi.org/10.1016/j.jmb.2006.11.084] [PMID: 17196616]

[164] Sawaya MR, Sambashivan S, Nelson R, *et al.* Atomic structures of amyloid cross-beta spines reveal varied steric zippers. Nature 2007; 447(7143): 453-7.
[http://dx.doi.org/10.1038/nature05695] [PMID: 17468747]

[165] Matthes D, Daebel V, Meyenberg K, *et al.* Spontaneous aggregation of the insulin-derived steric zipper peptide VEALYL results in different aggregation forms with common features. J Mol Biol 2014; 426(2): 362-76.
[http://dx.doi.org/10.1016/j.jmb.2013.10.020] [PMID: 24513105]

[166] Foderà V, Librizzi F, Groenning M, van de Weert M, Leone M. Secondary nucleation and accessible surface in insulin amyloid fibril formation. J Phys Chem B 2008; 112(12): 3853-8.
[http://dx.doi.org/10.1021/jp710131u] [PMID: 18311965]

[167] Masino L, Nicastro G, De Simone A, Calder L, Molloy J, Pastore A. The Josephin domain determines the morphological and mechanical properties of ataxin-3 fibrils. Biophys J 2011; 100(8): 2033-42.
[http://dx.doi.org/10.1016/j.bpj.2011.02.056] [PMID: 21504740]

CHAPTER 5

Surface-Enhanced Raman Scattering Based Rapid Pathogen Detection

Ujjal Kumar Sur[1,*] and **Amar Ghosh**[1]

[1] *Department of Chemistry, Behala College, University of Calcutta, Kolkata-60, India*

Abstract: During an epidemic, it is judgmentally vital to monitor the spread of the disease by proper diagnostic testing to detect pathogens. However, inadequate numbers of testing can hamper the detection of pathogens to prevent the spread. Consequently, healthcare systems require tests that are rapid, sensitive, inexpensive, and easy to use for diagnosing infections at earlier stages, even before symptoms become apparent, to reduce both the transmission and mortality rates. Among new and novel detection protocol, Surface-enhanced Raman scattering (SERS) has emerged as a versatile and popular surface sensitive spectroscopic analytical tool in consequence of the colossal amplification of weak Raman signal in the presence of plasmonic nanoparticles providing suitable detection of various chemical and biological systems and can be further utilized for rapid detection of pathogens. SERS technique has diverse applications ranging from plasmonics, sensing, catalysis to biomedical applications and diagnostics. This novel powerful analytical technique has been utilized to detect various biomolecules such as carbohydrates, proteins, amino acids both qualitative and quantitatively. SERS can even be used to detect pathogens including bacteria and viruses within a short timescale. There are few review articles written on the SERS based detection of various biomolecules. SERS technique has the great capability of a versatile bioanalytical tool for forefront implementations such as microorganism. In this book chapter, the exploitation of various SERS active substrates for the rapid identification of pathogens like viruses and bacteria has been discussed. The various studies involving the utilization of SERS technique to detect pathogens has been reviewed by us comprehensively in a comparative manner providing a novel approach in terms of diagnostics applications. Some examples have been illustrated from the recent studies on the detection of pathogens by SERS technique from our research group.

Keywords: Bacteria, Biosynthesis, Pathogen Detection, Raman Scattering, Silver Nanoparticles, Surface-enhanced Raman Scattering.

---

* **Corresponding author Ujjal Kumar Sur:** Department of Chemistry, Behala College, University of Calcutta, Kolkata-60, India; Tel:+919831445492; E-mail:uksur99@yahoo.co.in

Swati Jain and Sruti Chattopadhyay

5.1. INTRODUCTION

Various contagious ailments are brought about by parasites, viruses, bacteria and fungi, which are accountable for the death of 20 millions of people every year. Therefore, precise and prompt detection of various pathogens is essential for the rapid and effective investigation of contagious diseases with efficient disease management and epidemic preparedness. The most common technique for the detection of pathogens largely depends on highly sensitive polymerase chain reaction (PCR), which is methodologically demanding and expensive for extensive implementation, globally. Therefore, it is very important to develop new detection methods, which can bring inexpensive accurate, rapid and highly sensitive features to diagnostic tests. The recent outbreak of the Coronavirus SARS-CoV-2 (Covid-19) is growing severely and conventional methods of diagnosis such as polymerase chain reactions are not suitable owing to cost-effectiveness, accuracy and time duration of detection of this deadly virus. Therefore, a faster, cheaper and simpler method for the detection of viral infections in biofluids like saliva or blood can become a forefront device towards controlling the escalation of this deadly infection. Due to the plasmonic nanomaterials based high detection sensitivities, even up to single molecular levels, surface-enhanced Raman spectroscopy (SERS) has becoming as an emerging novel optical analytical tool for the rapid detection of pathogens providing an alternative to widely available PCR techniques.

Fleischmann and his research group from the University of Southampton, United Kingdom first showed the extremely large Raman signals attained from pyridine molecules adsorbed on a rough silver electrode in 1974 [1]. This exceptional scientific breakthrough brought noteworthy attention among researchers from various areas of science and technology. There is huge increase of the weak Raman scattering intensity by molecules in the vicinity of metallic nanostructured surfaces which is popularly known as Surface-enhanced Raman scattering (SERS) [1 - 4]. In SERS technique, generally the SERS enhancement factor is determined from the ratio between the Raman signals acquired from a certain number of molecules in the presence and absence of the metallic nanostructures. The enhancement factor in SERS phenomena is normally $\sim 10^6$, but it can be as high as 10^{10} [1 - 4].

The low sensitivity problem of traditional Raman spectroscopy has been solved as a result of the discovery of SERS and it also inspires the study of the interfacial processes relating boosted optical scattering from adsorbates on metal surfaces [5].

This book chapter demonstrates the SERS based detection of various biomolecules such as amino acids. In addition to this, we have also reviewed the fast identification of pathogens like viruses and bacteria employing the versatile SERS technique in this book chapter. We have reviewed the utilization of SERS technique to detect pathogens and provide a novel approach in terms of diagnostics application.

Owing to space restrictions, a full review of all up to date work on this new part of research is not possible. On the other hand, a small number of illustrations together with our own results had been shortened to exhibit the current development in SERS research to detect pathogens.

5.2. APPLICATIONS OF SERS

SERS is the most receptive analytical techniques accessible equally to surface science and nanoscience, which has been applied along with other surface sensitive techniques to investigate numerous fundamental and applied areas.

Nie and Emory [6] carried out the first study on single-molecule SERS by combining SERS technique together with the transmission electron microscopy (TEM) and scanning tunneling microscopy (STM) and detected Raman enhancement factor in the order of 10^{14} to 10^{15} for single Rhodamine 6G (R6G) molecule adsorbed on selected Ag nanoparticles.

Unstable reaction intermediates such as radical and radical ions on the electrode surface along with the overall reaction mechanism can be determined by the surface-enhanced Raman scattering spectroscopy SERS. Tian and his research group carried out the first *in-situ* electrochemical SERS (EC-SERS) experiment to study the electrochemical reduction of $PhCH_2Cl$ in organic solvent acetonitrile (CH_3CN) on Ag electrode [7] and benzyl radical anion and 3-phenylpropanenitrile were detected as the reaction intermediate and the main reaction product. The experimental SERS results were further established by theoretical quantum mechanical Density Functional Theory (DFT) calculations to identify the reaction intermediate and product as well as elucidate overall reaction mechanism.

Mulvihill *et al.* detected arsenate and arsenite ions in aqueous solutions with a detection limit of 1 ppb employing LB assemblies consist of various polyhedral Ag nanocrystals as SERS substrates [8]. Highly reproducible and highly transportable chemical sensor can be made from the SERS substrate, which could be effortlessly applied in field detection. Various molecules such as pesticides; herbicides; in water; chlorophenol derivatives and amino acids; chemical warfare agents; explosives; and a variety of organic pollutants can be detected employing SERS technique [9, 10].

SERS technique can be employed to detect glucose in human blood. Generally, glucose in blood can be checked by electrochemical-based sensors. A new method involving SERS substrates made-up by NSL technique had been applied to detect glucose in human blood [11].

5.3. SERS BASED DETECTION OF VARIOUS BIOMOLECULES

Biomolecules, such as nucleic acids, peptides/proteins and their building blocks can be demarcated as a chemical material created by a living organism. The detection of biomolecules is important in a variety of analytical, medical, biochemical and pharmaceutical application areas and techniques like immunological or molecular biological methods - enzyme-linked immunosorbent assay (ELISA) or polymerase chain reaction (PCR) can be applied to identify peptides and proteins as well as to augment nucleic acids for sequence-specific detection [12, 13]. Moreover, different optical and spectroscopic techniques such as fluorescence microscopy [14], UV-VIS absorption [15] and vibrational spectroscopy (IR absorption, Raman spectroscopy) [16] can be extensively employed for bioanalytical uses. Elements such as carbon, hydrogen, oxygen, nitrogen, phosphor and sulfur are the main molecular precursors in biomolecules which are the building blocks of life. Nucleic acids and proteins and its building blocks can be characterized and detected SERS as analytical tool which has been primarily demonstrated in this chapter. We have made an overview over the characterization and detection of biomolecules within this chapter.

5.3.1. Nucleotides and Nucleic Acids

Combination of sugar, phosphoryl group and nucleobases such as adenine (A), guanine (G) cytosine (C), thymine (T) and uracil (U) will produce nucleotides like DNA and RNA. Ribonucleic acid (RNA) and Deoxyribonucleic acid (DNA) are biopolymers of nucleotides, which can be named by the type of the sugar present in the backbone. D-ribose is the sugar component in RNA, while for DNA; it is 2-deoxy-D-ribose. RNA can act as a messenger (mRNA) by translating the intrinsic genetic information of the DNA into a defined peptide or protein sequence. Therefore, mRNA molecules can act as midway data storage, while the appearance of the DNA-coded information can be regulated at various stages into a given amino acid chain. Adenine related modes will dominate the SERS spectra of DNA and RNA molecules [17, 18]. Furthermore, adenine adsorbed on the metal nanoparticles can be employed as SERS active substrate The SERS spectra of adenine were observed [19] and a strong Raman peak at 730 cm^{-1} was observed corresponding to the breathing mode of vibration. The orientation of adenine and its derivatives polyadenine single stranded DNA (polyA) and adenosine monophosphate (AMP) were further studied at different pH values employing SERS along with surface enhanced IR absorption (SEIRA) on gold nanoshell particles [20]. It was proposed that the assumed orientation would be the 'end-on' attachment of adenine through the nitrogen atom N3 with the $C-NH_2$ bond perpendicular to the surface in comparison to the gold nanoshell surface.

The automatic SERS detection at various pH values was established [21] during several pH-sensitive SERS studies [22, 23]. The effect of the different experimental conditions, like type of colloid, adenine concentration and pH values on the SERS spectra was studied [18] and it was found that the SERS spectra were nicely corresponding with the N1-protonated form of adenine at acidic pH values applying both silver and gold colloids. In contradiction to earlier investigations, the SERS spectra at neutral and alkaline pH were assigned to the N9-deprotonated form rather than the neutral adenine molecule. Therefore, it was concluded that the 'typical' spectrum of adenine was attributed to the deprotonated form, similar to guanine and uracil. The variations in intensity with changing the adenine concentrations can be clarified by the high Raman cross section of the silver complexes.

5.3.2. Amino Acids and Peptides/Proteins

Amino acids are the most important of biomolecules which comprises of containing an amine and a carboxylic group as well as a side-chain stipulating each amino acid. They act as the building blocks of peptides and proteins. SERS technique had been applied to characterize amino acids, small peptides and proteins which have been described here. SERS technique can be applied as analytical tool along with quantum mechanical theoretical DFT (density functional theory) estimation to study the adsorption behavior of the zwitter ionic L-Cysteine on the silver surface *via* the carboxylate, ammonium and sulphydryl groups [24]. L-tryptophan was characterized by SERS technique and it was demonstrated that the Raman modes of carboxylate and amino groups became robust under experimental conditions [25] showed a unique spectral response, related It was concluded from the time-dependent SERS study that this amino acid can be preferentially attached *via* the carboxylate and amino groups of L-tryptophan [25, 26]. L-lysine attached to silver colloidal surfaces was studied using SERS spectroscopy indicating no exceptional conformation and orientation with respect to the metal surface [27]. The preferential interaction between the amino acid and the silver surface was verified by SERS and it was found that arginine will interact *via* the guanidinium moiety and not the carboxylate and amino groups [28]. It was observed that Ce^{3+} will reorient the structure of N-acetylalanine self-assembled at silver surfaces due to a interaction between Ce^{3+}

and the carbonyl as well as the amino moieties as monitored by SERS technique [29]. Addition of Pb^{2+} will regulate the structure of L-glutathione as observed by SERS suggesting the probable interaction between the Pb^{2+} and the carboxyl and amino groups of the amino acids [30]. SERS experiment was carried out at different pH values and electrode potentials to study the adsorption of enantiomeric and racemic forms of an amino acid [31]. Significant variances in the features of SERS spectra was observed by observing the SERS spectra of D-methionine and a racemic mixture of D- and L-methionine under acidic conditions with various electrode potentials. However, no differences in spectral response of observed SERS spectra were found in alkaline medium in contradiction of the observations under acidic conditions, as both the carboxylate and amino groups were responsible for the interaction with the electrode surface. Additionally, the occurrence of the stereoselective interaction between phenylalanine and monolayers of cysteine adsorbed on rough silver surfaces was also observed [32]. It was investigated that L-carnosine would interact mainly through the carboxylate group with the imidazole ring oriented perpendicular to the silver surface and the alanine moiety oriented parallel to the silver surface [33], was studied SERS along with DFT calculations were carried out to study Alafosfalin, the phosphono dipeptides of alanine [34]. Researchers had observed that these peptides will adsorb in the form of anionic species with the P-terminal acid group onto the silver surface Moreover, the adsorption mechanism of phosphono-dipeptides containing N-terminal glycine on silver surface was elucidated using SERS [35]. The SERS spectra observed for dipeptides containing dehydroalanine and dehydrophenylalanine [36] can be attributed to the symmetric stretching vibration of the carboxylate group, which suggests the preferential adsorption through the deprotonated carboxyl group. Homodipeptide diglycine on the silver surface was studied by Time-dependent and pH-dependent SERS [37]. SERS spectra of cysteine containing aromatic peptides were compared with the SERS spectra of the cell-penetrating peptide oligomer penetratin [38]. Adsorption of the phosphonate tripeptides on the metal surface took place through the phosphonate moiety causing SERS enhancement of the corresponding vibrational modes [39]. There were various studies on the characterization of A 14-amino acid peptide bombesin was investigated by SERS technique along with its various analogues [40 - 44]. It was concluded from these studies that the 8–14 bombesin portion had virtually a parallel orientation with respect to the metal surface [43]. The SERS study showed an irrelevant effect on the adsorption of human neurotensin and mutated analogues on silver surfaces due to the replacement of native amino acids in the investigated peptides [45]. Bombesin and five related peptides, along with neuromedin B were comparatively investigated and it was observed that the interaction between peptides and the metallic surface was dependent on the geometry of the tryptophan, amide bond, and S-C fragments of these molecules

[44]. Neuromedin B was studied employing several electrode materials and applied electrode potentials [46].

SERS based technique can be applied to characterize proteins by studying the interaction of bovine serum albumin (BSA) with gold nanoparticles. The presence of the S-S stretching vibration bands of disulfide bridges indicated that the disulfide bonds cannot be broken and the protein cannot be denatured at room temperature due to the attachment on the gold surface [47]. The adsorption of BSA to the metal surface was realized principally *via* the tryptophan residue. Therefore, BSA can be explicitly attached to the surface undergoing only minor conformational changes during the interaction with gold nanoparticles. The SERS spectrum of lysozyme, the main Raman modes within the SERS can be attributed to the amino acids tryptophan, tyrosine, phenylalanine and histidine, indicating the close vicinity of these moieties to the metal surface [48, 49]. The adsorption of the protein bovine pancreatic trypsin inhibitor (BPTI) on the metal surface was studied by employing cetyl trimethylammonium bromide (CTAB)-protected gold nanoparticles [50]. Cytochrome C immobilized on 2-mercaptoethanesulfonate (MES) monolayers adsorbed on silver nanostructures was investigated by SERS technique [51] and the heme orientation with respect to the metal surface was evaluated from the SERS spectrum. SERS was applied to investigate protein myeloperoxidase (MPO), its corresponding antibody and their immunocomplex [52]. SERS technique was applied to investigate the structure of amyloid beta peptide aggregates, which were responsible Alzheimer's disease [53]. Investigation by SERS technique showed the presence of the Amide III band providing information about the secondary structure (α-helix and β-sheet) of this protein at very low concentrations down to the femtomolar range. The cancer-promoting protein S100A4 was studied by employing silver colloids as SERS substrate [54]. The conformational changes of the photoactive yellow protein (PYP) were studied at single molecular level exhibiting the high potential of SERS as a bioanalytical device in detecting biomolecules [55].

5.4. APPLICATIONS OF SERS IN PATHOGEN DETECTION

SERS can be applied as a diagnostic analytical tool to differentiate pathogens for example bacteria and viruses [56, 57]. A number of research groups have reported the application of SERS-based assays for pathogen detection [56 - 59] from the time when Efrima *et al.* first studied a bacterial cell surface by the SERS technique [58, 59]. However, most of the SERS-based assays face a lot of problems on account of unnecessary large fluctuations of SERS signals, which take place basically from the heterogeneity in the SERS-active substrates and also owing to the irreconcilable binding between the bacterial cell surface and the

SERS substrate. Metal colloids and nanostructures, which are generally used as the SERS active substrates, have poor biocompatibility. Consequently, it is essential to develop new novel biocompatible SERS substrates for biological molecules especially for rapid identification of pathogens.

Antibiotic induced chemical changes in bacterial cell wall were monitored by [60] Liu *et al.* employing Ag/AAO nanostructured systems as SERS active substrates. Drug unaffected bacteria were detected within an hour from the vibrational data obtained from SERS spectrum of bacterial cell wall. The SERS based technique for pathogens detection was applied to a single bacterium. Clinical sample can be directly analyzed as a substitute to pure cultured sample by rapid SERS based method. Bacterial cell culture can be isolated followed by identification of isolates and monitoring the responses of isolates in presence of antibiotics in terms of viability can be carried out in the conventional protocols for bacterial diagnosis. Various PCR-based protocols have been developed for the detection of bacteria over the past decade. Mass spectrometry can also be potentially applied for culture-free bacterial diagnostics. Mass spectrometry is also reliant on the prevailing previous statistics on the pathogens just like PCR approach. Both PCR and mass spectrometric methods cannot be employed to live bacteria to monitor their reactions to antibiotics. On the other hand, SERS based diagnostic protocol can solve the limitations of PCR based methods. SERS substrates constructed on Ag/AAO system can be employed to investigate the fine changes in the bacterial cell wall during the different growth stages of bacterium and it's response to antibiotic treatment during early phase of exposure to antibiotics.

The current research activity of SERS technique points towards the direction of an early diagnosis of bacteremia as well as urinary tract infections. Bacteremia is a disease caused by the existence of bacteria in the patient's blood, caused by severe infections at locations in the body, surgical wounds, or filthy implanted devices. Urinary tract infections are common disease and almost all women will experience at least one in their generation. The rate of infection of this disease will lead to high monetary effect along with patient illness [61]. The existing standard testing protocol in most hospitals is based on analysis of bacterial culture, which typically consumes minimum three days [62]. This chapter which covers the SERS-based detection of various biomolecules, the SERS based detection of pathogens like bacteria and viruses has also been included.

Oligonucleotide-gold and oligonucleotide-silver nanoparticles had been applied for the rapid detection of *Staphylococcus aureus* [63]. Several *E. coli* strains had been detected by employing either different dye-labeled reporter probes or even Ag/Au-colloids [64, 65]. A *Staphylococcus aureus* and *Vibrio vulnificus* were detected by a multiplexed gold particle on-wire sensor based on SERS technique

[66] with a detection limit of 10 pM. The Hospital-acquired methicillin-resistant *Staphylococcus aureus* was also detected by SERS based multiplexed detection technique [67]. DNA molecule of bacteria *Staphylococcus aureus* DNA was identified specifically *via* the association of TaqMan assay with the aid of SERS [68]. *Mycoplasma mycoides*, was detected by employing SERS technique in combination of magnetic nanoparticles and DNA hybridization [69]. SERS was employed for fingerprint bacterial detection through the O-antigens as a portion of the lipopolysaccharides disclosing on the outer bacterial wall [70]. Bacteria *Salmonella typhimurium* and *E. coli* O16 were discriminated by the specialized vibrational modes in the range of 1250–1130 cm^{-1}, 1030–730 cm^{-1} and 650–430 cm^{-1} which corresponds to carbohydrates, particularly O-antigens. *Shewanella oneides* bacteria was identified by using SERS technique with the help of a liquid core photonic crystal fiber [71]. *Mycobacterium avium* subsp. *paratuberculosis* [72] was detected by SERS using gold nanoparticles. RNA genome virus such as theWest Nile virus (WNV), which is main pathogen responsible for West Nile fever and encephalitis was detected by SERS technique [73]. The human immunodeficiency virus (HIV) should be detected early to keep away from viral transmission in addition to slow down the growth of disease by proper medical involvement. So-called branched DNA technologies can be employed to estimate HIV virus in clinical laboratories [74]. SERS technique was applied to fabricate a molecular junction based biosensor with a detection limit as low as 10^{-19} M HIV-1 DNA [75]. SERS technique was also employed to detect deadly dengue RNA virus sequences, which is responsible mosquito-borne dengue fever [76]. Distinctive SERS peaks we observed at wavenumbers 1653 cm^{-1}, 1360 cm^{-1} and 1219 cm^{-1} by employing gold nanoparticles as SERS substrate [76]. SERS technique was employed for the reliable rapid detection of influenza viruses employing SERS based immunosensor [77]. The relevant SERS peaks were observed at 993 cm^{-1} and 1525 cm^{-1}, which correspond to phenylalanine and tryptophan peaks, respectively. The Vo-Dinh group at Duke University developed a bioassay-on-chip employing plasmonic bimetallic nanostructure as SERS substrate for the detection of dengue viral DNA [78]. Gold and silver nanoparticles were used to fabricate a highly sensitive SERS substrate which was subsequently functionalized with reporter probes attached with Raman tag for the rapid detection of dengue viral DNA. Table **1** illustrates the comparative data based on various SERS based pathogen detection.

Table 1. Comparative data for various SERS based pathogen detection.

S.No.	System	Name of the Detected Pathogen	Reference
1	Ag/AAO nanostructures	Drug resistant bacteria	Liu *et al.* [60]
2	Oligonucleotide-gold and oligonucleotide-silver nanoparticles	*Staphylococcus aureus*	Graham *et al.* [63]
3	Multiplexed gold particle on-wire sensor	*Staphylococcus aureus* and *Vibrio vulnificus*	Kang *et al.* [66]
4	SERS *via* combining magnetic nanoparticles and DNA hybridization	*Mycoplasma mycoides*	Strelau *et al.* [69]
5	Gold nanoparticles coated with DSNB and antibodies	*Mycobacterium avium* subsp. *paratuberculosis*	Yakes *et al.* [72]
6	Raman reporter tag conjugated gold nanoparticles	West Nile virus (WNV)	Neng *et al.* [73]
7	Nanoforest structured SERS active substrates	Influenza viruses	Seol *et al.* [77]
8	Plasmonic bimetallic nanostructure consisting of gold and silver nanoparticles	Dengue viral DNA	Vo-Dinh *et al.* [78]
9	Biosynthesized silver nanoparticles	*Staphylococcus aureus* and *Escherichia coli*, bacteria	Ankamwar *et al.* [79]
10	Biosynthesized silver nanoparticles	*Mycobacterium tuberculosis*	Sur *et al.* [80]
11	Citrate reduced silver nanoparticles	SARS-CoV-2 virus	Yacaman *et al.* [89]
12	Multilayer metal-molecule-metal nanojunctions consist of gold nanoparticles	HIV-1 DNA	Hu *et al.* [75]
13	DNA functionalized gold and silver nanoparticles based microfluidic SERS sensor	Dengue virus serotype 2	Huh *et al.* [76]

Highly stable and homogeneous SERS active substrates were made from biosynthesized silver nanoparticles (Ag NPs) employing *Neolamarckia cadamba* leaf extract for the rapid detection of two strains of bacteria, gram positive (*Staphylococcus aureus, S. aureus*) and gram negative (*Escherichia coli, E. coli*) bacteria by Ankamwar *et al.* [79]. (Fig. **1**) show the TEM image of the biosynthesized silver nanoparticles along with their UV-visible spectrum, SAED pattern and the resultant SERS spectra generated upon interaction with *S. aureus* and *E. coli*.

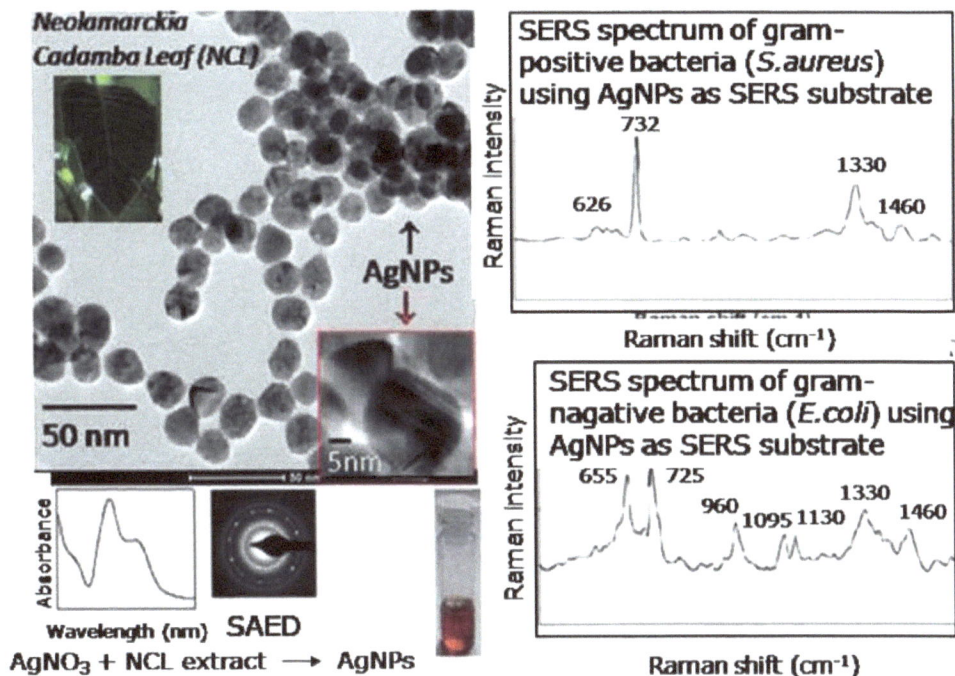

Fig. (1). The TEM image of the biosynthesized silver nanoparticles along with the UV- vis spectrum, SAED pattern and the resultant SERS spectra generated upon interaction with *S. aureus* and *E. coli* bacteria. Reproduced with permission from Ankamwar B, Sur U K., Das P. Anal Methods. 2016; **8**: 2335-2340. Copyright @ Royal Society of Chemistry, Inc.

The produced SERS substrates were tremendously stable even after three months. These almost consistent, stable SERS active substrates were utilized to differentiate Gram positive bacteria from Gram negative bacteria with. highly stable, uniform and reproducible SERS signal. Very low concentrations (10^3 CFU ml^{-1}) of *E. coli* can be detected by this biocompatible SERS active substrates with high sensitivity (See Fig. **2**). The SERS calibration curve acquired by plotting the SERS intensity of the peak at 1330 cm^{-1} (C–N stretching mode) against the concentration of bacteria *E. coli* is displayed in Fig. (**2**). The SERS intensity will rise exponentially with the increase of concentration of the bacterial solution between concentrations 10^3 CFU/ml to 10^8 CFU/ml. The main aim of this SERS based detection protocol employing biosynthesized Ag nanoparticles was to establish a rapid technique to detect bacteria especially *E. coli,* which is related to urinary tract infection (UTI), a common disease amongst people of all age groups in countries like India and China.

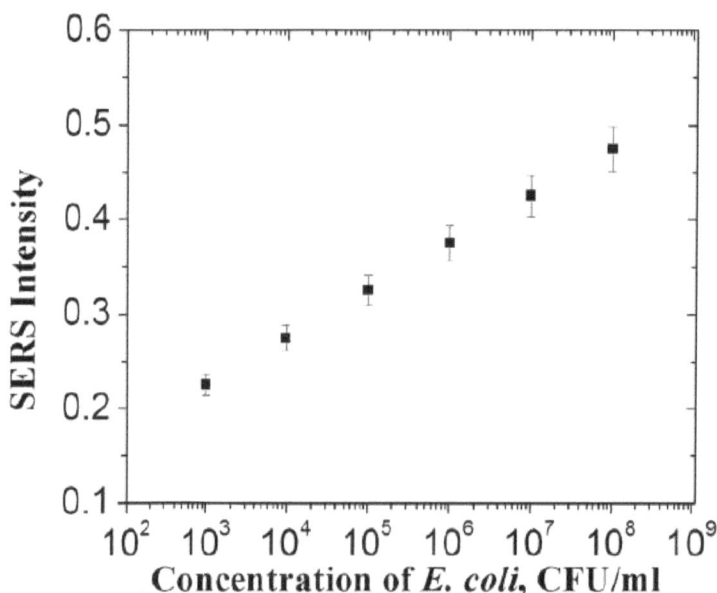

Fig. (2). The SERS calibration curve obtained with SERS peak area or SERS intensity of the peak at 1330 cm^{-1} (C–N stretching mode) as a function of concentration of bacteria *E. coli*. Reproduced with permission from Ankamwar B, Sur U K., Das P. Anal Methods. 2016; **8**: 2335-2340. Copyright @ Royal Society of Chemistry, Inc.

Biosynthesized Ag nanoparticles obtained from the plant extract of Reetha and Shikakai were employed as SERS substrate for fast identification of detrimental bacteria such as *Mycobacterium tuberculosis*, which is known as drug-resistive to most of the common drugs available commercially [80]. (Fig. **3**) illustrates the SERS spectrum of *Mycobacterium tuberculosis* on biosynthesized Ag nanoparticles and important peaks were observed at wavenumbers 437, 915, 1175 and 1390 cm^{-1} in the SERS spectrum.

SERS spectrum was monitored by exposing the whole bacterium *via* laser light during the interaction with the silver nanoparticles used as active substrate showing the molecular structure inside bacterial cell wall [81]. The calculated value of Raman enhancement factor for the biosynthesized Ag nanoparticles was found to be $(5 \pm 0.10) \times 10^9$. The corresponding SERS spectra observed for *Mycobacterium tuberculosis* showed characteristic features which are dissimilar with respect to both peak position and intensity from the Gram-positive and the Gram-negative bacteria as reported previously by the same research group [79]. The peak obtained at 730 cm^{-1} was very feeble and the peak observed at 1330 cm^{-1} was completely absent in the SERS spectra of *Mycobacteria*. This distinctive features may be due to the presence of long chain fatty acid mycolic acid in the outermost hydrophobic membrane of *Mycobacteria*, hindering the peptidoglycan layer from approaching towards the SERS substrate and therefore completely removing the 730 and 1330 cm^{-1} peaks [82, 83]. Subsequently, the compositions of the outermost membrane were comprising of biomolecules such as arabinogalactan, mycolic acids, lipids, which would contribute to the observed enormously complicated SERS spectra [83].

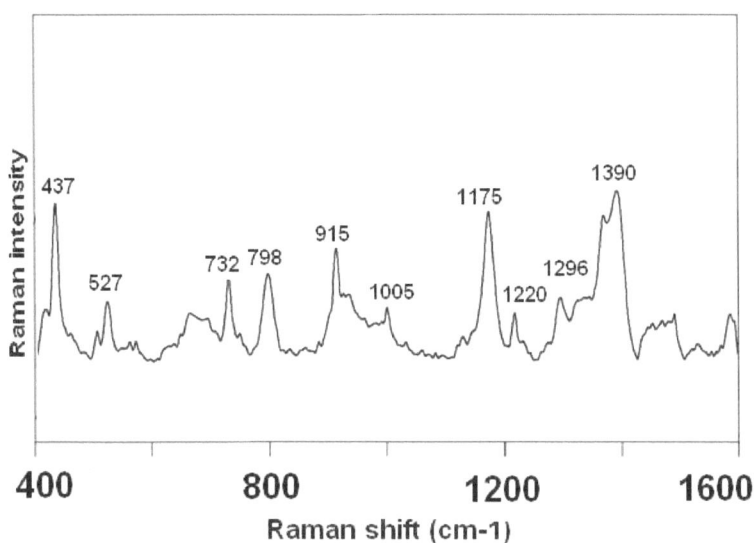

Fig. (3). The SERS spectrum of bacteria *Mycobacterium tuberculosis* on biosynthesized Ag nanoparticles. Reproduced with permission from Sur U K, Ankamwar B, Karmakar S, Halder A, Das P, *Materials Today: Proceedings,*5 (2018) 2321-2329. Copyright @ Elsevier Ltd.

Numerous transmittable diseases including lung tuberculosis (TB) in human being are caused by lethal *Mycobacterium tuberculosis* bacteria and it is responsible for over two million global deaths yearly [84]. Consequently, the perceptive discovery of pathogens is requisite for advance diagnosis, therapy, and control of this deadly disease. A few prevailing conventional diagnostic methods such as sputum smear microscopy, chest radiography and tuberculin skin testing are insensitive and also arduous, extended and results are inaccurate and normally unfocused [85, 86]. Numerous rapid current diagnostic protocols, which are commercially available in the market have been developed towards the improvement of the diagnostic accuracy for TB. PCR and other molecular amplification techniques are although both capable and prominent, none are more than adequate for the diagnosis of TB due to variable sensitivities of the test results [87]. Different rapid commercial techniques are now available for species identification of *M. tuberculosis* complex. However, these techniques are very expensive and limited to selected, frequently encountered species, as evident for available commercial techniques such as the reverse line blot assay, the Amplicor nucleic acid amplification test and the Gen-Probe Amplified *Mycobacterium tuberculosis* direct test [88]. Therefore, there is continuing demand for rapid, simplified choice owing to these limitations, which can be readily applied to cultured bacteria from clinical sample, facilitating the rapid detection of a wide spectrum of microorganisms.

5.5. RAMAN SPECTROSCOPY BASED DETECTION FOR COVID-19 TESTING

Professor Miguel Jose Yacaman and his research group at the Northern Arizona University, USA had developed a new protocol for the detection of SARS-CoV-2 virus using single molecule surface-enhanced Raman spectroscopy (SM-SERS) [89]. The researchers had applied the concept directly from areas such as nanotechnology, plasmonics and 2D materials to develop the protocol. This will provide non-traditional spectroscopic method to detect virus in infected patients. The researchers had used SM-SERS to detect S proteins of the SARS-CoV-2 virus. Yacaman had earlier applied SERS technique to identify biomolecules such as glycoproteins and sialic acid for investigating breast cancer [90]. This testing approach is now in the final approval stage for commercial use.

Research team led by Amit Dutt from the Mumbai based Tata Memorial centre has utilized conventional Raman Spectroscopy to detect RNA viruses present in saliva samples [91] by analysing the raw Raman data and compared the signals with both viral positive and negative samples. The signal set has 92.5% sensitivity and 88.8% specificity. The result was published in the Journal of Biophotonics [91].

5.6. LIMITATION OF SERS TECHNIQUE IN PATHOGEN DETECTION AND FUTURE TREND

It was demonstrated that the SERS technique can be useful for rapid detection of pathogens like bacteria and viruses based on the recently developed biosynthesized SERS-active substrates. SERS based pathogen detection is mostly useful for the investigation of slow-growing bacteria, which generally may take weeks during laboratory tests.

On the flip side, the SERS based detection protocol of bacteria cannot discriminate one strain from another within the same bacterial types. This is the utmost vital disadvantage of the SERS based detection of bacteria. The SERS based technique for the detection of pathogens are inferior compared to genome sequencing or mass spectrometric based proteomics analysis in terms of the molecular level specificity.

SERS technique can be employed to carry out clinical microbial diagnostics straight on a clinical specimen without the requirement of bacterial pure cultured samples, which are both time consuming and a little bit difficult to carry out. Clinical samples such as blood, urine, stool, saliva, sputum which are directly collected from the infected persons and patients in the hospitals and can be subsequently utilized to carry out SERS base detection technique of pathogens. (Fig. **4**) shows the schematic diagram explaining the SERS based pathogen detection along with its diagnostics applications from clinical samples.

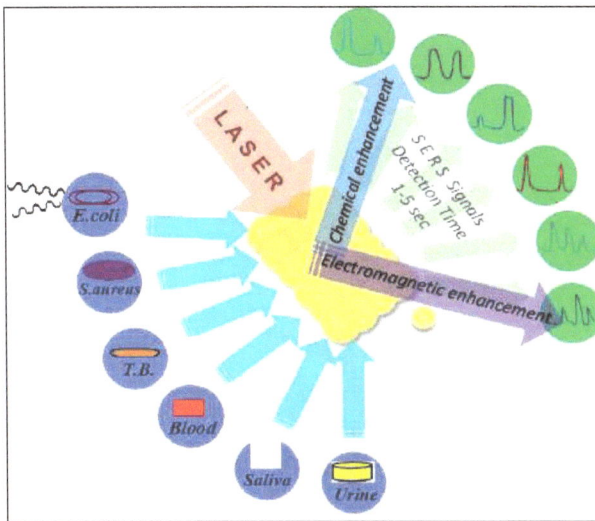

Fig. (4). The schematic diagram explaining the SERS based pathogen detection along with its diagnostics applications from clinical samples. Photo courtesy B Ankamwar.

It is expected that several new and unknown diseases based on viruses such as Swine Flu, Avian flu, Japanese Encephalitis, Ebola, Zika and Dengue which are very difficult to detect within a short period of time, can be qualitatively and quantitatively detected using the newly developed SERS based protocols. It is important to mention here that diseases like Swine Flu, Avian flu, Japanese Encephalitis and Dengue has emerged as life-threatening and harmful in different parts of the world in recent years. It is essential to develop new diagnostic tools and sensing devices to control these new diseases.

CONCLUSIONS

Surface-enhanced Raman scattering (SERS) is related to the large augmentation of the weak Raman signal by both organic and biological molecules in the presence of metallic nanostructured particularly, gold and silver nanoparticles. It has become a versatile analytical tool owing to the rapid development of nanoscience and nanotechnology for enormously sensitive and selective recognition of chemical and biological systems. This review article abridges the current results and progress of using SERS for the identification of biomolecules such as nucleotides and nucleic acids, amino acids and proteins.

The immense potential and versatility of the SERS technique as a bioanalytical technique to identify biological molecules has also been demonstrated by revolutionary applications in the fields of rapid pathogen detection and diagnosis. In the case of pathogen detection, the bacterial or viral contamination in culture free clinical sample is identified *via* their building blocks such as DNA and proteins or bacterial cell wall.

The sensitive and stable SERS profiles along with the "chemical features" obtained from SERS spectrum of bacterial cell wall facilitates prompt identification of pathogens like bacteria and viruses within a very short time scale.

We have demonstrated the recent use of various SERS active substrates for the rapid identification of pathogens like viruses and bacteria in this chapter. SERS can be ideal as well as suitable for the rapid, accurate and sensitive pathogens detection and it is expected that it will provide versatile diagnostic tool in the development of a rapid diagnostic system for detecting bacteria and viruses even novel Coronavirus. It is expected that proper and rapid diagnosis of various pathogens will facilitate the development of proper vaccines and antiviral medicines in the coming years.

CONSENT FOR PUBLICATION

Not applicable.

CONFLICT OF NTEREST

The authors declare no conflict of interest, financial or otherwise.

ACKNOWLEDGEMENTS

UKS would like to acknowledge financial support from the projects funded by the DHESTBT, Government of West Bengal (memo no. 161(sanc)/ST/P/S&T/9G-

50/2017 dated 8/2/2018). AG would like to acknowledge WBDST for providing JRF fellowship. The authors would like to grateful to all authors and publishers of various journals (Elsevier, RSC, Intech Inc, Techno Press, Indian Academy of Sciences) from which various figures and text portions has been reproduced in this paper. UKS would like to acknowledge Dr. B Ankamwar for valuable suggestions.

REFERENCES

[1]   Fleischmann M, Hendra PJ, McQuillan AJ. Raman spectra of pyridine adsorbed at a silver electrode. Chem Phys Lett 1974; 26: 163-6.
        [http://dx.doi.org/10.1016/0009-2614(74)85388-1]

[2]   Sur UK. Surface-Enhanced Raman Spectroscopy. Recent Advancement of Raman Spectroscopy Resonance 2010; 15: 154-64.

[3]   Sur UK, Chowdhury J. Surface-enhanced Raman scattering: overview of a versatile technique used in electrochemistry and nanoscience. Curr Sci 2013; 105: 923-39.

[4]   Sur UK. Surface-enhanced Raman scattering (SERS) spectroscopy: a versatile spectroscopic and analytical technique used in nanoscience and nanotechnology. Adv Nano Res 2013; 1: 111-24.
        [http://dx.doi.org/10.12989/anr.2013.1.2.111]

[5]   Cooney RP, Mahoney MR, McQuillan AJ. Advances of Infrared and Raman Spectroscopy. London: Heyden 1982; p. 188.

[6]   Nie S, Emory SR. Probing single molecules and single nanoparticles by surface enhanced Raman scattering. Science 1997; 275(5303): 1102-6.
        [http://dx.doi.org/10.1126/science.275.5303.1102] [PMID: 9027306]

[7]   Wang A, Huang YF, Sur UK, *et al. In situ* identification of intermediates of benzyl chloride reduction at a silver electrode by SERS coupled with DFT calculations. J Am Chem Soc 2010; 132(28): 9534-6.
        [http://dx.doi.org/10.1021/ja1024639] [PMID: 20575538]

[8]   Mulvihill M, Tao A, Benjauthrit K, Arnold J, Yang P. Surface-enhanced Raman spectroscopy for trace arsenic detection in contaminated water. Angew Chem Int Ed Engl 2008; 47(34): 6456-60.
        [http://dx.doi.org/10.1002/anie.200800776] [PMID: 18618882]

[9]   Liu SQ, Tang ZY. Nanoparticle assemblies for biological and chemical sensing. J Mater Chem 2010; 20: 24-35.
        [http://dx.doi.org/10.1039/B911328M]

[10]   Fan M, Andrade GFS, Brolo AG. A review on the fabrication of substrates for surface enhanced Raman spectroscopy and their applications in analytical chemistry. Anal Chim Acta 2011; 693(1-2): 7-25.
        [http://dx.doi.org/10.1016/j.aca.2011.03.002] [PMID: 21504806]

[11]   Shafer-Peltier KE, Haynes CL, Glucksberg MR, Van Duyne RP. Toward a glucose biosensor based on surface-enhanced Raman scattering. J Am Chem Soc 2003; 125(2): 588-93.
        [http://dx.doi.org/10.1021/ja028255v] [PMID: 12517176]

[12]   Ma L-n., Zhang J, Chen H-t, Zhou Jh, Ding YZ, Liu YS. An overview on ELISA techniquesfor FMD. Virol J 2011; 8: 9-18.

[13]   Malou N, Raoult D. Immuno-PCR: a promising ultrasensitive diagnostic method to detect antigens and antibodies. Trends Microbiol 2011; 19(6): 295-302.
        [http://dx.doi.org/10.1016/j.tim.2011.03.004] [PMID: 21478019]

[14]   Seibel J, König S, Göhler A, *et al.* Investigating infection processes with a workflow from organic chemistry to biophysics: the combination of metabolic glycoengineering, super-resolution fluorescence

imaging and proteomics. Expert Rev Proteomics 2013; 10(1): 25-31.
[http://dx.doi.org/10.1586/epr.12.72] [PMID: 23414357]

[15] Nicklas JA, Buel E. Quantification of DNA in forensic samples. Anal Bioanal Chem 2003; 376(8): 1160-7.
[http://dx.doi.org/10.1007/s00216-003-1924-z] [PMID: 12739098]

[16] Ataka K, Kottke T, Heberle J. Thinner, smaller, faster: IR techniques to probe the functionality of biological and biomimetic systems. Angew Chem Int Ed Engl 2010; 49(32): 5416-24.
[http://dx.doi.org/10.1002/anie.200907114] [PMID: 20818765]

[17] Barhoumi A, Halas NJ. Label-free detection of DNA hybridization using surface enhanced Raman spectroscopy. J Am Chem Soc 2010; 132(37): 12792-3.
[http://dx.doi.org/10.1021/ja105678z] [PMID: 20738091]

[18] Papadopoulou E, Bell SEJ. Structure of adenine on metal nanoparticles: pH equilibria and formationof Ag^+ complexes detected by surface-enhanced Raman spectroscopy. J Phys Chem C 2010; 114: 22644-51.
[http://dx.doi.org/10.1021/jp1092256]

[19] Feng F, Zhi G, Jia HS, Cheng L. SERS detection of low-concentration adenine by a patterned silver structure immersion plated on a silicon nanoporous pillar array. Nanotechnol. 2009; 20: p. 6.

[20] Kundu J, Neumann O, Janesko BG, Zhang D, *et al.* Adenine- and Adenosine Monophosphate (AMP)-gold binding interactions studied by surface-enhanced Raman and infrared spectroscopies. J Phys Chem C 2009; 113: 14390-7.
[http://dx.doi.org/10.1021/jp903126f]

[21] Carrillo-Carrión C, Armenta S, Simonet BM, Valcárcel M, Lendl B. Determination of pyrimidine and purine bases by reversed-phase capillary liquid chromatography with at-line surface-enhanced Raman spectroscopic detection employing a novel SERS substrate based on ZnS/CdSe silver-quantum dots. Anal Chem 2011; 83(24): 9391-8.
[http://dx.doi.org/10.1021/ac201821q] [PMID: 22047639]

[22] Primera-Pedrozo OM, Rodríguez GdelM, Castellanos J, Felix-Rivera H, Resto O, Hernández-Rivera SP. Increasing surface enhanced Raman spectroscopy effect of RNA and DNA components by changing the pH of silver colloidal suspensions. Spectrochim Acta A Mol Biomol Spectrosc 2012; 87: 77-85.
[http://dx.doi.org/10.1016/j.saa.2011.11.012] [PMID: 22169024]

[23] Papadopoulou E, Bell SEJ. Surface-enhanced Raman evidence of protonation, reorientation, and Ag^+ complexation of Deoxyadenosine and Deoxyadenosine-5 '-Monophosphate (dAMP) on Ag and Au surfaces. J Phys Chem C 2011; 115: 14228-35.
[http://dx.doi.org/10.1021/jp204369f]

[24] Diaz Fleming G, Finnerty JJ, Campos-Vallette M, *et al.* Experimental and theoretical Raman and surface-enhanced Raman scattering study of cysteine. J Raman Spectrosc 2009; 40: 632-8.
[http://dx.doi.org/10.1002/jrs.2175]

[25] Chuang C-H, Chen Y-T. Raman scattering of L-tryptophan enhanced by surface plasmon of silver nanoparticles: vibrational assignment and structural determination. J Raman Spectrosc 2009; 40: 150-6.
[http://dx.doi.org/10.1002/jrs.2097]

[26] Aliaga AE, Osorio-Roman I, Leyton P, Garrido C, Carcamo J, *et al.* Surface-enhanced Raman scattering study of L-tryptophan. J Raman Spectrosc 2009; 40: 164-9.
[http://dx.doi.org/10.1002/jrs.2099]

[27] Aliaga AE, Osorio-Roman I, Garrido C, Leyton P, *et al.* Surface enhanced Raman scattering study of L-lysine. Vib Spectrosc 2009; 50: 131-5.
[http://dx.doi.org/10.1016/j.vibspec.2008.09.018]

[28] Aliaga AE, Garrido C, Leyton P, *et al.* SERS and theoretical studies of arginine. Spectrochim Acta A Mol Biomol Spectrosc 2010; 76(5): 458-63.
[http://dx.doi.org/10.1016/j.saa.2010.01.007] [PMID: 20471905]

[29] Yang H, Zhu X, Song W, *et al.* N-acetylalanine monolayers at the silver surface investigated by surface-enhanced Raman scattering spectroscopy and X-ray photoelectronspectroscopy: effect of metallic ions. J Phys Chem C 2008; 112: 15022-7.
[http://dx.doi.org/10.1021/jp8042544]

[30] Sheng C, Zhao H, Gu F, Yang H. Effect of Pb^{2+} on L-glutathione monolayers on a silver surface investigated by surface-enhanced Raman scattering spectroscopy. J Raman Spectrosc 2009; 40: 1274-8.
[http://dx.doi.org/10.1002/jrs.2277]

[31] Graff M, Bukowska J. Surface-enhanced Raman scattering (SERS) spectroscopy of enantiomeric and racemic methionine on a silver electrode-evidence for chiral discrimination in interactions between adsorbed molecules. Chem Phys Lett 2011; 509: 58-61.
[http://dx.doi.org/10.1016/j.cplett.2011.04.089]

[32] Graff M, Bukowska J. Enantiomeric recognition of phenylalanine by self-assembled monolayers of cysteine: Surface-enhanced Raman scattering evidence. Vib Spectrosc 2010; 52: 103-7.
[http://dx.doi.org/10.1016/j.vibspec.2009.11.003]

[33] Thomas S, Biswas N, Malkar VV, Mukherjee T, Kapoor S. Studies on adsorption of carnosine on silver nanoparticles by SERS. Chem Phys Lett 2010; 491: 59-64.
[http://dx.doi.org/10.1016/j.cplett.2010.03.059]

[34] Podstawka E, Andrzejak M, Kafarski P, Proniewicz LM. Comparison of adsorption mechanism on colloidal silver surface of alafosfalin and its analogs. J Raman Spectrosc 2008; 39: 1238-49.
[http://dx.doi.org/10.1002/jrs.1977]

[35] Podstawka E, Kafarski P, Proniewicz LM. Effect of an aliphatic spacer group on the adsorption mechanism of phosphonodipeptides containing N-terminal glycine on the colloidal silver surface. J Raman Spectrosc 2008; 39: 1396-407.
[http://dx.doi.org/10.1002/jrs.2010]

[36] Malek K, Makowski M, Królikowska A, Bukowska J. Comparative studies on IR, Raman, and surface enhanced Raman scattering spectroscopy of dipeptides containing ΔAla and ΔPhe. J Phys Chem B 2012; 116(4): 1414-25.
[http://dx.doi.org/10.1021/jp208586j] [PMID: 22208201]

[37] Yuan X, Gu H, Wu J. Surface-enhanced Raman spectrum of Gly-Gly adsorbed on the silver colloidal surface. J Mol Struct 2010; 977: 56-61.
[http://dx.doi.org/10.1016/j.molstruc.2010.05.009]

[38] Wei F, Zhang D, Halas NJ, Hartgerink JD. Aromatic amino acids providing characteristic motifs in the Raman and SERS spectroscopy of peptides. J Phys Chem B 2008; 112(30): 9158-64.
[http://dx.doi.org/10.1021/jp8025732] [PMID: 18610961]

[39] Podstawka E, Kafarski P, Proniewicz LM. Structural properties of L-X-L-Met-L-Ala phosphonate tripeptides: a combined FT-IR, FT-RS, and SERS spectroscopy studies and DFT calculations. J Phys Chem A 2008; 112(46): 11744-55.
[http://dx.doi.org/10.1021/jp803674q] [PMID: 18942819]

[40] Podstawka E. Effect of amino acid modifications on the molecular structure of adsorbed and nonadsorbed bombesin 6-14 fragments on an electrochemically roughened silver surface. J Raman Spectrosc 2008; 39: 1290-305.
[http://dx.doi.org/10.1002/jrs.1996]

[41] Podstawka E, Niaura G, Proniewicz LM. Potential-dependent studies on the interaction between phenylalanine-substituted bombesin fragments and roughened Ag, Au, and Cu electrode surfaces. J

Phys Chem B 2010; 114(2): 1010-29.
[http://dx.doi.org/10.1021/jp909268c] [PMID: 20025214]

[42] Podstawka E, Ozaki Y, Proniewicz LM. Structures and bonding on a colloidal silver surface of the various length carboxyl terminal fragments of bombesin. Langmuir 2008; 24(19): 10807-16.
[http://dx.doi.org/10.1021/la8012415] [PMID: 18759412]

[43] Podstawka-Proniewicz E, Ozaki Y, Kim Y, Xu Y, Proniewicz LM. Surface-enhanced Ramanscattering studies on bombesin, its selected fragments and related peptides adsorbed at the silver colloidal surface. Appl Surf Sci 2011; 257: 8246-52.
[http://dx.doi.org/10.1016/j.apsusc.2011.02.012]

[44] Podstawka E, Proniewicz LM. The orientation of BN-related peptides adsorbed on SERS-active silver nanoparticles: comparison with a silver electrode surface. J Phys Chem B 2009; 113(14): 4978-85.
[http://dx.doi.org/10.1021/jp8110716] [PMID: 19296643]

[45] Podstawka-Proniewicz E, Kudelski A, Kim Y, Proniewicz LM. Structure and binding of specifically mutated neurotensin fragments on a silver substrate: vibrational studies. J Phys Chem B 2011; 115(21): 7097-108.
[http://dx.doi.org/10.1021/jp201316n] [PMID: 21548565]

[46] Ignatjev I, Podstawka-Proniewicz E, Niaura G, Lombardi JR, Proniewicz LM. Potential induced changes in neuromedin B adsorption on Ag, Au, and Cu electrodes monitored by surface-enhanced Raman scattering. J Phys Chem B 2011; 115(35): 10525-36.
[http://dx.doi.org/10.1021/jp2026863] [PMID: 21812441]

[47] Iosin M, Toderas F, Baldeck PL, Astilean S. Study of protein-gold nanoparticle conjugates by fluorescence and surface-enhanced Raman scattering. J Mol Struct 2009; 924–926: 196-200.
[http://dx.doi.org/10.1016/j.molstruc.2009.02.004]

[48] Das R, Jagannathan R, Sharan C, Kumar U, Poddar P. Mechanistic study of surface functionalization of enzyme lysozyme synthesized Ag and Au nanoparticles using surface enhanced Raman spectroscopy. J Phys Chem C 2009; 113: 21493-500.
[http://dx.doi.org/10.1021/jp905806t]

[49] Chandra G, Ghosh KS, Dasgupta S, Roy A. Evidence of conformational changes in adsorbed lysozyme molecule on silver colloids. Int J Biol Macromol 2010; 47(3): 361-5.
[http://dx.doi.org/10.1016/j.ijbiomac.2010.05.020] [PMID: 20685371]

[50] Kaminska A, Forster RJ, Keyes TE. The impact of adsorption of bovine pancreatic trypsin inhibitor on CTAB-protected gold nanoparticle arrays: a Raman spectroscopic comparison with solution denaturation. J Raman Spectrosc 2010; 41: 130-5.

[51] Krolikowska A, Bukowska J. Surface-enhanced resonance Raman spectroscopic characterization of cytochrome c immobilized on 2-mercaptoethanesulfonate monolayers on silver. J Raman Spectrosc 2010; 41: 1621-31.
[http://dx.doi.org/10.1002/jrs.2618]

[52] Papazoglou ES, Babu S, Hansberry DR, Mohapatra S, Patel C. SERS study on myeloperoxidase and its immunocomplex: Identification of binding interactions. Spectros Int J 2010; 24: 183-90.
[http://dx.doi.org/10.1155/2010/169292]

[53] Choi I, Huh YS, Erickson D. Ultra-sensitive, label-free probing of the conformational characteristicsof amyloid beta aggregates with a SERS active nanofluidic device. Microfluid Nanofluidics 2012; 12: 663-9.
[http://dx.doi.org/10.1007/s10404-011-0879-1]

[54] Abdali S, De Laere B, Poulsen M, Grigorian M, Lukanidin E, Klingelhofer J. Towardmethodology for detection of cancerpromoting S100A4 protein conformations in subnanomolar concentrations using Raman and SERS. J Phys Chem C 2010; 114: 7274-9.
[http://dx.doi.org/10.1021/jp908335z]

[55]   Singhal K, Kalkan AK. Surface-enhanced Raman scattering captures conformational changes of single photoactive yellow protein molecules under photoexcitation. J Am Chem Soc 2010; 132(2): 429-31.
[http://dx.doi.org/10.1021/ja9028704] [PMID: 19788179]

[56]   Jarvis RM, Goodacre R. Discrimination of bacteria using surface-enhanced Raman spectroscopy. Anal Chem 2004; 76(1): 40-7.
[http://dx.doi.org/10.1021/ac034689c] [PMID: 14697030]

[57]   Sengupta A, Laucks ML, Davis EJ. Surface-enhanced Raman spectroscopy of bacteria and pollen. Appl Spectrosc 2005; 59(8): 1016-23.
[http://dx.doi.org/10.1366/0003702054615124] [PMID: 16105210]

[58]   Efrima S, Bronk BV. Silver colloids impregnating or coating bacteria. J Phys Chem B 1998; 102: 5947-50.
[http://dx.doi.org/10.1021/jp9813903]

[59]   Efrima S, Zeiri L. Understanding SERS of bacteria. J Raman Spectrosc 2009; 40: 277-88.
[http://dx.doi.org/10.1002/jrs.2121]

[60]   Liu TT, Lin YH, Hung CS, *et al.* A high speed detection platform based on surface-enhanced Raman scattering for monitoring antibiotic-induced chemical changes in bacteria cell wall. PLoS One 2009; 4(5): e5470.
[http://dx.doi.org/10.1371/journal.pone.0005470] [PMID: 19421405]

[61]   Brumbaugh AR, Mobley HLT. Preventing urinary tract infection: progress toward an effective Escherichia coli vaccine. Expert Rev Vaccines 2012; 11(6): 663-76.
[http://dx.doi.org/10.1586/erv.12.36] [PMID: 22873125]

[62]   Wertheim H, Verbrugh HA, van Pelt C, de Man P, van Belkum A, Vos MC. Improved detection of methicillin-resistant Staphylococcus aureus using phenyl mannitol broth containing aztreonam and ceftizoxime. J Clin Microbiol 2001; 39(7): 2660-2.
[http://dx.doi.org/10.1128/JCM.39.7.2660-2662.2001] [PMID: 11427589]

[63]   Graham D, Stevenson R, Thompson DG, Barrett L, Dalton C, Faulds K. Combining functionalised nanoparticles and SERS for the detection of DNA relating to disease. Faraday Discuss 2011; 149: 291-9.
[http://dx.doi.org/10.1039/C005397J] [PMID: 21413187]

[64]   Faulds K, Jarvis R, Smith WE, Graham D, Goodacre R. Multiplexed detection of six labelled oligonucleotides using surface enhanced resonance Raman scattering (SERRS). Analyst (Lond) 2008; 133(11): 1505-12.
[http://dx.doi.org/10.1039/b800506k] [PMID: 18936827]

[65]   Papadopoulou E, Bell SEJ. Label-free detection of nanomolar unmodified single- and double-stranded DNA by using surface-enhanced Raman spectroscopy on Ag and Au colloids. Chemistry 2012; 18(17): 5394-400.
[http://dx.doi.org/10.1002/chem.201103520] [PMID: 22434729]

[66]   Kang T, Yoo SM, Yoon I, Lee SY, Kim B. Patterned multiplex pathogen DNA detection by Au particle-on-wire SERS sensor. Nano Lett 2010; 10(4): 1189-93.
[http://dx.doi.org/10.1021/nl1000086] [PMID: 20222740]

[67]   MacAskill A, Crawford D, Graham D, Faulds K. DNA sequence detection using surface-enhanced resonance Raman spectroscopy in a homogeneous multiplexed assay. Anal Chem 2009; 81(19): 8134-40.
[http://dx.doi.org/10.1021/ac901361b] [PMID: 19743872]

[68]   Harper MM, Robertson B, Ricketts A, Faulds K. Specific detection of DNA through coupling of a TaqMan assay with surface enhanced Raman scattering (SERS). Chem Commun (Camb) 2012; 48(75): 9412-4.
[http://dx.doi.org/10.1039/c2cc34859d] [PMID: 22889872]

[69] Strelau KK, Brinker A, Schnee C, Weber K, Moller R, Popp J. Detection of PCR products amplified from DNA of epizootic pathogens using magnetic nanoparticles and SERS. J Raman Spectrosc 2011; 42(3): 243-50.
[http://dx.doi.org/10.1002/jrs.2730]

[70] Osorio-Román IO, Aroca RF, Astudillo J, Matsuhiro B, Vásquez C, Pérez JM. Characterization of bacteria using its O-antigen with surface-enhanced Raman scattering. Analyst (Lond) 2010; 135(8): 1997-2001.
[http://dx.doi.org/10.1039/c0an00061b] [PMID: 20532346]

[71] Yang X, Gu C, Qian F, Li Y, Zhang JZ. Highly sensitive detection of proteins and bacteria in aqueous solution using surface-enhanced Raman scattering and optical fibers. Anal Chem 2011; 83(15): 5888-94.
[http://dx.doi.org/10.1021/ac200707t] [PMID: 21692506]

[72] Yakes BJ, Lipert RJ, Bannantine JP, Porter MD. Detection of Mycobacterium avium subsp. paratuberculosis by a sonicate immunoassay based on surface-enhanced Raman scattering. Clin Vaccine Immunol 2008; 15(2): 227-34.
[http://dx.doi.org/10.1128/CVI.00334-07] [PMID: 18077613]

[73] Neng J, Harpster MH, Wilson WC, Johnson PA. Surface-enhanced Raman scattering (SERS) detection of multiple viral antigens using magnetic capture of SERS-active nanoparticles. Biosens Bioelectron 2013; 41: 316-21.
[http://dx.doi.org/10.1016/j.bios.2012.08.048] [PMID: 23021841]

[74] Tsongalis GJ. Branched DNA technology in molecular diagnostics. Am J Clin Pathol 2006; 126(3): 448-53.
[http://dx.doi.org/10.1309/90BU6KDXANFLN4RJ] [PMID: 16880139]

[75] Hu J, Zheng P-C, Jiang J-H, Shen G-L, Yu RQ, Liu GK. Sub-attomolar HIV-1 DNA detection using surface-enhanced Raman spectroscopy. Analyst (Lond) 2010; 135(5): 1084-9.
[http://dx.doi.org/10.1039/b920358c] [PMID: 20419260]

[76] Huh YS, Chung AJ, Cordovez B, Erickson D. Enhanced on-chip SERS based biomolecular detection using electrokinetically active microwells. Lab Chip 2009; 9(3): 433-9.
[http://dx.doi.org/10.1039/B809702J] [PMID: 19156293]

[77] Seol M-L, Choi S-J, Baek DJ, et al. A nanoforest structure for practical surface-enhanced Raman scattering substrates. Nanotechnology. 2012.
[http://dx.doi.org/10.1088/0957-4484/23/9/095301]

[78] Ngo HT, Wang HN, Fales AM, Nicholson BP, Woods CW, Vo-Dinh T. DNA bioassay-on-chip using SERS detection for dengue diagnosis. Analyst (Lond) 2014; 139(22): 5655-9.
[http://dx.doi.org/10.1039/C4AN01077A] [PMID: 25248522]

[79] Ankamwar B, Sur UK, Das P. SERS study of bacteria using biosynthesized silver nanoparticles as SERS substrate. Anal Methods 2016; 8: 2335-40.
[http://dx.doi.org/10.1039/C5AY03014E]

[80] Sur UK, Ankamwar B, Karmakar S, Halder A, Das P. Green synthesis of Silver nanoparticles using the plant extract of Shikakai and Reetha. Mater Today Proc 2018; 5: 2321-9.
[http://dx.doi.org/10.1016/j.matpr.2017.09.236]

[81] Kahraman M, Yazici MM, Sahin F, Bayrak OF, Culha M. Reproducible surface-enhanced Raman scattering spectra of bacteria on aggregated silver nanoparticles. Appl Spectrosc 2007; 61(5): 479-85.
[http://dx.doi.org/10.1366/000370207780807731] [PMID: 17555616]

[82] Hoffmann C, Leis A, Niederweis M, Plitzko JM, Engelhardt H. Disclosure of the mycobacterial outer membrane: cryo-electron tomography and vitreous sections reveal the lipid bilayer structure. Proc Natl Acad Sci USA 2008; 105(10): 3963-7.
[http://dx.doi.org/10.1073/pnas.0709530105] [PMID: 18316738]

[83] Buijtels PCAM, Willemse-Erix HFM, Petit PLC, *et al.* Rapid identification of mycobacteria by Raman spectroscopy. J Clin Microbiol 2008; 46(3): 961-5.
[http://dx.doi.org/10.1128/JCM.01763-07] [PMID: 18174303]

[84] WHO report 2006. Geneva, Switzerland: World Health Organization 2006.

[85] Rapid diagnostic tests for tuberculosis: what is the appropriate use? American Thoracic Society Workshop. Am J Respir Crit Care Med 1997; 155(5): 1804-14.
[http://dx.doi.org/10.1164/ajrccm.155.5.9154896] [PMID: 9154896]

[86] Buijtels PC, Petit PL, Verbrugh HA, van Belkum A, van Soolingen D. Isolation of nontuberculous mycobacteria in Zambia: eight case reports. J Clin Microbiol 2005; 43(12): 6020-6.
[http://dx.doi.org/10.1128/JCM.43.12.6020-6026.2005] [PMID: 16333092]

[87] Nahid P, Pai M, Hopewell PC. Advances in the diagnosis and treatment of tuberculosis. Proc Am Thorac Soc 2006; 3(1): 103-10.
[http://dx.doi.org/10.1513/pats.200511-119JH] [PMID: 16493157]

[88] Kirschner P, Böttger EC. Species identification of mycobacteria using rDNA sequencing. Methods Mol Biol 1998; 101: 349-61.
[http://dx.doi.org/10.1385/0-89603-471-2:349] [PMID: 9921490]

CHAPTER 6

From Cells to Clinic - Direct Biomolecule Quantification of Clinically Relevant Biomolecules

Swati Jain[1,2,*], **Harsimran Singh Bindra**[3] and **Sruti Chattopadhyay**[4,*]

[1] *Amity Institute of Nanotechnology, Amity University, Noida, UP, India*

[2] *Department of Science & Technology, Technology Bhavan, New Mehrauli Road, New Delhi, India*

[3] *School of Biotechnology, Sher-e-Kashmir University of Agricultural Sciences and Technology of Jammu, Jammu and Kashmir, India*

[4] *Center for Biomedical Engineering, Indian Institute of Technology Delhi (IITD), New Delhi, India*

Abstract: Translation of investigative cellular analysis into reliable clinical settings is a challenge and surface enhanced Raman spectroscopic (SERS) technique has the potential to move beyond laboratorial examinations. Quantitative and qualitative measurement of cellular components and their properties is essential indicative of healthy or disease state. Pre-emptive analysis of diseased state offers synchronization with early diagnosis of certain medical conditions and is requisite for initiation of therapeutic interventions. High sensitivity and capacity for multiplexing renders SERS suitable for biochemical analysis for disease diagnosis a critical step towards formulation of therapeutic regime. SERS assists in a deeper understanding of cellular processes and its micro environment without disturbing and damaging the cellular milieu in 3 dimensional (3D) set up without invading the tissues. Fabrication of novel nanostructures with enhanced plasmonic effects has also propelled the growth of SERS based analysis of cellular structure in normal and abnormal circumstances. Thus, SERS is operating as diagnostic tool for *in-vitro*, ex-*vivo* and *in-vivo* investigations for assessing onset of disease as well as prognosis of the therapy in hospitals and clinics.

Keywords: Continuous Glucose Monitoring, Neurotransmitters, Surface-enhanced Spatially-offset Raman Spectroscopy (SESORS).

* **Corresponding author(s) Swati Jain:**Amity Institute of Nanotechnology, Amity University, Noida, UP, India; Current affiliation: Department of Science & Technology, Technology Bhavan, New Mehrauli Road, New Delhi, India; and **Sruti Chattopadhyay:** Center for Biomedical Engineering, Indian Institute of Technology Delhi (IITD), New Delhi, India ; Tel:+9101126596360; E-mail:swatijain.iitd@gmail.com, sruticiitd@gmail.com

6.1. INTRODUCTION

Monitoring metabolites with high specificity and reliability to systematically understand the cellular dynamics is the established viewpoint of healthcare R&D professionals. This knowledge of cell species in their native micro-environment, transport mechanism of molecules across cells and concentration of various proteinaceous, nucleic acids and other species is essential in measuring the extent of deviation from desired state. Therefore, it is imperative to analyse biomolecules in their complex biological surroundings along with methods that enable fast, accurate and precise detection of biomolecules. The ever-evolving medical science has gained significant aid from the development of new and improved materials and instrumentations that offer potential to detect minute changes in analyte concentration. For instance, luminescence based assays are routinely engaged in clinical settings for analysis but the need of tracking disease progression at early stages is still desired that facilitates therapeutic intermediation. Spectroscopic techniques have also been employed using UV-Visible absorption, infra-red spectroscopy, near-infrared absorption, X-ray diffraction and absorption. Presently, surface enhanced Raman spectroscopy (SERS) is a well-established spectrometric technique with superior qualities than these traditionally used spectroscopic techniques. The highly sensitive molecular fingerprinting of biomolecules has propelled the applicability of SERS in medical and biological domains. Furthermore, production of field deployable small hand-held SERS devices has further accelerated their use at clinical level. Offering an eased and less specific sample preparative technique, recording SERS data of either tissues or cells can be performed by avoiding labour intensive procedures.

When any electromagnetic (EM) radiation bombards a molecule, its enables elastic or inelastic interactions which eventually denotes the transition in different motions of any molecule. In case of in-elastic interactions of EM waves with the molecules, they undergo either gain or loss of photons or energy by accepting or donating energy to the vibrational and rotational motions of the analyte molecule. The spectra generated with this inelastic interaction yields a stokes or anti-stokes band in Raman spectroscopy. These resultant bands corresponding to the vibrational or rotational transitions are specific to the molecular structure and called as fingerprints of the molecule that enables the chemist to identify and characterize molecules. For a long period of time, Raman spectroscopy was employed for structural identification and analysis of solvated molecules, however, limited exploration was possible owing to feeble nature of these bands. Independent studies in late 1970's by group of two scientists noticed a significant increase in the intensity of weak Raman signals when samples were absorbed on a rough silver metal electrode [1]. These postulates were put to further experimentation and thus began the journey of surface enhanced Raman spectroscopy. Data gathered from Raman spectroscopy and SERS was similar, however, SERS provides signals with high magnitude and distinct 'fingerprint' peaks of the molecules. Thus, has been exploited to much larger range of molecular structures than Raman spectroscopy. These advantages propelled the growth of SERS analysis in biological and biomedical applications.

On interaction of electromagnetic radiations with metallic nano-patterns or nanoparticles which have lower dimensions than the wavelength of incident light a displacement of electronic clouds occurs inducing polarization within the metallic nanostructures. The enhancement appears as a maxima when plasmon frequency is in resonance with the incident radiation. This excitation of local plasmon resonance owe to resonant electron dynamics, plasmon resonance conditions and dielectric constant of the surrounding medium. This excitation of localized surface plasmons which are associated with strong field enhancement is known as electromagnetic mechanism of SERS. The dipolar effect is strongly pronounced, this leads to significant increase in the intensity of peaks. The field enhancement intensifies the incident radiation in the exciting the Raman mode of the target molecule. In addition, the surface on which the molecule is absorbed further amplifies the signal owing to the surface characteristics of roughness. Scattering of light occurs when plasmon oscillations are perpendicular to the surface of the nanomaterials because if they are in-plane no scattering will occur. Hence, the surfaces have rough morphology attained by either nanomaterials or nano-patterns by nanofabrication process contributing to dipolar effects. EM factor sharply decreases as the distance between the molecule and nanomaterial increases.

Apart from EM mechanism, chemical theory has been envisaged for molecules which were not previously described by electromagnetic enhancement and polarization effects of nanomaterials. The chemical mechanism contributes to the overall SERS response enhancing the signal to about 10^1 to 10^3 times; thus, SERS shows roughly increase of 10^{14} times intensified signal than Raman spectra. Therefore, in condition when SERS scattering light is in resonance with the frequency of localized surface plasmon resonance (LSPR) generated by nanostructures, the SERS signal is maximum resulting to enhancement factor of E^4 Hence, the excitation light and the Raman scattering can gain the same EM enhancement in a close proximity. The SERS signal intensity is theoretically calculated under the effects of both EM and CE:

$$PSERS \ \alpha \ \{\text{Chemical enhancement} \times \text{EM enhancement}\} \qquad \textbf{(1)}$$

For maximum enhancement to occur, the molecule of interest has to be in close vicinity of SERS material or nanostructure or 'hot-spots' and enhancement factor of about 10^{11} has been attained. The signalling attributes are directly correlated with size, shape and nature of nanomaterials as well as relies on the orientation of the molecule towards the metallic surface which impacts the spectral shape. As noted above, Raman modes of target molecules placed normal to the metal nanostructure are preferentially enhanced and more pronounced in SERS spectra. In addition, Raman modes in close vicinity get enhanced to larger extent owing to evanescent character of the EM waves on the metal surface. Hence, hot-spots are created to gain maximum insight into molecular structure of the material of interest using nanomaterials as solid support. For biomolecular investigation, this feature is only attained when they are coupled to nanomaterials chemically. This type of binding preferentially occurs as a covalent attachment which is facilitated *via* oxygen, nitrogenous and sulphur containing moieties present on the biomolecules.

In general, two different modes for biomolecular analysis are carried out as depicted in Fig. (**1**). This leads to complexation between various participating groups in intrinsic and extrinsic formats that eventually fates the bioanalysis process. These are also called as direct or label-free or indirect approaches. When sample and SERS active nanomaterials are in direct interaction, enhancement in Raman modes of spectra prevails. This provides intrinsic fingerprints of the biomolecule with high sensitivity and has been particularly very useful for analysing small molecules. The resultant spectrum contains the conformation and orientation information on adsorbed biomolecules [2]. Investigation with direct SERS biosensing is highly recommended for small biomolecules such as endo-exogenous antioxidants, certain lipids, oligonucleotides, glucose *etc.*

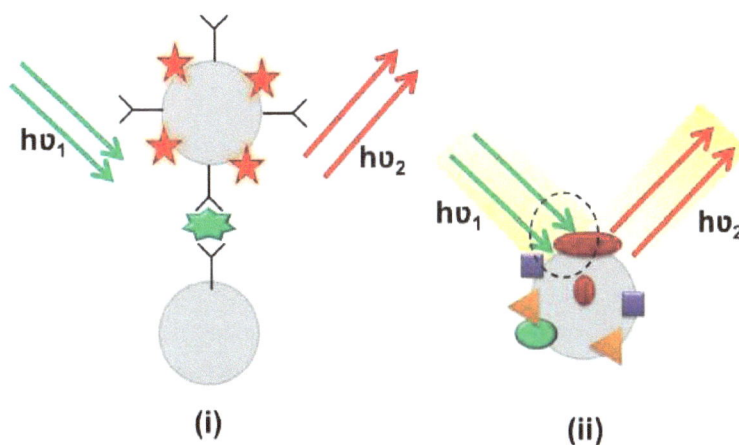

Fig. (1). Schematic representation of (i) indirect and (ii) direct SERS detection mode.

Indirect or extrinsic format employs certain Raman tags namely Raman reporter molecules (RRMs) along with nanomaterials to greatly enhance the signal. Here, the optical activity of plasmonic nanomaterials is combined with inherently SERS active messenger or reporter molecules which are in resonance with wide range of excitation lasers. The target analyte on binding with bio-receptor attached nanostructure is indirectly measured through RRM. These RRMs have distinct Raman spectra owing to high chemical enhancement arising from adsorption on SERS active nanostructures. Most of sulphur or nitrogen containing dyes such as DTTC, Rhodamine 6G, crystal violet. and thiolated small molecules are preferred as Raman reporters due to their high affinity for Au and Ag [3 - 6]. It has been experimentally observed that molecules with large Raman scattering cross-section, perform phenomenally as RRMs for biosensing and bioanalysis. Herein, each nanoparticle is coated with large number of RRMs (10^3-10^4) and hence, the response of an individual binding event is significantly enhanced. Therefore, consecrated efforts are directed by synthetic chemists to synthesize improved RRMs which not only enhance the Raman signal intensity but also render multiplexing capabilities to SERS.

To attain further insight in the concept, this chapter delivers a brief review of SERS as diagnostic tool for the detection of biological moieties for assessing medical ailments which have the potential to be translated as clinical procedures. Many of the proof-of concepts are highlighted along with methods which are currently in clinical phase trials. Besides, discussion on the investigation of cells and tissues related to cancer diagnostics as well as prognosis, neurodegenerative disorders, lifestyle related conditions such as diabetic and cardiac issues and other hormonal imbalances is also presented.

6.2. DIAGNOSTIC SERS: KEY FEATURES AND ADVANTAGES

Diagnosing diseases using SERS analysis is highly lucrative. Since it takes advantage from the synergistic combinatorial approach of integral structural specificity and experimental setup of Raman spectroscopy technique. High sensitivity leading to molecule detection has been attempted with SERS using colloidal silver (Ag) nanomaterials that are randomly absorbed on glass substrates. It operates at very low molecular concentration typically less than 10^{-8} M that statistically allows no more than one molecule per colloid. Thus, the Raman signal generated from this sample could be considered from a single molecule. The key consideration is to provide large number of "hot spots" or boosting active-SERS sites on colloidal nanoparticles. Extensive research is underway to fabricate surfaces with high number of hot-spots either in colloidal formulation such as Ag heterodimers [7], star-shaped nanoparticles [8], and oxide shell-isolated nanoparticles [9]. In addition, nanofabrication techniques like lithography have been beneficial in getting either silver arrays [10], or bowtie-nano antennas [11].

The synthesis of novel nanoacrchitectures of SERS active materials has significantly increased the efforts towards estimating and analysing the molecular, cellular and intracellular events by tracking biomolecules *in-vitro*, *ex-vivo* and *in-vivo* set-up. SERS diseases diagnosis relying on active nanomaterials offers high sensitivity, large spectral resolution capacity and reliable signals in a non-invasive mode without damaging biological sample. In addition, unlike fluorescence microscopy and spectroscopy which greatly suffers from the problem of photobleaching of fluorophores, SERS active nanomaterials and strong RRMs does not get photobleached or quench under laser. Moreover, existence of water does not affect Raman spectra. Therefore, with little interference from autofluorescence of other molecules and water, SERS spectra is relatively clean devoid from background noise which is considered as an extremely beneficial feature of SERS in disease diagnosis. Furthermore, different RRMs with distinct Raman activity infer possibilities of multiplexing for checking different analytes for same excitation of light. Enhancement in chemometrics and multivariant analogues for analysis has also been specially found useful in SERS based disease diagnosis.

Many of the procedural designs have been taken up by researchers to create methods for detecting clinically relevant molecules *in-vitro*, stimulated conditions as well as *in-vivo* conditions in order to confer as additional benefit towards biosensing of neurological diseases, glucose, cardiovascular diseases and discreet cellular entities which regulate metabolic pathways during normal as well as in disease conditions.

6.3. NEUROLOGICAL DISEASES

6.3.1. General Features

Neurons are fundamental units of the nervous system. The adult human brain is estimated to contain in excess of 80×10^9 neurons establishing several trillion synaptic connections. Diseases stemming from central and peripheral nervous system which includes brain, spinal cord and nerves connecting them impair the normal functioning of an organism. Neurological diseases can occur owing to injury to brain or spine, impaired blood supply leading to stroke, faulty genes such as muscular dystrophy, microbial infections like meningitis, nerve degenerative diseases like Parkinson's or Alzheimer's or Seizure disorders – epilepsy. In most cases these structural, biochemical or electrical abnormalities often leads to irreparable damages in a person's ability to respond, coordinate, memorize, sense or even possess altered consciousness [12]. In many cases, losses of motor function go undetected till the symptoms aggravate causing irrevocable damages, thus instigating the research for the development of rapid procedures for neurological pathways and systems which govern the functioning of motor nerves. Multitudes of investigative methods have been developed in order to delve deeper into the inner workings of the brain matter and thus neurochemistry is designed spanning the realms of molecular biology, genetics, chemistry, optics, and engineering. This conformed nexus of discoveries in neurosciences has accelerated our understanding of the brain functions for motor control, learning and behaviour.

Neuronal interaction with nerve cells and other cellular components regulated *via* chemo-electrical messenger system manifest as behavioural perception [13]. Chemical messengers or neurochemicals include neurotransmitters, neuropeptides and certain psychopharmaceuticals participate in neural activities and influence the physiology of nervous system. Qualitative and quantitative changes in neurotransmitters manifest as behavioural disorders and experimentally it is proved that there is positive correction of psychological events with distribution and function of neurotransmitters and neurotrophic peptides. Therefore, neurochemicals serve as possible markers for neurological diseases particularly during early stages of development of diseases and their detection becomes paramount in underlining appropriate therapeutic regime.

Till today, the brain's neurochemical state is being interrogated using classical tools borrowed from analytical chemistry, such as liquid chromatography and amperometry or classic electrophysiology. By definition electrophysiology is the branch of neuroscience devoted to understand the electrical activity of living neurons while performing investigative analysis of molecular and cellular processes that govern the signalling pathway. Electrophysiology techniques estimate these signals by monitoring electrical activity, allowing scientists to decode intercellular and intracellular messages. In most of these techniques, host of miniature electrodes or sensors are used which allow the detection of membrane currents caused by the activation of ionotropic membrane receptors or neurotransmitters themselves at the single neuron level, in acute brain slices or in living brain tissue [14, 15]. The matrix marked as Table **1** sketches a detailed outlook of various illustrative neurotransmitters.

Table 1. Classes of different neurotransmitters, their function and possible implications in mental health

Neurotransmitter (NT)	Designation of NT	Molecular identity	Releasing organ	Functions	Type of function	Neuropsychiatric disease
Acetylcholine (Ach)	Learning NT	Cholinergic	Basal ganglia, parasympathetic post-synaptic nerve terminals of Cerebral cortex, limbic structures, motor neurons, hippocampus, presynaptic nerve terminals	Regulates the sleep cycle, essential for muscle functioning, pain perception, memory	EY except in the heart (IY)	Alzheimer's disease
Serotonin	Mood	Monoamine	Spinal cord, Serotonin Hypothalamus, thalamus, limbic system, cerebral cortex, cerebellum,	Feelings of well-being and happiness, regulates mood, food intake, and social behaviour like aggression	IY	Depression Migraine aging Attention deficit disorder Anxiety
Dopamine	Pleasure	Catechol amine	Substantia nigra, Frontal cortex, limbic system, basal ganglia, thalamus, posterior pituitary and spinal cord.	Regulates bodily movement, involved in feelings of pleasure, motivation/reward, wakefulness, directs attention to specific tasks or activity	Both EY and IY	High Schizophrenia Low Parkinsonism
GABA	Calming	Amino acid	Hypothalamus, hippocampus, cortex, cerebellum, basal ganglia, spinal cord, retina	Controls fear anxiety, contributes to motor control and vision, counterbalance the action of the excitatory neurotransmitter, regulates apetite and metabolism	IY	Epileptic Seizers Low Gaba concentration-Anxiety, Chronic stress, Depression, Difficulty, concentrating and memory problems, Muscle pain and headaches, Insomnia and other sleep problems

(Table 1) cont.....

Neurotransmitter (NT)	Designation of NT	Molecular identity	Releasing organ	Functions	Type of function	Neuropsychiatric disease
Glutamate	Memory	Amino acid	Pyramidal cells of the cortex, cerebellum and the primary sensory afferent systems, hippocampus. thalamus, hypothalamus, spinal cord	Long-term potentiation (LTP)-relay of sensory information Shape learning and memory, promotes neuronal excitation, mediates cognition, emotions, sensory information, and motor coordination	EY	Migraine Stroke Autism
Adrenaline (NT and Hormone)	Fight or flight	Catecholamine (Monoamine)	-	Secreted in stressed or excited condition, increases heart rate and blood flow, leading to heightened awareness and physical boost	EY	High blood pressure, anxiety, sweating, headaches
Noradrenaline (NT and Hormone)	Concentration	Catecholamine (Monoamine)	Sympathetic post-synaptic nerve terminals. CNS – Thalamus, hypothalamus, limbic system, hippocampus, cerebellum,cerebral cortex	Increases the level of alertness and wakefulness, stimulates various processes of the body-locomotion, mood, cardiovascular functioning	EY	Mood disorders-anxiety and depression, impaired sleep cycle, S chizophrenia
Endomorphins	Euphoria	Neuropeptides	thalamus, limbic structures, mid brain and brain stem	Not completely understood, associated with pain relief experiences and reduced peristalsis	EY	autism, depression, and depersonalization disorder and activities such as laughter and vigorous aerobic exercise

EY: Excretory; IY: Inhibitory

Name	Structure	Involved With
Melatonin		Regulates the sleep-wake cycle; free radical scavenger
Serotonin		Regulation of mood, appetite and sleep; cognitive functions
Epinephrine		Mobilizes brain and body for action, sleep and wake cycle
g-aminobutyric acid		Inhibitory action at receptors
Dopamine		Motor control, motivation, cognition and reward
Glutamate		Neural communication, memory, formation, learning regulation

● Carbon	● Nitrogen	● Oxygen	○ Hydrogen

Fig. (2). Characteristic of essential neurotransmitters.

Typically, for *in-vivo* analysis the detector methods including micro-dialysis and cyclic voltammetry are invasive by nature involving surgical procedures to remove certain parts of skull for implantation of devices for continual monitoring of concentration of neurotransmitters. Most of the deep brain stimulation (DBS) neurosurgical techniques are only done in rare extreme cases. SERS has captured attention of the medical professionals and research scientists for the detection of these neurochemicals in minimally intrusive manner. Chiefly, the metallic plasmonic nanoparticles have been identified as signal enhancing materials in SERS for the detection of such neurochemicals. Electrochemistry and electrophysiology techniques though have grown tremendously over the last decade providing unprecedented data for understanding neuronal activity, however most of the work is devoted towards selective neurochemicals. Glutamate, GABA, serotonin and dopamine constitute a small albeit paramount group of neurotransmitters which are linked with functioning of brain chemistry. SERS gives an edge over other methods to analyse neurotransmitters as it provides multiplexing capabilities to detect large number of neurotransmitters which are in close contact with neurons and neuronal junctions. Silver (Ag) and gold (Au) nanoparticles (Nps) have both been explored for SERS based analysis however, it has been reported that Au Nps are good intracellular probes reporting local cellular environment during *in-vivo* SERS detection and imaging [16].

6.3.2. Detection of Neurotransmitters by SERS

Amongst the various classes of neurotransmitters, the detection of monoamines such as dopamine and serotonin are well studied and documented. Deficiency of dopamine leads to numerous psychological behavioural anomalies and other disease pathological condition. In Parkinson's disease (PD) the dopamine (DA) concentration is majorly depleted and much significant correlation has been observed between low levels of dopamine with depression, anxiety, and memory related complications in patients [17]. DA is excitatory catecholamine neurotransmitter and its deficiency develops into anhedonia and motivational loss. Thus, reversal of depression symptoms in PD patients has been achieved with dopamine precursor Levodopa, bringing noticeable improvement in the patients' attitude [18, 19]. Serotonin on the other hand is an inhibitory monoamine implying opponency exists between contrasting functions of these two neurotransmitters. Serotonin or 5-hydroxytryptamine (5-HT) is generated from L-Tryptophan which converts into 5-Hydroxy-L-tryptophan (5- HTP) which in turn yields Serotonin through the activity of specific decarboxylase. Serotonin or 5-HT regulates mood and emotions involved in decision making process and its balanced levels have been known to be associated with several neurological and psychological behaviours. Analysis of these clinically effective neurotransmitters is important in understanding neuronal systems as well as efficacy of drug administration in the treatment of mood and behavioural disorders. SERS analysis of dopamine and serotonin has been attempted independently and in combination with one another to understand their complex interconnection. Gold/Silver or gold/silver/gold dimer and heterodimer were used for the detection of low concentration of DA [20, 21]. Rapid investigation of DA levels in cerebrospinal fluid and mouse striatum was investigated by Ranc *et al.* using nanocomposite of magnetite and silver nanoparticles [22]. They used a DA selective reporter compound - iron nitriloacetic acid enabling target selection; magnetite core for simple sample preparation, easy separation of targeted molecules and silver shell on nanoparticles for enhanced SERS signal. Selective capture of DA can be achieved in sandwich assay setup involving Au nanoplates attached to DA specific antibodies consisting of one layer and Au Nps with DNA and antibodies constituting another layer [23]. Likewise, serotonin concentration specific to physiological levels can be monitored through SERS investigation especially it's unwanted oxidized form. Computational density functional theory (DFT) calculations and SERS algorithmic analysis revealed ultrasensitive detection of serotonin compounds as low as 10-11 molar using Ag NPs. These results indicated the presence of different serotonin molecular forms which includes ionic, neutral and oxidized compounds. It was observed that simultaneous detection of these neurotransmitters posse problems for their accurate detection at physiological levels owing to overlapping bands in SERS. Similarity of their chemical structures both containing typical hydroxyl groups induce overlapping SERS characteristics bands complicating their analysis particularly in direct measurement without labelling these neurotransmitters. In stimulated body fluids single molecule ultrasensitive identification of DA and serotonin was done using Graphene-Au nanopyramids heterostructure platform. The strategy allows to attain enhancement factor of 10^{10} as a function of quasi-periodic Au structures propelling high-density and high-homogeneity hotspots, assisting to achieve distinct spectral analysis of DA and serotonin in 1s at 10^{-9}M level [24]. In a more detailed research, SERS based biosensing of 7 neurotransmitters is reported by Bhavna Sharma lab (Moody *et al.*) where detection is extended to other families of NTs *i.e.* melatonin, serotonin, glutamate, dopamine, GABA, norepinephrine, and epinephrine through two metal nanoparticles [25]. They highlighted that for catecholamine NTs, Au Nps gave worthy results at an excitation wavelength of 785 nm, while for amino acid chain neurotransmitters, the best detection was achieved with Ag Nps at 633 nm. SERS biosensing and bioanalysis has been extended for other types of neurotransmitters such as glutamate and -amino butyric acid (GABA), important amino acid neurotransmitters that are essential for neuroendocrine control as well as are associated with epilepsy. These were detected using silver nanoparticles with limit of detection (LOD) ~ 10^{-7} and 10^{-4} M for glutamate and GABA respectively. Table **2** compiles details of documented reports of some of the illustrative NTs, their role in neuronal communication and analysis through SERS.

Table 2. Matrix showing NT assisted SERS analysis.

Neurotransmitter (NT)	Neuronal Communication with Binding Molecule	SERS Response	Reference
Glutamate and γ-Amino Butyric Acid (GABA)	AgNPs	Limits of detection (LODs) of 10^{-7} M for glutamate and 10^{-4} M for GABA	[26]

Neurotransmitter (NT)	Neuronal Communication with Binding Molecule	SERS Response	Reference
Dopamine, serotonin, acetylcholine, γ-aminobutyric acid, and glutamate without Raman reporters	Au nanoislands on quartz glasses spread spectrum SERS (ss-SERS)	1.9, 3.0, 1.0, 99, 39 aM LOD for Dopamine, Serotonin, Acetylcholine, GABA, Glutamate resp.	[27]
Dopamine	Ag Np in presence of NaCl	specific adsorption on AgNp surface, low concentration detection in presence of interfering ascorbic and uric acid	[28]
Dopamine	Au core-Ag shell NP-Au nanorod heterodimers	0.02 nM	[29]
Dopamine	AuNR dimer coated with an Ag shell	0.006 pM LODs	[20]
Glutamate and GABA	SERS with a partial least squares (PLS) analysis	LOD of 8 µM for both in serum	[30]
Choline and Catecholamine neurotransmitters including acetylcholine, dopamine, and epinephrine	AgNP electrodeposition onto tin-doped indium oxide (ITO)	Detection limits of 2 µM for choline, 4 µM for acetylcholine, 10 µM for dopamine, and 0.7 µM for epinephrine	[31]

In another report, Félix Lussier and Jean-François Masson have demonstrated detection of ATP, glutamate (glu), acetylcholine (ACh), GABA and dopamine (DA), among other neurotransmitters in one experiment itself using Dynamic SERS technique [32]. They have proposed the use of pulled glass electrodes to facilitate adoption of SERS based measurement by neuroscience community. These electrodes are similar to carbon fibre nanoelectrodes and to patch-clamp nanopipettes, regularly used by scientists and medical professionals working in neuroscience. The proposed SERS based nanosensor chiefly comprise (a) silicon functionalized pulled glass capillaries having morphology of nanopipettes of nearly 500 nm diameter and 13° cone angle and (b) gold nano scale raspberries attached onto pulled capillaries. This sensor was used to detect panel of highly relevant neurotransmitters in physiological conditions near neurons. Fluorescence imaging was coupled with the SERS design to validate the position of nanosensor near the expected sources of neurotransmitter secretion and cultured neurons. Chemometric data procession algorithm was optimized by their group for assessing neurotransmitters in a multiplexed arrangement while improving selectivity and specificity of the detection.

Optimization of series of parameters in the bar-coding algorithm was implemented to maximize selectivity and sensitivity using a combinatorial computational constructed map of time-wise levels of neurotransmitters. The DA-

SERS nanosensor was then located near cultured mouse dopaminergic neurons particularly in a region containing a large number of axonal varicosities and dendrites to maximize the probability of sensing DA. The same group has also non-destructively monitored metabolite secretion near the living cells using the similar technology [33]. Here the nanosensor was fabricated from borosilicate nanopipettes decorated with Au nanoparticles analogous to the patch clamp and detected multiple metabolites, such as pyruvate, lactate, ATP, and urea simultaneously near to Madin-Darby canine kidney (MDCKII) epithelial cells.

The NT capturing surfaces particularly - deposition of dense and well dispersed nanoparticles on highly curved surfaces, are continuously being advanced to improve their sensitivity. Zhu *et al.* have use a polystyrene-block-poly-4-vinylpyridine) block copolymer (BCP) and plasmonic gold nanoparticles (AuNPs, 52 nm diameter) for the 200nm pulled fibres nanosensor fabrication [34]. This template was able to detect co-releasing dopamine and glutamate from living mouse brain dopaminergic neurons with high sensitivity.

Certain biomolecules do not have high affinity for Au/Ag NPs, to overcome this limitation, Prof. Van Duyne group has very recently proposed capture agent-free method for detection of NTs in solution phase [35]. In this attempt, the analyte is physiochemically trapped *via* interactions between the substrate, surrounding media, and molecules within the sensing volume (< 1 nm from the surface) of a SERS substrate. Au Nps were aggregated with salty buffer in presence of target analyte (5 neurotransmitters - dopamine, epinephrine, norepinephrine, serotonin, and histamine) and then polyvinylpyrrolidone (PVP) is added to halt the aggregation. This close proximity of analyse near SERS active nanomaterial leads to a high signal to noise ratio allowing them to achieve high sensitivity of detection in nM range for all the NTs, spanning from 5.7×10^{-4} M to 1.7×10^{-10} M.

6.3.3. New Technique: Surface Enhanced Spatially Offset Raman Spectroscopy (SESORS)

As an upgradation in SERS technique, a spectroscopic technique has been developed which chiefly involves combined activity of SERS with the non-invasive, subsurface sampling capabilities of spatially offset Raman spectroscopy (SORS) collectively called as Surface enhanced spatially-offset Raman spectroscopy (SESORS) and is depicted in schematic representation of (Fig. **3**). This technique shows great promise during *in-vivo* imaging and is highly acclaimed method as it has the potential to provide highly selective, label free, and minimally invasive ultrasensitive detection of low levels of biomolecules in rapid sampling pattern. SESORS could lead to earlier detection of neurological diseases and provide critical knowledge about their progression in non-invasive method. SORS is being employed for food, health supplements and pharmaceutical sample analysis of encapsulated, coated or sealed packages in non-or semi - transparent containers.

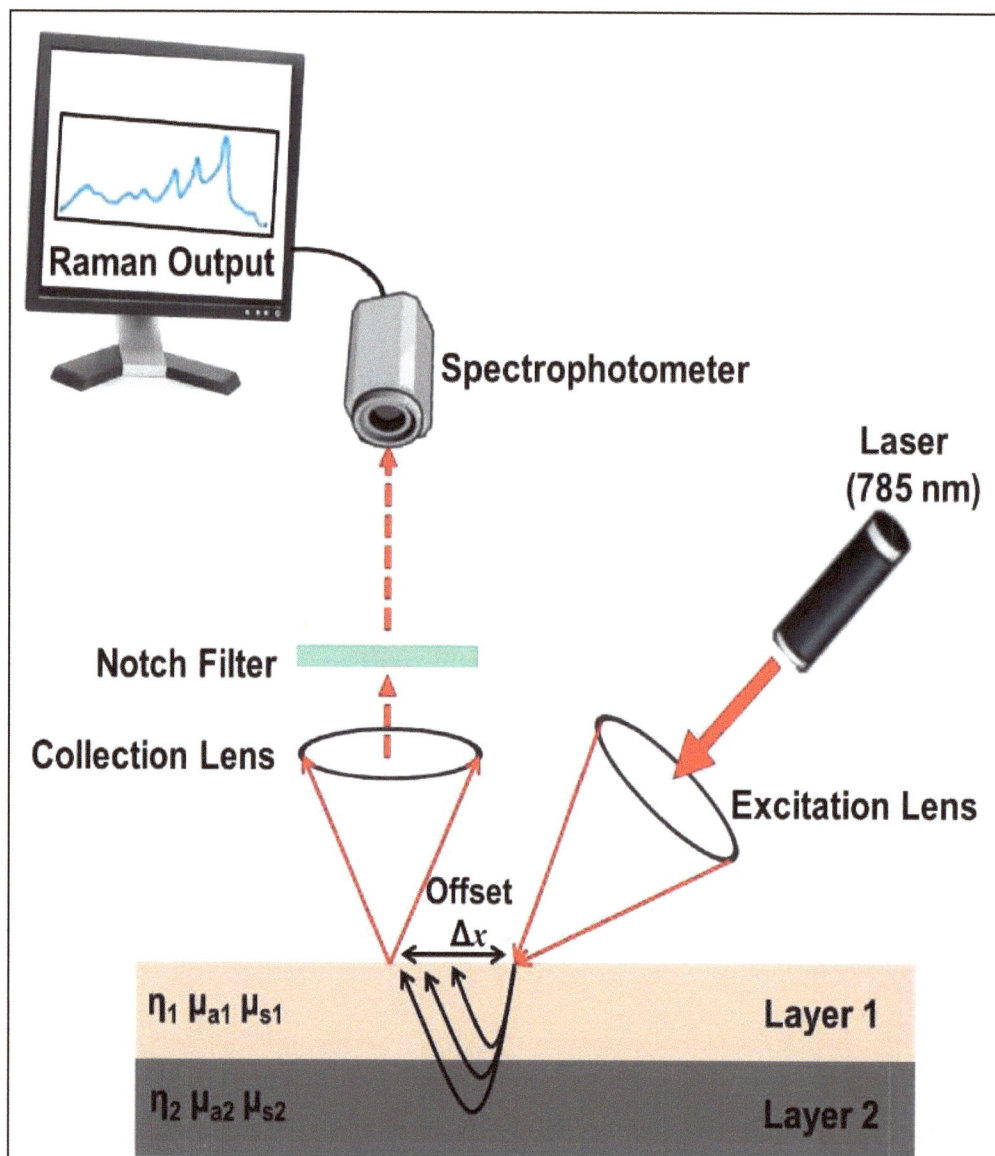

Fig. (3). Illustration for depiction of Surface enhanced spatially-offset Raman spectroscopy SESORS.

For SORS, Raman signals are acquired at multiple points which are spatially offset from the incident illumination. Upon optimization of distance between point of illumination and sample surface Raman signals which are gathered from different deeper/internal layers within the sample can be retrieved. Incident photons having large offset deeply penetrate inside the sample obtaining Raman signal from further beneath the surface of the sample.

Raman signals in spatially offset resolved state was employed for determining concentration of neurotransmitters through intact skull highlighting the concept of non-destructive bioimaging and analysis [36]. Melatonin, serotonin and epinephrine were detected at 100 μM range through cat skull using SESORS technique. The reported detection method is unique as it is unsupervised multivariant principle component analysis of PCA which significantly reduced the large data sets or clusters of individual neurotransmitters. This paves the way for potential *in-vivo* and real time analysis of neurotransmitter concentration though skull with damaging or destroying any of its part unlike DBS.

6.4. DIABETES-GLUCOSE MONITORING

6.4.1. The Rising Phenomenon Called Diabetes

The chronic disease of Diabetes is a major cause of kidney failure, blindness, strokes, hearth disorders, and lower limb amputation. Diabetes and hyperglycaemia or rise in blood sugar levels are chief concern in underdeveloped countries which see escalated increase in the number of people suffering from this disease and related medical condition. It is reported that the both types of diabetic conditions 1 and 2 will have increased prevalence to more than 54.9 million Americans between 2015 and 2030 while deaths owing to diabetes will jump by 38% to 385,800 [37]. Highly staggering statistical data is present analysing diabetes for the entire global population presented by international diabetes federation (IDF). It is estimated that every 7 seconds someone is expected to die from diabetes or related complications.

The International Diabetes Federation (IDF) Diabetes Atlas Ninth edition 2019 reports that approximately 463 million adults aged between 20-79 years are currently living with diabetes and by 2045 this will rise to 700 million. Most of the world's population below 60 years of age is diagnosed with diabetes conditions and the situation is more precarious in underdeveloped countries with 50% of the cases being undiagnosed. In fact, one in every six cases report gestational diabetes in live births or during pregnancy thus, risking both mother and child. The serious spreading and increase in diabetes incidences are a cumulative effect of global obesity epidemic and successful revascularisation therapy of people with diabetes worldwide.

In addition to upsurge in type 1 and 2 disease, there is momentous manifestation in adult population having pre-stage of diabetes, called Impaired Glucose Tolerance (IGT) as well as Impaired Fasting Tolerance (IFT) forming an intermediate stage in the natural history of diabetes mellitus. This state arises when glucose concentration comes 140 to 199 mg/mL during 75-g oral glucose tolerance test and 100 to 125 mg/mL (5.6 to 6.9 mmol/L) in fasting patients respectively [38]. The blood sugar concentration vis-à-vis diabetic or normal condition is detailed in Fig. (**4**). These glucose levels are even though lower than the level that is diagnostic for diabetes but are above normal and have high risk of developing diabetes. People with IGT also have higher peril of developing cardiovascular problems thus, necessitating adoption of primary preventive methods.

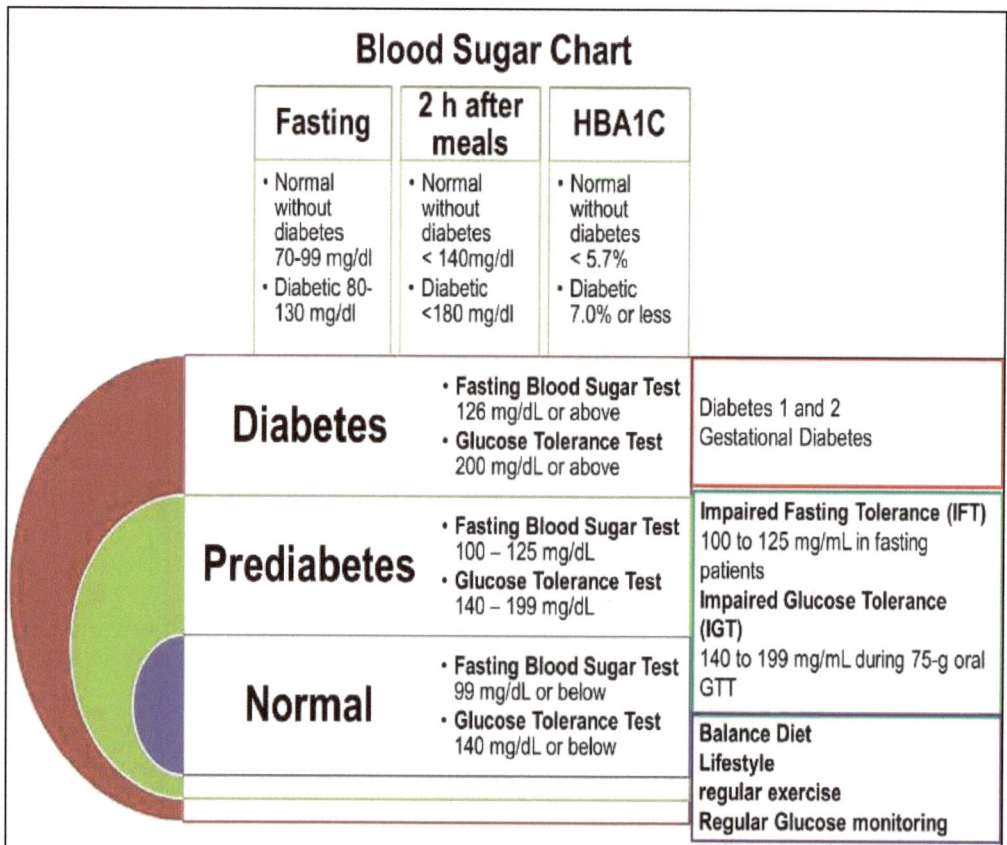

Fig. (**4**). Blood sugar concentration chart vis-à-vis their.

Furthermore, in gestational condition (during pregnancy), IGT and diabetes type 1/2 necessitates tightly controlling glycaemic concentration through diet and exercise as well as frequent blood glucose measurement. As per the recommendation of American diabetologists, in conditions of pre-gestational diabetes, blood sugar levels should be monitored after 6 weeks of delivery and after every 3 years [39].

Regular blood glucose monitoring aids in making informed decisions regarding therapeutic regime and instigating changes in diet, exercise and lifestyle. Moreover, secondary health complications such as cardiovascular issues owing to diabetes can be directly avoided through glucose monitoring plan. This can mitigate long-term hyperglycaemic conditions as well as avoid dangerous nocturnal hypoglycaemia avoiding disease related complications. Glucose tests employed in hospitals and clinics check amount of glycated haemoglobin which is representative of the average blood glucose levels over the last 120 days using redox reactions or enzyme mediated reactions with about 1 ml of blood. Enzymatic methods using glucose oxidase and hexokinase have been adopted in making point-of care tests strips which use one drop of blood [40]. However, in both cases blood needs to be drawn which is a painful procedure especially in emergency situation, post meal in type 2 diabetes, pregnant females. Accurate determination of blood glucose concentration is pre-requisite for developing state-of the art diabetes management technology.

Another limiting factor in diabetes management is low glucose concentration or the condition of hypoglycaemia which can be attained post insulin shot translating into temporal fluctuation of glucose concentration. This rapidly changing glucose levels are hazardous and need careful monitoring to ensure optimized levels. In addition, physicians can be made more confident about the effective dosage administration decisions if sensors were able to detect low levels of glucose. This would act as an asset to numerous sufferers who are not able to accomplish BGL targets despite testing their blood glucose multiple times daily.

6.4.2. Continuous Glucose Monitoring

At present no technology is capable enough to detect very low blood glucose levels (VBGL) and in spite of 40 years of innovation in improved diabetes management, the development of a continuous glucose monitor (CGM) remains a worldwide goal. CGM is proved to be useful for daily managing glucose levels of patients and is much better than every day multiple capillary blood glucose determinations which is not only cumbersome but expensive and inconvenient as well. CMG is recommended to be very effective as real time data can be gathered about current or momentarily BGL through interstitial fluids. In addition, it is possible to get short term feedback of the therapeutic intervention dosing and frequency of insulin administration and it gives warnings when glucose concentrations become dangerously high or low. For this reason, the creation of a non-invasive approach for blood glucose measurement has long been regarded as the holy grail.

In a typical CGM sensor, electrodes are inserted subcutaneously under the skin of either arm or abdomen (Fig. **5**). The device consists of a disposable sensor which checks glucose levels; a transmitter attached to the sensor, and a receiver that displays and stores data. The stored data or information is converted into estimated mean values of glucose standardized to capillary BGL calculated during calibration.

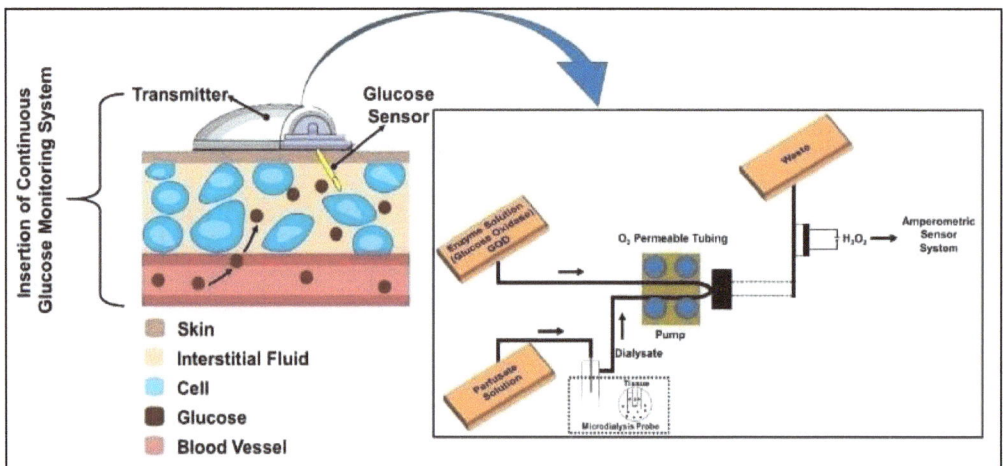

Fig. (5). Schematic representation of working of continuous glucose monitoring (CGM) system. The transmitter is placed transdermally and gives signal for glucose concentration from blood capillaries.

Optical methods are termed accurate in sensing sensitive glucose levels in a painless manner with techniques such as fluorescence, vibrational spectroscopy, microwave spectroscopy, optical coherence tomography and lastly Raman spectroscopy. Raman scattering owing to 'molecular fingerprinting' resulting through inelastic interactions between incoming photons and molecular vibrations, has shown superiority over other spectroscopic technique in estimating clear and concise intelligent information for human skin. In recent years, much progress has been made towards the development of Raman based sensors to detect specific biomarkers like glycated albumin, leucine, and isoleucine [41, 42]. Progress in Raman analysis for glucose monitoring is spurred with the development of specific tissue modulators working at the interface of tissue and sensor tip, introduction of tailored optical probes that facilitate in collecting

Raman photon scattering from blood analytes after interactions. Mathematical interpretation relying on non-linear calibration model altogether incurs modelling of curved data in representation of the spectral-concentration profile.

6.4.3. Development of SERS Sensors for CGM

SERS is exploited for developing CGM, since Raman cross-section of glucose molecule is small while SERS provides manifested signal, thus combating the problem of weak scattering signal. In addition, SERS is compatible with aqueous solutions easily discriminating interferants by spectral characteristics and thus is an attractive method for glucose monitoring in complex biological fluids. Currently, SERS prototypes demonstrate *in-vitro* analysis of glucose in body fluids and implantable devices used in animal subjects. The new versions of CGMs as well as SERS based monitoring methods rely on redox reactions of glucose which are mediated by glucose oxidase enzyme (GOx) producing gluconic acid and hydrogen peroxide. SERS based 'turn-off' device has been proposed by Qi *et al* using Glucose oxidase (GOx) having a Raman-active chromophore (flavin adenine dinucleotide) working as signalling molecule [43]. They have optimized electrostatic assembly consisting of enzyme GOx over silver nanoparticles linked though specific polyelectrolyte maintaining pH of 6.86. When redox reaction occurs, it leads to the generation of gluconic acid, which further lowers pH of the serum allowing bonding between GOx and polyelectrolyte to get weak resulting in chipping off of GOx from the surface. Thus, the SERS intensity of enzyme decreases with increasing concentration of glucose. The overall method is non-toxic as well as non-invasive enabling *in-vivo* estimation of glucose feasible with high sensitivity [43]. However, it was pointed out that glucose does not bind efficiently with roughened metallic surfaces, a pre-requisite for electrochemical enhancement to occur for signal amplification. This results in using of large target concentration with longer signal acquisition time as well as higher laser power. Another demoralising factor for SERS based glucose analysis is very small Raman scattering cross section owing to weak interaction between excitation photon and glucose molecule yielding weak signals.

To overcome these challenges, two solutions were devised and most of the work in this field is credited to Prof. Van Duyne's group. For SERS based analysis, **Film Over Nanospheres Technology** called as **FON** is developed which work as SERS active substrate [44]. FON are fabricated by drop-coating an aqueous suspension of preformed polystyrene or poly silica nanospheres onto supporting glass substrate. These nanoparticles are allowed to self-assemble into close packed array on drying and a metallic layer of approximately 200 nm thickness (or half of sphere diameter) is deposited over these spheres through vapour deposition system. Nanosphere lithography techniques employed to generate silver film over nanospheres AgFON exhibit enhancement of about 10^7. In addition, there is marked increase in active-surface area of the substrate for interacting with analyte molecule. Ag is also in contact with supporting glass surface and these contact work as heat sinks to prevent photothermal damages.

The poor adsorption over nanoparticle surfaces can be overcome by modulating surfaces of nanoparticles to introduce capturing feature on them by linking molecular recognition moieties onto SERS active substrate. Strategically chemistry of nanoparticle surface is functionalized with self-assembled monolayers (SAM) to improve affinity of glucose molecule. These SAM layers trap glucose within their alkane structures protruding out from the surface of nano particles reducing the distance between analyte and nanoparticle surface. These SAM structure operates as portioning layers and, in many cases, they work in reversible manner reflecting the fluctuations in glucose concentration. The portioning layer and ultimately sensor does not get saturated within physiological concentration range.

In initial studies, metal surfaces were modified by simple straight chain alkane thiols like 1-decanethiol [$HS(CH_2)_9CH_3$] triethylene glycol terminated alkane thiols (EG_3) which fashion themselves according to polarity arrangement of the environment. When further characterized it was seen that both these molecules lack in giving proper orientation in water owing to their inherent hydrophobicity and difficult synthesis procedure. To overcome this challenge recently mixed portioning layers or two component systems composed of decanethiol [$HS(CH_2)_9CH_3$] and 6-mercapto-1-hexanol [$HS(CH_2)_6OH$] have been created, with dual hydrophobic/hydrophilic capability for glucose portioning in water phase making them appropriate for *in-vivo* use. This DT/MH monolayer was fabricated over AgFON which shows stability up to 10 days towards sensing glucose in real time for 25 seconds with temporal response <30 seconds in a flow cell setup. The sensing system work in physiological concentrations of glucose (0 – 450 mg/dL) in a complex biological milieu, bovine plasma demonstrating significant improvement from EG_3 SAM based AgFON system. In addition, DT/MH-functionalized AgFON in phosphate-buffered saline demonstrate reversible partitioning and de-partitioning. *In-vivo* studies done in rats with the help of partial least-squares chemometric analysis are successful highlighting the further possibilities of extending these work to suit human applications [45].

Torul *et al.* reported a paper-membrane based SERS platform as well as two component self-assembled monolayer functionalized substrates for glucose detection [46, 47]. In these studies, they were able to detect glucose concentration up to 5mM and 0.5mM respectively. For improving implantable continuous monitoring technology, a novel system composed of AgFON coated with mixed SAM of DT/MH to detect glucose is designed based on SESORS [48]. Targeting to develop an *in-vivo* injectable CGM sensor, functionalized AgFON substrate was surgically implanted under the skin of a rat and the sensor was in contact with the interstitial fluid. A glass window placed along the midline of the rat's back was used as optical window for analysis. The SERS spectra from glucose were acquired at 785 nm through a window using Ti:Sapphire laser with a power of 50 mW for 2 minutes. The approach of SESORS has unleashed new pathways for in *vivo*, continuous sensing of metabolic analytes. This biosensor works directly by sensing glucose giving amplified signal as DT and MH create a dynamic pocket to fill similarly sized glucose molecule enabling them to come close to SERS active surface for fast and effective sensing when injected transcutaneous in rats [48]. The signal acquired using SESORS device through skin of rats, has also enabled an accurate detection of hypoglycaemic levels of glucose working for a minimum of 17 days [49].

In order to facilitate multiplexed analysis in timely fashion, by using chemometric analysis, gold layers were sought to be deposited over nanospheres [50]. AuFON operate in near-IR region where Raman scattering is lowest for water molecules thus decreasing the noise percentages in the signal. AuFON with stabler SAM have been developed using shorter chain length version of EG_3 as the partition layer. These AuFON have red-shifted the SERS resonance, reducing the biological autofluorescence through water molecules and permitting greater biological tissue depth penetration. This sensor has accurate glucose detection with larger glucose concentration range (10 – 800 mg/dL, 0.5 – 44 mM). Thus, the overall strategy is useful for large scale analysis of diabetes patients. Few other amphiphilic molecules have also been studies as portioning layers. Boronic acid and its derivatives show better binding capacity and selectivity for glucose molecule [51 - 53]. AuFON modified with bisboronic acid receptors selectively captures and senses glucose in presence of large amounts of fructose directly *via* SERS [46]. Using two 4-amino-3-fluorophenylboronic acid molecules over AuFON, Sharma *et al.* have shown direct and selective sensing of glucose using computational modelling assigning normal modes and vibrational frequencies for bisboronic acids, glucose and fructose [54]. Applying multivariate analysis, they were able to differentiate between normal (4–8 mM), hypoglycaemic (<4 mM), and hyperglycaemic (>8 mM) glucose levels *in-vitro*.

Bidentate complexes have also been explored which sandwich glucose between boronic acid labelling them as distinct Raman peaks [55]. Kong *et al.* have used alkyne functionalized boronic acid attached to bi-metallic film over nanosphere (BMFON) working as SERS active substrates for sensing glucose. They have used alkyne Raman signal at 1996 cm^{-1} as fingerprint peak which changes with glucose molar concentration. This sensor is cost effective due to its reusability after being washed with a mildly acidic solution. In an earlier work, they have used Au and Ag coated substrate functionalized with boronic acid to capture glucose [56]. The metal carbonyl group gives intense signal in correlation with glucose concentration making accurate quantification of 5 mM glucose in urine possible. Recently, a more direct sensing has been attempted using 2-Thienylboronic acid as a linker molecule which gets coupled directly to AgFONs as well as to a glucose molecule [57]. This molecule works as linker or bridge molecule that attached to the silver surface on one side and to the glucose on the other side and the unique peak arising from interaction with glucose molecule was used for the detection purpose. Their results seem promising in developing a sensor for non-invasive detection of glucose in diabetic patients using saliva samples.

Promising approaches to counter the persistent challenges that hamper adoption of the otherwise transformative non-invasive glucose sensing method and an outlook of its bench-to-bedside translation is also presented. A typical nanoprobes for *in-vivo* sensing comprises of metallic nanostructures functionalized with some Raman reporter molecules (RRMs) coated with an antifouling surface layer such as PEG, silica *etc* along with a biorecognition element which is capable of binding with target site of interest [58]. In case of non-targeted approach SERS probe does not have biorecognition element attached with the nanoprobes. Multiplexed bioassays or the detection of multiple analytes are designed by using variety of RRMs and specific biomolecules each encoding for different target analyte.

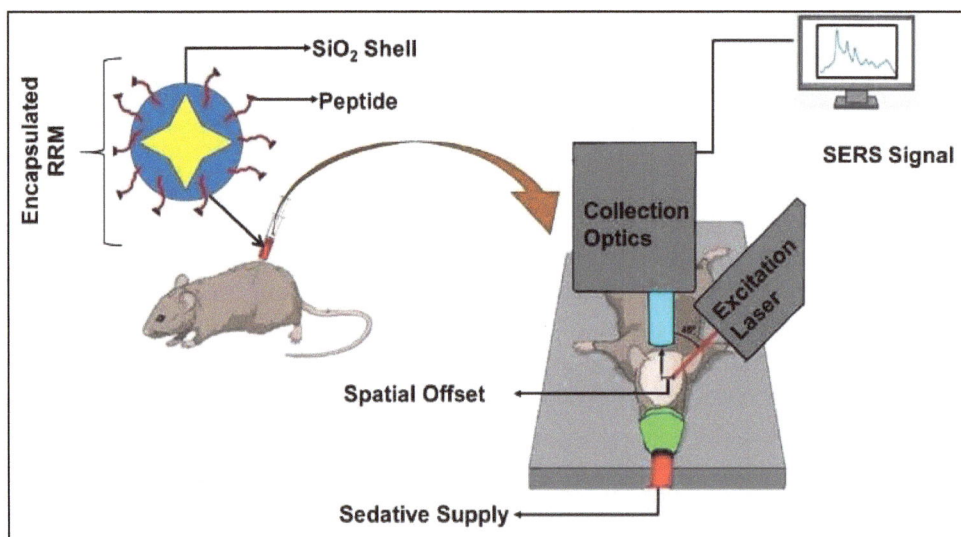

Fig. (6). SESORS in CGM/ Glucose Monitoring.

6.5. DISCREET CELLULAR MOLECULES AND CELLULAR MICROENVIRONMENT

Currently many reports seek SERS for direct quantitation of chemical and biologically relevant molecules in *in-vivo* conditions. The gathered spectral data is utilized to map the cellular environment to differentiate normal and unhealthy state of being. Reactive oxygen species (ROS) a cumulative index of oxygenated radical species derived from oxygen metabolism is an important physiological parameter for knowing cellular microenvironment. The term ROS is used for the collection of hydroxyl, oxy, super-hydroxy, hydrogen peroxide, singlet oxygen, hypochlorous acid species which govern multiple regulatory procedures of cell. ROS are involved in cell transport, communication, gene regulation and immune reactions, thus, their optimal concentration is necessary for cell. In case of overproduction of ROS cell structures like lipidic and proteinaceous membranes, nucleic acids and protein molecules undergo oxidative damage *via* destruction of chemical structure leading to diseased state. Hydrogen Peroxide detection has been taken up using SERS technology where sensing system composed of gold nanoparticles functionalized with 4-carboxyphenylboronic acid (4-CA) was able to selectively detect the molecule [59]. The working of this nanosensor was based on specific transformation into phenol of arylboronate in presence of H_2O_2, leading to SERS spectral changes of AuNPs/4-CA. The sensor showed good stability and biocompatibility with cellular environment. Other attempts have been made for detecting H_2O_2 in presence of other oxygenated ROS species using boronate nanoprobes [60]. This sensor actively consisting of gold nanoparticles functionalized with 3-mercaptophenylboronic acid (3-MPBA) had been employed to check exogenous and endogenous peroxide concentration in living cells.

6.6. MEDICAL IMAGING THROUGH SERS

Healthcare system has been completely revolutionized by the rapid development in the field of medical imaging technologies enabling healthcare experts for quick diagnosis leading to improved decision making while preparing course of action for treatment. Imaging systems aid in early disease diagnosis, reducing the requirements for cumbersome and painful invasive, in-patient procedures, thus, facilitating shorter recovery times. These cost-effective technologies have taken over the concept of traditional exploratory surgeries and are mandatorily used as outpatient procedure for detecting, monitoring and treating possible disease, injury, or similar sort of anomaly. In addition, imaging is also done to assess the effectiveness of medical treatment in certain diseases such as cancer, neurodegenerative complications *etc*. Imaging is carried out by scanning the cellular and molecular targets in living species *via* imaging agents which get accumulated or bonded to these specific targets reporting in the form of a measurable signal. Many imaging modalities have been developed catering to demands of healthcare world such as optical, ultrasound, magnetic resonance imaging; computed tomography; single-photon emission computed tomography (SPECT), and positron emission tomography (PET) along with coupled techniques such as CT, PET/CT, and PET/MRI. PET and SPECT are the workhorses of molecular imaging since being the most evolved modalities for anatomical visualization and physiological mapping as well as monitoring of cellular analytes. Nanoparticle mediated signal acquisition has been attempted in all of these techniques which enhance their imaging capabilities. However, no single modality work as perfect fit and has their own merits and demerits.

As discussed in previous sections SERS provides high sensitivity and its nanoprobes offering higher specificity than other imaging modalities like fluorophores. Moreover, multiplexing capabilities are highly favourable in SERS with unique fingerprint region allowing gathering of much more information about the cellular and sub-cellular components. In addition, low background signal and high spatial resolution make SERS competitive contender in medical imaging arena for monitoring of intracellular microenvironments and tracking of the cellular distribution of extrinsic molecules. As a result, there is tremendous rise in the number of articles published demonstrating SERS as prominent tool in biomedical applications such as diagnostics, sequencing, clinical translation and sensing microarrays particularly in cancer diagnostics. SERS imaging has also up-surged with the innovative SERS substrate preparative techniques like - nanofabrication techniques, biofunctionalization, nanosphere lithography *etc.* several groups are currently working on SERS based bioimaging of molecular targets in animal models of human disease [61, 62]. SERS imaging system comprises of nanoprobe or tag consisting of (i) plasmonically active metallic nanostructures (ii) Raman reporter molecule with unique and strong Raman fingerprint and (iii) a bio-interface providing the tags with a specific signature for identification of target in recognition event. Such particles are further encapsulated by polymer or silica as controlled aggregates to protect them from protein corona formation inside cell [63]. On addition of bioligands with cell-surface receptor, specific affinity of these nanoparticles can be used in microscopy and *in-vivo* imaging of cells, tissues and organs. Au/Ag nanoparticles attached with Monoclonal antibodies using spacer poly(ethylene glycol) (PEG) for imaging of the expression of phospholipase Cg1 (PLCg1) on the surface of HEK293 cells [64]. In another study, non-invasive SERS-based tumor detection was achieved in animal models using thiolated PEG stabilized RRM attached Au nanoparticles [65]. These SERS nanoparticles target specifically to epidermal growth factor (EGF) through tumor-targeting single-chain variable fragment (ScFv). These tags were found be brighter than popular semiconductor quantum dots (QDs). Since Qds have disadvantages of photo-blinking and toxicity, SERS nanotags can serve as an important alternate to these nanomaterials. Novel nanomaterial architectures are also explored which possess desirable properties to sustain microenvironment of the cell. Self-assembled monolayers (SAMs) of Raman reporter molecules created *via* adsorption onto the surface of metal nanoparticles have been reported for the detection of biomarker PSA [66]. SAM renders the metallic nanoparticles hydrophilic in nature which were further stabilized by two different PEG linkers to ensure aqueous dispersibility of colloidal nanoparticles as independent of RRMs. Anti-PSA antibodies were hooked onto these colloidal nanoparticles enabling immune-SERS microscopy to image PSA in prostrate cancerous tissues. Design of nanomaterial construct along with their rational for imaging and detection of significant biomarkers is discussed in Table **3**.

Table 3. Detailed overview on nanomaterial synthesis and their potential to detect specific biomarkers.

No.	Nanotag (NP-RRM-Spacer)				Imaging Target	Model	Remark	Ref.
	NP	RRM	Spacer	Bio-ligand	-	-	-	-
1.	Hollow Au nanospheres (HGNs)	DSNB coated gold nanoparticles with R6G	-	Ab	HER2 receptors in MCF7 breast cancer cells	-	HGNs have homogeneous scattering properties than AgNPs Work as multimodal agents for both dark-field and SERS detection,useful as highly sensitive and homogeneous sensing probes for the detection of cancer markers in cells.	[66]
2.	Au Np	Cyanine and Triphenylmethine	Lipoic acid (LA)	-	HER2 and EGF receptor (EGFR)	different cancer cell lines	Multiplex SERS nanotags (B2LA and Cy3LA) are developed for the concurrent detection of related cancer cells.	[67]
3.	Gold nanorods	Mercaptopyridine	Poly(sodium 4-styrene-sulfonate) and anti-rabbit IgGs	Conjugated Ab	HER2 in cancer cells	MCF7 cancer cells	Highly sensitive targeting and imaging of cancer cell surface markers.	[68]

(Table 3) cont.....

No.	Nanotag (NP-RRM-Spacer)			Imaging Target		Model	Remark	Ref.
4.	Gold/silver core–shell nanoparticles	R6G	-	mAbs	Live HEK293 cells expressing PLCgamma1	PLCγ1-expressing HEK293 cells	Novel metal nanoprobe, conjugated with specific antibodies is used for the sensitive imaging of HEK293 cells expressing PLCγ1 cancer markers.	[64]
5.	Self-assembled monolayers (SAMs) on	5,5'-dithiobis(2-nitrobenzoic acid) (DTNB	A short monoethylene glycol (MEG-OH) and a longer triethylene glycol (TEG-COOH)	Ab	-	Epithelium of biopsies from patients with prostate cancer	Dual SAM SERS labels are used in immuno-SERS microscopy for selective imaging of prostate-specific antigen Uniform molecular orientation Minimal co-adsorption of other molecules from the surrounding Water solubility and stability Controlled bioconjugation	[69]
6.	Au@Ag core-shell nanorods	4-mercaptobenzoic acid	Silica	Folic acid	HeLa cervical cells	-	Fabrication of a SERS and fluorescence dual mode cancer cell targeting probe based on silica coated Au@Ag NRs conjugated with CdTe QDs and FA	[70]
7.	Gold cores	Methylene blue	Silica shell	-	BT549 breast cancer	-	SERS detection (diagnostic) and singlet-oxygen generation (therapeutic) in a single Theragnostic platform	[71]
8.	Au NPs	3,30-diethylthiatricarbocyanine (DTTC)	Thiol-modified polyethylene glycols (PEG)	Single-chain variable fragment (ScFv) antibodies	Epidermal growth factor receptor (EGFR),	Nude mice bearing human head-and-neck squamous cell carcinoma (Tu686)	More sensitive than fluorescent NIR quantum dots xenograft tumor	[65]

Combinatorial studies are need of the hour providing large data set and more information which are useful in analysing multiple parameters or dual applications. Application of SERS nanotags useful in imaging has been extended to study pharmacokinetics and theragnostic. Maiti *et al* have studied the *in-vivo* kinetics of tumour and liver site for a week [72]. They have analysed physical nature of three nanotags in target site for more than a week. These three nanotags namely- Anti-EGFR conjugated CyNAMLA-381, Cy7LA nanotags (EGFR positive), and anti-HER2-conjugated Cy7.5LA nanotags (EGFR negative), were iv injected together into the mice xenograft model. They recorded SERS spectra from the tumour and liver sites daily and deciphered that there was no significant change in the signal intensity of multiplexed SERS nanotags even after 8 days of injection at the tumour site. In its comparison signal intensity at the liver site decreased rapidly falling to almost zero after 8 days on injection. These results highlight the applicability of nanotags in *in-vivo* tumour detection and pharmacokinetic modelling in tandem.

Multimodal probes are emerging with merging of SERS technologies with other imaging systems to widen the intrinsic capabilities of molecular imaging. Bi-functional nanomaterials constituted of superparamagnetic material and a plasmonic metal has properties for measuring signals for both magnetic resonance imaging (MRI) and SERS. MRI-SERS hybrid probes were synthesized by reduction and growth of Au onto the surface of dextran-coated superparamagnetic iron oxide NPs [73]. Sequential steps of seeding and growth were adopted where citrate reduction of $HAuCl_4$ doped small Au Np seeds on the iron oxide particle surface followed by hydroxylamine reduction of $HAuCl_4$ in the suspension enlarging preformed Au seeds. Then RRM 3,30-diethylthiatricarbocyanine (DTTC) was adsorbed on the Np surface as the SERS reporter molecule followed by coating of protective PEG to avoid aggregation and destabilization of Nps. The results are good indicator for simulants performing MRI and SERS detection as T2-weighted magnetic resonance images showed that the signal intensity of the obtained bifunctional hybrid particles was comparable to iron oxide NPs.

Triple modality of MRI-photoacoustic SERS nanotags have been developed for carrying out herculean task of Delineation of brain tumour margins. This translation clinical research has long term positive implications in for preoperative planning and intraoperative resection of brain tumours.

The nano construct with multiple facet constitutes 60 nm Au core with trans-1,--bis(4-pyridyl)-ethylene RRM and 30 nm thick protective silica shell. Further MRI agent 1,4,7,10-tetraazacyclododecane-1,4,7,10-tetraacetic acid (DOTA)-Gd^{3+} was linked on the surface creating RRM-Au-silica-Gd^{3+} nanoparticle [74]. These nanotags were intravenous injected in glioblastoma-bearing mice, and it was recorded that the probes accumulated in tumour areas only instead of healthy tissues facilitating accurate delineation of brain tumour margins. Thus, these multimodal probes permit the use of single imaging agent for, whole-brain tumour localization (MRI), 3D imaging with high spatial resolution (photoacoustic), and high-resolution surface imaging of tumor margins (SERS). These results indicate the use of SERS as effective next generation molecular imaging tool for diagnosis as well as guided surgeries.

Theragnostic modalities are under intense analysis with considerable amount of research focussed on developing SERS nanoprobes integrated with selective cell targeting. The action of cytotoxic agent is stemming *via* photothermal therapy, photoacoustics, singlet oxygen generation or drug based therapeutics. Photothermal therapy where tumour cells are destroyed owing to hyperthermia or rise in local temperature *via* accumulated Au Nps of different shapes and sizes when irradiated with light in NIR range. Ray *et al* have designed a popcorn-shaped Au Nps and attached two biomolecules- RRM Rhodamine 6G(Rh6G) modified A9 RNA anti-prostate specific membrane antigen (PSMA) aptamer and monoclonal anti-PSMA antibody, to increase the binding efficiency of the NPs on the LNCaP human prostate cancer cells [75]. When irradiated with 785 nm laser photothermal therapy was done while at 670 nm SERS detection was carried out. Subsequent studies have involved the hybridization of popcorn AuNPs and other types of nanomaterials like single walled carbon nanotubes SWNTs with aptamers targeting breast cancer SKBR-3 cells [76].

Recently, a new kind of bio-orthogonal Raman reporter and aptamer functionalized SERS nanotags are fabricated for a precise and effective strategy for cancer theragnostic. The authors report that SERS nanotags have strong Raman signal in the biologically Raman-silent range at 2205 cm–1 these tags on irradiation with laser light demonstrated excellent photothermal capabilities restraining tumour growth with 99% inhibition rate [77].

CONCLUDING REMARKS

As the world is shrinking and moving towards biological catastrophes, world of medical diagnostics is fast expanding in volume incurring greater demands for faster, sensitive and reliable tools or kits for detecting as well as monitoring biomolecules. The common forum of scientists is required to work trans - boundaries and inter-governmental ensuring protection against diseases, empowerment of human resources and enhancement of knowledge regarding biomolecules and pathogens. Evolution of technologies and knowledge regarding their judicial usage is defining the new age era empowering medical and healthcare resource professionals.

SERS is one of the most fascinating technologies of the new decade for both material scientists and biologists who seek application of SERS in various domain of medical and paramedical field. Unique molecular fingerprinting with ultrahigh sensitivity for probe molecule under aqueous conditions is the key feature of SERS putting it above the pedestal amongst other spectroscopic techniques in world of medical diagnostic. This chapter reviewed progress in SERS as molecule indicative technique in biomedical and bioanalytical field. Studies on application of SERS in diagnosis are reviewed focusing on investigation of some of the highly relevant biomolecules while cancer cells and CTC are analysed in proceeding chapter. Finally, some the multimodal theragnostic platforms are discussed illustrating imaging theragnostic capabilities of SERS.

SERS has been successfully applied for the sensitive detection of cancer cells, pathogens, discreet biomolecules and assessment of cellular pathways. Some reports have even emphasized fruitful single molecule detection using SERS opening plethora of activities in the multidimensional space of SERS diagnosis. However, more work is needed in the area of SERS in all aspects before its complete implementation in healthcare and medical settings moving from laboratories to practical field to answer significant biomedical questions. Consistency of SERS in diagnosis relies on rational approach on SERS substrate, sample preparation and data analytics. Some of the key challenges and their prospective solutions are summarized in Table **4**.

Table 4. Some of the key challenges and their prospective solution for implementation of SERS as biosensing tool in practical settings of clinics and hospitals.

S. No.	Challenge	Reason	Prospective Solution	Tips for biomedical analysis
1.	Selective detection	Cells are extremely complex and theoretical all molecules in cells can give SERS spectral signal	Data analysis-chemometrics	1. Creation of accurate spectral database for biomolecules 2. Proper data processing and analysis 3. Usage of internal standards for quantitative investigation 4. Proper calibration in stimulated solutions and buffer

(Table 4) cont.....

S. No.	Challenge	Reason	Prospective Solution	Tips for biomedical analysis
2.	Non-Reproducible SERS substrate	No proportionality between intensity and concentration of analyte	Fabrication of novel clean SERS-active substrates	1. Innovations in fabricating technologies to synthesize nanomaterials in uniformity an d batch consistency 2. Introduction of highly biocompatible protective layer for desorption of unspecific proteins 3. Functionalization of nanomaterials with target specific biomolecules 4. Capability to target specific analyte of interest 5. Trapping the analytes in the "hot spots" with the functionalized molecules
	-	-	Introduction of innovative internal standards	1. Introduction of Raman reporter molecules for indirect detection. 2. RRMs with simple and sharp bands; large Raman cross-section and high affinity for the metallic surface
3.	Assessment in living cells	Cell is an extreme dynamic system with respect to time and space and it is very difficult to obtain information about cells in their native state reflecting real physiological activities	Creation of multimodal approach-integration of *in vivo* cell culturing microscopy with spectroscopy	1. Innovative design of cell culturing chambers to keep cells and tissues under controlled external parameters 2. Spectral assessment of biomolecules in their native state 3. Multifunctional nanoprobes for rapid, highly temporal and spatial detection. 4. Future trends highlight the concept of microfluidic systems for high throughput measurements with increased automation, necessary in clinics and hospitals

S. No.	Challenge	Reason	Prospective Solution	Tips for biomedical analysis
4.	Long term activity of biomolecule	Biomolecules especially Abs lose efficacy on immobilization	Effective sample preparation	1. Assessment of suitable biofunctionalization to retain activity of biomolecules in their native state 2. Storage of biomolecules in ambient conditions to avoid their degradation and loss of activity

CONSENT FOR PUBLICATION

Not Applicable.

CONFLICT OF INTEREST

The author declares no conflict of interest, financial or otherwise.

ACKNOWLEDGEMENTS

Author, Sruti Chattopadhyay acknowledges financial support given by Department of Health Research (DHR), Govt. of India (12013/15/2019-HR).

REFERENCES

[1]   Fleischmann M, Hendra PJ, McQuillan AJ. Raman spectra of pyridine adsorbed at a silver electrode. Chem Phys Lett 1974; 26: 163-6.
[http://dx.doi.org/10.1016/0009-2614(74)85388-1]

[2]   Cotton TM, Schultz SG, Van Duyne RP. Surface-enhanced resonance Raman scattering from cytochrome c and myoglobin adsorbed on a silver electrode. J Am Chem Soc 1980; 102: 7960-2.
[http://dx.doi.org/10.1021/ja00547a036]

[3]   Cao YWC, Jin R, Mirkin CA. Nanoparticles with Raman spectroscopic fingerprints for DNA and RNA detection. Science (80-) 2002; 297: 1536-40.
[http://dx.doi.org/10.1126/science.297.5586.1536]

[4]   Yang Y, Zhong XL, Zhang Q, *et al.* The role of etching in the formation of Ag nanoplates with straight, curved and wavy edges and comparison of their SERS properties. Small 2014; 10(7): 1430-7.
[http://dx.doi.org/10.1002/smll.201302877] [PMID: 24339345]

[5]   Zeman EJ, Schatz GC. An accurate electromagnetic theory study of surface enhancement factors for silver, gold, copper, lithium, sodium, aluminum, gallium, indium, zinc, and cadmium. J Phys Chem 1987; 91: 634-43.
[http://dx.doi.org/10.1021/j100287a028]

[6]   Reguera J, Langer J, Jiménez de Aberasturi D, Liz-Marzán LM. Anisotropic metal nanoparticles for surface enhanced Raman scattering. Chem Soc Rev 2017; 46(13): 3866-85.
[http://dx.doi.org/10.1039/C7CS00158D] [PMID: 28447698]

[7]   Kleinman SL, Ringe E, Valley N, *et al.* Single-molecule surface-enhanced Raman spectroscopy of crystal violet isotopologues: theory and experiment. J Am Chem Soc 2011; 133(11): 4115-22.
[http://dx.doi.org/10.1021/ja110964d] [PMID: 21348518]

[8] Rodríguez-Lorenzo L, Álvarez-Puebla RA, Pastoriza-Santos I, *et al.* Zeptomol detection through controlled ultrasensitive surface-enhanced Raman scattering. J Am Chem Soc 2009; 131(13): 4616-8.
[http://dx.doi.org/10.1021/ja809418t] [PMID: 19292448]

[9] Li JF, Huang YF, Ding Y, *et al.* Shell-isolated nanoparticle-enhanced Raman spectroscopy. Nature 2010; 464(7287): 392-5.
[http://dx.doi.org/10.1038/nature08907] [PMID: 20237566]

[10] Sawai Y, Takimoto B, Nabika H, Ajito K, Murakoshi K. Observation of a small number of molecules at a metal nanogap arrayed on a solid surface using surface-enhanced Raman scattering. J Am Chem Soc 2007; 129(6): 1658-62.
[http://dx.doi.org/10.1021/ja067034c] [PMID: 17284005]

[11] Hatab NA, Hsueh C-H, Gaddis AL, *et al.* Free-standing optical gold bowtie nanoantenna with variable gap size for enhanced Raman spectroscopy. Nano Lett 2010; 10(12): 4952-5.
[http://dx.doi.org/10.1021/nl102963g] [PMID: 21090585]

[12] Kumar H, Gupta N. Neurological disorders and barriers for neurological rehabilitation in rural areas in Uttar Pradesh: A cross-sectional study. J Neurosci Rural Pract 2012; 3(1): 12-6.
[http://dx.doi.org/10.4103/0976-3147.91923] [PMID: 22346183]

[13] Basbaum AI, Bautista DM, Scherrer G, Julius D. Cellular and molecular mechanisms of pain. Cell 2009; 139(2): 267-84.
[http://dx.doi.org/10.1016/j.cell.2009.09.028] [PMID: 19837031]

[14] Evanko D. Primer: spying on exocytosis with amperometry. Nat Methods 2005; 2(9): 650-0.
[http://dx.doi.org/10.1038/nmeth0905-650] [PMID: 16118634]

[15] Robinson DL, Hermans A, Seipel AT, Wightman RM. Monitoring rapid chemical communication in the brain. Chem Rev 2008; 108(7): 2554-84.
[http://dx.doi.org/10.1021/cr068081q] [PMID: 18576692]

[16] Huefner A, Septiadi D, Wilts BD, *et al.* Gold nanoparticles explore cells: cellular uptake and their use as intracellular probes. Methods 2014; 68(2): 354-63.
[http://dx.doi.org/10.1016/j.ymeth.2014.02.006] [PMID: 24583117]

[17] Segura-Aguilar J, Paris I, Muñoz P, Ferrari E, Zecca L, Zucca FA. Protective and toxic roles of dopamine in Parkinson's disease. J Neurochem 2014; 129(6): 898-915.
[http://dx.doi.org/10.1111/jnc.12686] [PMID: 24548101]

[18] Maricle RA, Valentine RJ, Carter J, Nutt JG. Mood response to levodopa infusion in early Parkinson's disease. Neurology 1998; 50(6): 1890-2.
[http://dx.doi.org/10.1212/WNL.50.6.1890] [PMID: 9633754]

[19] Baronti F, Davis TL, Boldry RC, Mouradian MM, Chase TN. Deprenyl effects on levodopa pharmacodynamics, mood, and free radical scavenging. Neurology 1992; 42(3 Pt 1): 541-4.
[http://dx.doi.org/10.1212/WNL.42.3.541] [PMID: 1549214]

[20] Tang L, Li S, Han F, *et al.* SERS-active Au@Ag nanorod dimers for ultrasensitive dopamine detection. Biosens Bioelectron 2015; 71: 7-12.
[http://dx.doi.org/10.1016/j.bios.2015.04.013] [PMID: 25880832]

[21] Luo Y, Ma L, Zhang X, Liang A, Jiang Z. SERS Detection of Dopamine Using Label-Free Acridine Red as Molecular Probe in Reduced Graphene Oxide/Silver Nanotriangle Sol Substrate. Nanoscale Res Lett 2015; 10(1): 937.
[http://dx.doi.org/10.1186/s11671-015-0937-9] [PMID: 26055475]

[22] Ranc V, Markova Z, Hajduch M, *et al.* Magnetically assisted surface-enhanced raman scattering selective determination of dopamine in an artificial cerebrospinal fluid and a mouse striatum using Fe(3)O(4)/Ag nanocomposite. Anal Chem 2014; 86(6): 2939-46.
[http://dx.doi.org/10.1021/ac500394g] [PMID: 24555681]

[23]   An JH, Choi D-K, Lee K-J, Choi JW. Surface-enhanced Raman spectroscopy detection of dopamine by DNA Targeting amplification assay in Parkisons's model. Biosens Bioelectron 2015; 67: 739-46.
[http://dx.doi.org/10.1016/j.bios.2014.10.049] [PMID: 25465795]

[24]   Wang P, Xia M, Liang O, *et al.* Label-Free SERS Selective Detection of Dopamine and Serotonin Using Graphene-Au Nanopyramid Heterostructure. Anal Chem 2015; 87(20): 10255-61.
[http://dx.doi.org/10.1021/acs.analchem.5b01560] [PMID: 26382549]

[25]   Moody AS, Sharma B. Multi-metal, Multi-wavelength Surface-Enhanced Raman Spectroscopy Detection of Neurotransmitters. ACS Chem Neurosci 2018; 9(6): 1380-7.
[http://dx.doi.org/10.1021/acschemneuro.8b00020] [PMID: 29601719]

[26]   Tiwari VS, Khetani A, Monfared AMT, *et al.* Detection of amino acid neurotransmitters by surface enhanced Raman scattering and hollow core photonic crystal fiber. In: Achilefu S, Raghavachari R, Eds. Reporters, Markers, Dyes, Nanoparticles, and Molecular Probes for Biomedical Applications IV. 82330Q.
[http://dx.doi.org/10.1117/12.907754]

[27]   Lee W, Kang B-H, Yang H, *et al.* Spread spectrum SERS allows label-free detection of attomolar neurotransmitters. Nat Commun 2021; 12(1): 159.
[http://dx.doi.org/10.1038/s41467-020-20413-8] [PMID: 33420035]

[28]   Figueiredo MLB, Martin CS, Furini LN, Rafael JG. Surface-enhanced Raman scattering for dopamine in Ag colloid: Adsorption mechanism and detection in the presence of interfering species 2020; 522: 146466.

[29]   Gao F, Liu L, Cui G, *et al.* Regioselective plasmonic nano-assemblies for bimodal sub-femtomolar dopamine detection. Nanoscale 2017; 9(1): 223-9.
[http://dx.doi.org/10.1039/C6NR08264E] [PMID: 27906395]

[30]   Monfared AMT, Tiwari VS, Trudeau VL, *et al.* Surface-enhanced raman scattering spectroscopy for the detection of glutamate and γ-Aminobutyric acid in serum by partial least squares analysis. IEEE Photonics J 2015; 7: 1. Epub ahead of print.
[http://dx.doi.org/10.1109/JPHOT.2015.2423284]

[31]   Siek M, Kaminska A, Kelm A, *et al.* Electrodeposition for preparation of efficient surface-enhanced Raman scattering-active silver nanoparticle substrates for neurotransmitter detection. Electrochim Acta 2013; 89: 284-91.
[http://dx.doi.org/10.1016/j.electacta.2012.11.037]

[32]   Lussier F, Brulé T, Bourque M-J, Ducrot C, Trudeau LÉ, Masson JF. Dynamic SERS nanosensor for neurotransmitter sensing near neurons. Faraday Discuss 2017; 205: 387-407.
[http://dx.doi.org/10.1039/C7FD00131B] [PMID: 28895964]

[33]   Lussier F, Brulé T, Vishwakarma M, Das T, Spatz JP, Masson JF. Dynamic-SERS optophysiology: A nanosensor for monitoring cell secretion events. Nano Lett 2016; 16(6): 3866-71.
[http://dx.doi.org/10.1021/acs.nanolett.6b01371] [PMID: 27172291]

[34]   Zhu H, Lussier F, Ducrot C, *et al.* Block Copolymer Brush Layer-Templated Gold Nanoparticles on Nanofibers for Surface-Enhanced Raman Scattering Optophysiology. ACS Appl Mater Interfaces 2019; 11(4): 4373-84.
[http://dx.doi.org/10.1021/acsami.8b19161] [PMID: 30615826]

[35]   Vander Ende E, Bourgeois MR, Henry AI, *et al.* Physicochemical trapping of neurotransmitters in polymer-mediated gold nanoparticle aggregates for surface-enhanced raman spectroscopy. Anal Chem 2019; 91(15): 9554-62.
[http://dx.doi.org/10.1021/acs.analchem.9b00773] [PMID: 31283189]

[36]   Moody AS, Baghernejad PC, Webb KR, Sharma B. Surface Enhanced Spatially Offset Raman Spectroscopy Detection of Neurochemicals Through the Skull. Anal Chem 2017; 89(11): 5688-92.
[http://dx.doi.org/10.1021/acs.analchem.7b00985] [PMID: 28493674]

[37]   Rowley WR, Bezold C, Arikan Y, Byrne E, Krohe S. Diabetes 2030: Insights from Yesterday, Today, and Future Trends. Popul Health Manag 2017; 20(1): 6-12.
[http://dx.doi.org/10.1089/pop.2015.0181] [PMID: 27124621]

[38]   Madziarska K, Zmonarski S, Penar J, *et al.* Glucose challenge test (50-g GCT) in detection of glucose metabolism disorders in peritoneal dialysis patients: preliminary study. Int Urol Nephrol 2015; 47(4): 695-700.
[http://dx.doi.org/10.1007/s11255-014-0900-1] [PMID: 25539618]

[39]   Gestational diabetes mellitus. Diabetes Care 2003; 26 (Suppl. 1): S103-5.
[http://dx.doi.org/10.2337/diacare.26.2007.S103] [PMID: 12502631]

[40]   Ekun OA, Ogunyemi GA, Azenabor A, *et al.* A comparative analysis of glucose oxidase method and three point-of-care measuring devices for glucose determination. Ife J Sci 2018; 20: 43.
[http://dx.doi.org/10.4314/ijs.v20i1.4]

[41]   Dingari NC, Horowitz GL, Kang JW, Dasari RR, Barman I. Raman spectroscopy provides a powerful diagnostic tool for accurate determination of albumin glycation. PLoS One 2012; 7(2): e32406.
[http://dx.doi.org/10.1371/journal.pone.0032406] [PMID: 22393405]

[42]   Birech Z, Mwangi PW, Bukachi F, Mandela KM. Application of Raman spectroscopy in type 2 diabetes screening in blood using leucine and isoleucine amino-acids as biomarkers and in comparative anti-diabetic drugs efficacy studies. PLoS One 2017; 12(9): e0185130.
[http://dx.doi.org/10.1371/journal.pone.0185130] [PMID: 28926628]

[43]   Qi G, Wang Y, Zhang B, *et al.* Glucose oxidase probe as a surface-enhanced Raman scattering sensor for glucose. Anal Bioanal Chem 2016; 408(26): 7513-20.
[http://dx.doi.org/10.1007/s00216-016-9849-5] [PMID: 27518716]

[44]   Stuart DA, Yuen JM, Shah N, *et al. In vivo* glucose measurement by surface-enhanced Raman spectroscopy. Anal Chem 2006; 78(20): 7211-5.
[http://dx.doi.org/10.1021/ac061238u] [PMID: 17037923]

[45]   Lyandres O, Shah NC, Yonzon CR, Walsh JT Jr, Glucksberg MR, Van Duyne RP. Real-time glucose sensing by surface-enhanced Raman spectroscopy in bovine plasma facilitated by a mixed decanethiol/mercaptohexanol partition layer. Anal Chem 2005; 77(19): 6134-9.
[http://dx.doi.org/10.1021/ac051357u] [PMID: 16194070]

[46]   Torul H, Çiftçi H, Çetin D, Suludere Z, Boyacı IH, Tamer U. Paper membrane-based SERS platform for the determination of glucose in blood samples. Anal Bioanal Chem 2015; 407(27): 8243-51.
[http://dx.doi.org/10.1007/s00216-015-8966-x] [PMID: 26363778]

[47]   Torul H, Çiftçi H, Dudak FC, *et al.* Glucose determination based on a two component self-assembled monolayer functionalized surface-enhanced Raman spectroscopy (SERS) probe. Anal Methods 2014; 6: 5097-104.
[http://dx.doi.org/10.1039/C4AY00559G]

[48]   Yuen JM, Shah NC, Walsh JT Jr, Glucksberg MR, Van Duyne RP. Transcutaneous glucose sensing by surface-enhanced spatially offset Raman spectroscopy in a rat model. Anal Chem 2010; 82(20): 8382-5.
[http://dx.doi.org/10.1021/ac101951j] [PMID: 20845919]

[49]   Ma K, Yuen JM, Shah NC, Walsh JT Jr, Glucksberg MR, Van Duyne RP. *In vivo*, transcutaneous glucose sensing using surface-enhanced spatially offset Raman spectroscopy: multiple rats, improved hypoglycemic accuracy, low incident power, and continuous monitoring for greater than 17 days. Anal Chem 2011; 83(23): 9146-52.
[http://dx.doi.org/10.1021/ac202343e] [PMID: 22007689]

[50]   Stuart DA, Yonzon CR, Zhang X, *et al.* Glucose sensing using near-infrared surface-enhanced Raman spectroscopy: gold surfaces, 10-day stability, and improved accuracy. Anal Chem 2005; 77(13): 4013-9.

[http://dx.doi.org/10.1021/ac0501238] [PMID: 15987105]

[51]   Kanayama N, Kitano H. Interfacial recognition of sugars by boronic acid-carrying self-assembled monolayer. Langmuir 2000; 16: 577-83.
[http://dx.doi.org/10.1021/la990182e]

[52]   Barriet D, Yam CM, Shmakova OE, Jamison AC, Lee TR. 4-Mercaptophenylboronic acid SAMs on gold: comparison with SAMs derived from thiophenol, 4-mercaptophenol, and 4-mercaptobenzoic acid. Langmuir 2007; 23(17): 8866-75.
[http://dx.doi.org/10.1021/la7007733] [PMID: 17636994]

[53]   Piergies N, Proniewicz E, Ozaki Y, Kim Y, Proniewicz LM. Influence of substituent type and position on the adsorption mechanism of phenylboronic acids: infrared, Raman, and surface-enhanced Raman spectroscopy studies. J Phys Chem A 2013; 117(27): 5693-705.
[http://dx.doi.org/10.1021/jp404184x] [PMID: 23758215]

[54]   Sharma B, Bugga P, Madison LR, *et al.* Bisboronic Acids for Selective, Physiologically Relevant Direct Glucose Sensing with Surface-Enhanced Raman Spectroscopy. J Am Chem Soc 2016; 138(42): 13952-9.
[http://dx.doi.org/10.1021/jacs.6b07331] [PMID: 27668444]

[55]   Kong KV, Ho CJH, Gong T, Lau WK, Olivo M. Sensitive SERS glucose sensing in biological media using alkyne functionalized boronic acid on planar substrates. Biosens Bioelectron 2014; 56: 186-91.
[http://dx.doi.org/10.1016/j.bios.2013.12.062] [PMID: 24487255]

[56]   Kong KV, Lam Z, Lau WKO, Leong WK, Olivo M. A transition metal carbonyl probe for use in a highly specific and sensitive SERS-based assay for glucose. J Am Chem Soc 2013; 135(48): 18028-31.
[http://dx.doi.org/10.1021/ja409230g] [PMID: 24168766]

[57]   Botta R, Rajanikanth A, Bansal C. Silver nanocluster films for glucose sensing by Surface Enhanced Raman Scattering (SERS). Sens Biosensing Res 2016; 9: 13-6.
[http://dx.doi.org/10.1016/j.sbsr.2016.05.001]

[58]   Mauriz E. Low-Fouling Substrates for Plasmonic Sensing of Circulating Biomarkers in Biological Fluids. Biosensors (Basel) 2020; 10(6): 10. Epub ahead of print.
[http://dx.doi.org/10.3390/bios10060063] [PMID: 32531908]

[59]   Qu L-L, Liu Y-Y, He S-H, Chen JQ, Liang Y, Li HT. Highly selective and sensitive surface enhanced Raman scattering nanosensors for detection of hydrogen peroxide in living cells. Biosens Bioelectron 2016; 77: 292-8.
[http://dx.doi.org/10.1016/j.bios.2015.09.039] [PMID: 26414026]

[60]   Gu X, Wang H, Schultz ZD, Camden JP. Sensing Glucose in Urine and Serum and Hydrogen Peroxide in Living Cells by Use of a Novel Boronate Nanoprobe Based on Surface-Enhanced Raman Spectroscopy. Anal Chem 2016; 88(14): 7191-7.
[http://dx.doi.org/10.1021/acs.analchem.6b01378] [PMID: 27356266]

[61]   Cervo S, Mansutti E, Del Mistro G, *et al.* SERS analysis of serum for detection of early and locally advanced breast cancer. Anal Bioanal Chem 2015; 407(24): 7503-9.
[http://dx.doi.org/10.1007/s00216-015-8923-8] [PMID: 26255294]

[62]   Feng S, Chen R, Lin J, *et al.* Nasopharyngeal cancer detection based on blood plasma surface-enhanced Raman spectroscopy and multivariate analysis. Biosens Bioelectron 2010; 25(11): 2414-9.
[http://dx.doi.org/10.1016/j.bios.2010.03.033] [PMID: 20427174]

[63]   Abalde-Cela S, Aldeanueva-Potel P, Mateo-Mateo C, Rodríguez-Lorenzo L, Alvarez-Puebla RA, Liz-Marzán LM. Surface-enhanced Raman scattering biomedical applications of plasmonic colloidal particles. J R Soc Interface 2010; 7 (Suppl. 4): S435-50. Epub ahead of print.
[http://dx.doi.org/10.1098/rsif.2010.0125.focus] [PMID: 20462878]

[64]   Lee S, Kim S, Choo J, *et al.* Biological imaging of HEK293 cells expressing PLCgamma1 using

surface-enhanced Raman microscopy. Anal Chem 2007; 79(3): 916-22.
[http://dx.doi.org/10.1021/ac061246a] [PMID: 17263316]

[65]   Qian X, Peng XH, Ansari DO, *et al. In vivo* tumor targeting and spectroscopic detection with surface-enhanced Raman nanoparticle tags. Nat Biotechnol 2008; 26(1): 83-90.
       [http://dx.doi.org/10.1038/nbt1377] [PMID: 18157119]

[66]   Yoon KJ, Seo HK, Hwang H, *et al.* Bioanalytical Application of SERS Immunoassay for Detection of Prostate-Specific Antigen †. Bioanal Appl SERS Immunoass Bull Korean Chem Soc 2010; 31: 1215.
       [http://dx.doi.org/10.5012/bkcs.2010.31.5.1215]

[67]   Maiti K, Samanta A, Vendrell M, *et al.* Multiplex cancer cell detection by SERS nanotags with cyanine and triphenylmethine Raman reporters. pubs.rsc.org, https://pubs.rsc.org/ko/content/articlehtml/2011/cc/c0cc05265e
       [http://dx.doi.org/10.1039/c0cc05265e]

[68]   Park H, Lee S, Chen L, *et al.* SERS imaging of HER2-overexpressed MCF7 cells using antibody-conjugated gold nanorods. Phys Chem Chem Phys 2009; 11(34): 7444-9.
       [http://dx.doi.org/10.1039/b904592a] [PMID: 19690717]

[69]   Jehn C, Küstner B, Adam P, *et al.* Water soluble SERS labels comprising a SAM with dual spacers for controlled bioconjugation. Phys Chem Chem Phys 2009; 11(34): 7499-504.
       [http://dx.doi.org/10.1039/b905092b] [PMID: 19690725]

[70]   Zong S, Wang Z, Yang J, *et al.* A SERS and fluorescence dual mode cancer cell targeting probe based on silica coated Au@ Ag core–shell nanorods. Elsevier, https://www.sciencedirect.com/science/article/pii/S0039914012003426

[71]   Fales AM, Yuan H, Vo-Dinh T. Silica-coated gold nanostars for combined surface-enhanced Raman scattering (SERS) detection and singlet-oxygen generation: a potential nanoplatform for theranostics. Langmuir 2011; 27(19): 12186-90.
       [http://dx.doi.org/10.1021/la202602q] [PMID: 21859159]

[72]   Maiti KK, Dinish US, Samanta A, *et al.* Multiplex targeted *in vivo* cancer detection using sensitive near-infrared SERS nanotags. Nano Today 2012; 7: 85-93.
       [http://dx.doi.org/10.1016/j.nantod.2012.02.008]

[73]   Osborne EA, Atkins TM, Gilbert DA, *et al.* Rapid Microwave-Assisted Synthesis of Dextran-Coated Iron Oxide Nanoparticles for Magnetic Resonance Imaging https://iopscience.iop.org/article/10.1088/0957-4484/23/21/215602/meta
       [http://dx.doi.org/10.1088/0957-4484/23/21/215602]

[74]   Kircher M, La Zerda AD, Jokerst J, *et al.* A brain tumor molecular imaging strategy using a new triple-modality MRI-photoacoustic-Raman nanoparticle https://www.nature.com/articles/nm.2721

[75]   Lu W, Singh AK, Khan SA, Senapati D, Yu H, Ray PC. Gold nano-popcorn-based targeted diagnosis, nanotherapy treatment, and *in situ* monitoring of photothermal therapy response of prostate cancer cells using surface-enhanced Raman spectroscopy. J Am Chem Soc 2010; 132(51): 18103-14.
       [http://dx.doi.org/10.1021/ja104924b] [PMID: 21128627]

[76]   Beqa L, Fan Z, Singh AK, Senapati D, Ray PC. Gold nano-popcorn attached SWCNT hybrid nanomaterial for targeted diagnosis and photothermal therapy of human breast cancer cells. ACS Appl Mater Interfaces 2011; 3(9): 3316-24.
       [http://dx.doi.org/10.1021/am2004366] [PMID: 21842867]

[77]   Wang J, Liang D, Jin Q, Feng J, Tang X. Bioorthogonal SERS Nanotags as a Precision Theranostic Platform for *in Vivo* SERS Imaging and Cancer Photothermal Therapy. Bioconjug Chem 2020; 31(2): 182-93.
       [http://dx.doi.org/10.1021/acs.bioconjchem.0c00022] [PMID: 31940174]

CHAPTER 7

Smartphone Based Biosensors on Lateral Flow Assay Coupled to SERS: Point-of-Care Applications

Rajasekhar Chokkareddy[2], Suvardhan Kanchi[1,3,*], Surendra Thakur[4] and **Venkatasubba Naidu Nuthalapati[5,*]**

[1] *Department of Chemistry, Durban University of Technology, Durban, South Africa*

[2] *Department of Chemistry, Aditya College of Engineering and Technology, Andhra Pradesh, India*

[3] *Department of Chemistry, Sambhram Institute of Technology, Jalahalli East, Bengaluru, India*

[4] *SkillsCoLab, Durban University of Technology, Durban, South Africa*

[5] *Department of Chemistry, Sri Venkateswara University, Tirupati, Andhra Pradesh, India*

Abstract: Point-of-care (POC) diagnostic analysis is a fast rising arena that goals to improve the rapidness, low-cost, sensitive and selective *in-vitro* diagnostic analysis platforms that are independent, moveable, and can be used everywhere from current clinics to isolated and low resource regions. In addition, surface enhanced Raman spectroscopy (SERS) offers a suitable sensory stage whereby objective particles at smaller concentration are recognized, potentially identifying a particular substances. There are some analytical methods which can induce motivated Raman scattering, significantly attractive the possibility of Raman scattering, but most of these procedures need ultra-fast tunable lasers and microscopy set-ups making them impossible for most users. SERS exploits surface plasmons to improve Raman scattering by numerous orders of magnitude, without lacking various equipment than traditional Raman spectroscopy. As a result, SERS has been extensively useful in a selection of research activities using various substrates, one of which is a smartphone based biosensors. In outlining the development of SERS methods over the past few years joined with new expansions in smartphone based biosensors, metal and metal oxide nanomaterials, low-cost paper diagnostics, and high-quantity of microfluidics, a wide-ranging number of novel potentials display the possible for decoding SERS biosensors to the POC.

* **Corresponding author(s) S Kanchi:** Department of Chemistry, Sambhram Institute of Technology, M.S. Palya, Jalahalli East, Bengaluru 560097, India; and Department of Chemistry, Durban University of Technology, Durban, South Africa; and **Venkatasubba Naidu Nuthalapati:** Department of Chemistry, Sri Venkateswara University, Tirupati, Andhra Pradesh, India; E-mail: nvsn69@gmail.com

Keywords: Lateral Flow-assay, Point-of-care Application, Smartphone, Surface Enhanced Raman Spectroscopy.

7.1. INTRODUCTION

Point-of-care (POC) investigations have altered the distribution and influence of health care by simplifying a relocation of diagnostic analysis from consolidated laboratories to the medical site of patient care, or even a person's homemade [1]. Investigation and improvement of POC investigations has gradually over the last 30 years, and the worldwide POC investigative market is estimated to exceed US $40 billion in the year 2020 [2]. POC investigations have been established for usage of few therapeutic situations, an extensive range of medicinal areas and subspecialties, and for equally contagious and non-infectious sicknesses. New POC methodologies are becoming growing more classy and complex with lab-o--a-chip abilities, and the up-to-date group of POC investigations offer quick study of humanoid and pathogen genomic information to accelerate more suitable and modified treatments [3, 4]. Any positive POC methodologies will need a matching of rapidness, precision, user-friendliness, and low-cost. Although an important objective is to be clinically perfect, POC assessments that can be rapidly and simply did by medical health care workers and at low working costs will be the maximum positive. The main potential influence will be in resource-partial situations, where exclusive laboratory arrangement is also out-of-the-way or totally engaged [5, 6]. Though, numerous investigations of POC analysis have confirmed only nominal influence on the burden of infection in resource-partial situations, signifying their combination in health organizations has also been lacking or poorly instigated. While national orientation laboratories will continue a serious role in providing positive analysis, upcoming healthcare distribution is possible to be more dependent on POC methodologies [7, 8]. Diagnostics, since of their planned position at the connection among patients and their clinically actionable data, straight distress the patient familiarity and the superiority of attention that persons obtain (Fig. **1a**). They also provide valued tools for medical analysis. Diagnostics allow workers to expand upon "one-size-fits-all" action plans and in its place offer modified care on the source of influences such as comorbidities, hereditary makeup, actual serologic calculations, and responses to treatment [9, 10]. Moreover, the POC is ability to deliver a high-quality biomarker measurements enhanced for the superior limitations of various medical locations with critical care, casualty hospitals, homes, medical investigation centers, rural regions, and the increasing world (Fig. **1b**). In critical care situations such as the operational area, emergency room, or intensive care unit, cardiac catheterization group, doctors pursue real-time response to enhance attention and modify treatments to the active situations they confront [11, 12].

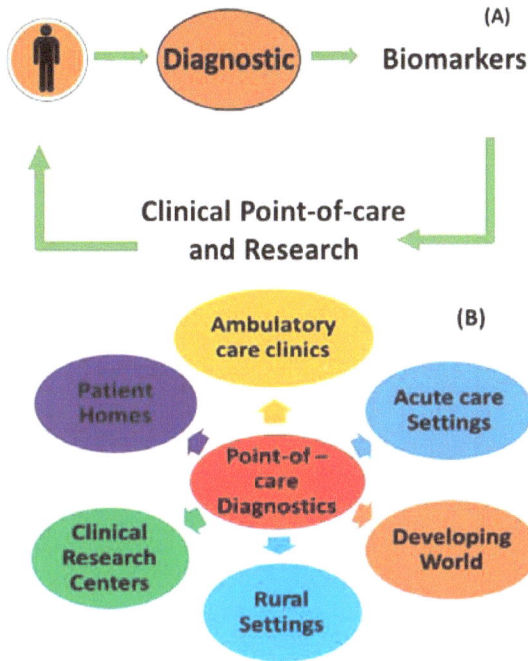

Fig. (1).   (A) Diagnostics attain quantities from a patient of related biological sample and produce imaginative or measurable productions know as biomarkers, which assist medical POC, patient communication and medical investigations. (B) Point-of-care methodologies.

7.1.1. What is POC Procedure

Have you always remained at a hospital or been to the alternative room? If you have, then you know there are a lot of analysis and processes that go on throughout your appointment. Approximately of the analysis you leave the room for, but some of the investigations and actions can be done accurate at your bedside by a surgeon, nurse or medicinal associate. POC, also identified as point of care analysis, is precisely what it sounds like: examinations that are done right where the patient is. For example, a patient's blood glucose level is measured a point of care analysis. It can be done where the patient is at the correct time, which is directly before the patient eats breakfast, lunch, or dinner, and based on the results, nutritional or medicine adjustments can take place directly. Moreover, the World Health Organization (WHO) has well-defined a usual of limits that plan the conditions for POC process in rising regions. This plan was initially generated for the improvement of portable devices for acquired immune deficiency syndrome (AIDS), human immunodeficiency virus (HIV) analysis. General point of care technology (POCT) strategies includes:

- Urine dipsticks
- Blood glucose monitors
- Hemoglobin level monitoring
- Pregnancy investigations
- Fast HIV tests
- Fast strep tests
- Cardiac markers
- Cholesterol/lipids
- Food pathogens

7.1.2. Recent POC Technologies

There are POC procedures that commercially exist. The best familiar example is the glucose meter for dealing diabetes, a long-lasting disease causing 30 million people in the America only [13]. It is well-defined as a set of metabolic sicknesses where finally the body's pancreas ensures not produce sufficient insulin or does not appropriately react to insulin formed, ensuing in high blood sugar heights over a protracted era. Glucose meters and other POC procedures develop a variety of approaches for identifying and detecting biomarkers with electrochemical biosensors [14], optical [15], label-free spectroscopic investigations [16], magnetic [17], plasmonic nanoparticle based electrochemical sensors [18] and colorimetric sensors [19]. Normally, electrochemical determination uses amperometric, potentiometric, and impedance extents in combination with electroactive labels or free fluid electro-active analytes [17 - 20]. Several of the commercial glucose determination monitors Arkray (Edina, MN, USA), One Touch Ultra (Johnson & Johnson, Wayne, PA, USA), BD Test Strip (Becton Dickinson, Franklin Lakes, NJ, USA), Ascensia Contour (Ascensia, Parsippany, NJ, USA) [21], are examples of colorimetric and electrochemical procedures that use investigation strips. The m type of Chips, is a device that analysis the whole blood for AIDS/HIV using a united microfluidic and protein-based immunoassay method [58], and later on the Ativa Micro Lab were introduced a execute a number of different blood investigations with the usage of a distinct device [22]. In addition, Ativa's portable Micro Lab was established as the potential to do up to 25 different blood investigations using its several one-use test passes for each investigation. The Micro Lab system uses flow cytometry, microfluidics, electrochemistry, and colorimetric statistics in arrangement with conventional imaging methods to conduct and communicate significant diagnostic material. The lateral flow assays (LFAs) is the one of the most powerful raising POC methods, it contains the colorimetric barcodes that can be observed with the bare eye. The best LFA is the pregnancy investigation it was introduced by Church & Dwight Co., Inc. US and Clearblue Swiss Precision Diagnostics in Switzerland

[23]. There are numerous POC strategies commercially existing, they are mainly intended to produce semi measureable or quantitative evidence for a one analyte per investigation. As significance, investigation and manufacturing determinations are still essential to report these issues to expand POC methods.

7.1.3. Biosensors

The electrochemical biosensor is a bioanalytical scheme which has the capability to mimic somewhat biological substances, for example nucleic acids, antigens/antibodies, enzymes. Those biological substances are combined into a transducing microsystem. The transducer might be thermometric, electrochemical, piezoelectric, or magnetic, optical, and this is named as label-free determination, while label-based determination contains fluorescent immunoassays, and quantum dots. In addition, the electrochemical biosensors utility with the bio-recognition of specific basics for exacts targeted analytes and the preservation of sensitivity and selectivity in the presence of other interference substances. In the clinical arena, electrochemical biosensor uses are developing quickly, such as for the analysis of diabetes, urinary tract infections, identification of end-stage heart failure, and serious leukemia (Table **1**).

Table 1. Explanation of POC procedures for determination of sicknesses over exact salivary biomarkers.

Salivary Biomarkers	Infections/Situations	Established POC	References
Hep C	Hepatitis	OraQuick	[24]
HIV	AIDS	Oraquick, tablet-based kiosks	[25]
A-amylase	Medical decision for stress-prompted sickness	Salivary α-Amylase electrochemical biosensor method	[26]
Cortisol	Stress stages	Label-free chemiresistor immuno-electrochemical sensor	[27]
HPV	HPV-related cancers, sexually spread illnesses	simple fluorescent and colorimetric analyze that facilitates DNA and RNA determination	[28]
C-reactive protein, myeloperoxidase and myoglobin	Acute myocardial infarction	lab-on-a-chip procedures, Luminex	[29]
DNA, proteins (Dipeptidyl peptidase *etc.*), metabolites	Periodontitis	Lab-on-a-chip, microfluidic platform for oral diagnostics	[30]
IL-8, IL-8mRNA	Cancer	Electrochemical biosensors	[31]
Cytokines	Chronic disruptive pulmonary sickness and asthma	Fiber-optic microsphere-based antibody array	[32]

Salivary Biomarkers	Infections/Situations	Established POC	References
Salivary nicotine metabolites	Smoking and tobacco use	POC investigation for salivary nicotine metabolites	[33]
Pulmonary and inflammation biomarkers (NO$_2$– and uric acid)	Asthma and chronic obstructive pulmonary sickness patients end-stage renal illness	Optical fibre microarrays	[34]
Chlamydia and gonorrhoea	Sexually spread diseases	POC	[24]
Salivary glucose	Diabetes	Glucose checking using saliva nanostructured electrochemical bio-sensor	[35]
Salivary anti-Ro60 and anti-Ro52 Antibody Outlines	Sjogren's Syndrome	Luciferase Immunoprecipitation Systems	[36]
Psa (free and complexed), Her2-neu, cea, ca125	Cancer	2D type nanomaterials, programmable bio-nano-chip method	[37]
ctnl,cRP, Myo, MPo,cK-MB, apoa1, apoB, BnP, d-dimerMcP-1, scd40l, adiponectin, nt-proBnP,	Cardiovascular disease	Programmable bio-nano-chip method	[37]

7.1.4. Smartphone-Based Biosensors

Smartphones function also to small computers, acting as low cost, moveable analytical laboratory procedures. They are useful for the determination, and analysis, of several sicknesses, such as tuberculosis, cancer, and glucose. (Fig. **2**) demonstrates the limited shapes and schemes of currently-used POC methodologies while Table **1** shows all the utilities, methods, and medical values of these procedures in a brief way. Furthermore, Lee and co-workers have developed the smartphone utility as a solid optical microscope in which ambient light as a light basis was existence used in its place of a chip-scale technique. This lens-less imaging structure tolerates sub-micron determination and the integrated android use conveys robustness, easiness, and employability for numerous arena uses. Practical approaches complicated in electrochemical biosensor improvement are nano- technically constructed, which be contingent upon also the label-based determination or label-free determination. Similarly, nanotechnology facilitates the influence of materials at the nanoscale and has exposed possible to improve selectivity, sensitivity, and low cost of an analysis.

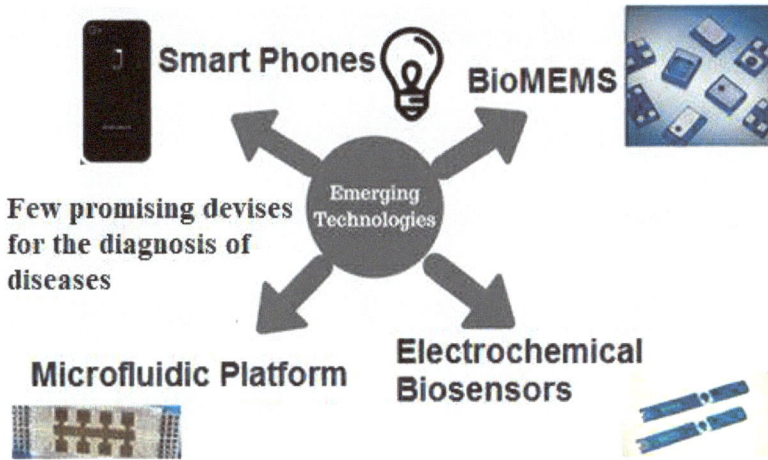

Fig. (2). Schematic representation of a variety of favourable developing procedures of PCC methodologies.

7.1.5. Challenges of Accepting POC in Emergency Department

It is always significant to regulate the unmet medical necessity that will be addressed by the use of some investigation, and in what way the assessment will really be used in the medical trail. In addition, this is similarly true for POC as it is for laboratory analysis. Then the entire determination of POC is typically to reduce the time to choice creating and thus of the complete medical path, then attention must be specified to in what way this will influence on the action of the emergency department. This fact is highlighted by Schilling and Rooney who in their conversation of the active use of POC in the emergency department highlight the circumstance that medical paths and emergency logistics might need extensive alteration to exploit the medical and financial assistances of quick TATs delivered [38]. Furthermore, the quality administration of POC is a significant feature of the extensive quality administration of facilities delivered by the emergency department. The situation can be a stimulating situation for POC with an extensive range of possible operatives and the essential to validate capability [39]. For example, the variety of investigations engaged in the emergency department enlarges, the necessity to preserve the apparatus might also become more exciting. The outline of more quick valuation, with the co-position of a chief care doctor in the emergency department, maintained by POC can expand patient movement in the emergency department [40]. The real tasks of improving patient movement can be addressed with the use of lean intelligent, which also has the improvement of providing evidence on reserve operation, as well as helping to escort where inefficiencies happen and properties might be protected [41]. Also, there are regularly seeming economic tests with the overview of POC due to an obsession on just the cost of the investigation (almost all the POC test are expensive compare to the laboratory equivalent) somewhat the attentions of the price of the complete path which POC can decrease [42]. All these challenges can be overawed and that care paths and related procedures can be different in the emergency department to optimally use the facility of more quick investigation effects from POC, then a quicker challenging can be influence on medical, working and financial consequences.

7.1.6. Influence of POC on Clinical Results in the Emergency Department

Table 1 shows the some of the uses of POC in the emergency department. This entry makes the implied statement that blood electrolyte, glucose and gas extents have a well-known effectiveness in the emergency situation to evaluate patient position at the time of admittance in a specific range of patients. Specified that assembling in the emergency department can lead to a rise in time to action [43], it is sensible to imagine that the overview of POC in the emergency department will decrease the time to action and thus expand medical consequences. Rooney and co-workers determined that POC when used successfully, it might be ease the destructive influences of overfilling on the protection, efficiency, and person-centeredness of carefulness in the emergency department. In addition, the development it goes without saying that, in somewhat medical situation, choice creation in the emergency department can only increase efficiency and competence if the consequences are accessed and replaced upon. However, there are instances described where laboratory outcomes have not been retrieved. The absence of continuation of assessment results for patients treated in the emergency department extended from 1-75% when measured as a quantity of assessments in an analysis of the occurrence of misplaced test results [44].

7.1.7. Essential Features of POC Procedures

Inventers of POC procedures began with the requirements of their operators and these necessities will to some point depend on the medical venue. Though some features are collective to essentially all operators in all situations. Some of the important requirements contain:

- Consumables and components are strong in storage and treatment.
- Easy to use.
- Outcomes must be concordant with a well-known laboratory process.
- Device together with related consumables and substances are nontoxic to usage.

Through the increasing potential for POC to advance healthcare in the evolving world, mainly over timely determination of contagious sicknesses, inventors of such policies have been directed by more detailed design principles. These are to guarantee that the method can address the requirements of the consumer in a clinically and low cost manner and avoid the overview of probably expensive procedures which fail to bring the essential results. Therefore, the World Health Organization has delivered some strategies for those evolving POC methods for the determination of sexually transferred diseases, a key health problem in the evolving world, and in the established world for sicknesses such as HIV/AIDS and Chlamydia. In recent times a novel challenge has risen, specifically the capability to simultaneously measure numerous analytes on the similar unit, or multiplexing as it is identified. Multiplexing is a somewhat wide-ranging and approximate time but, with the allowance of strategies used in dangerous care such as blood gas analyzers, the number of multi-analyte POC policies is comparatively insufficient. Though as new healthcare transfers away from the clinic into the public, the request produces for multi-analyte POC platforms meanwhile these avoid the essential for numerous procedures, all with the associated necessities of numerous training, superiority administration and interfacing procedures and the improved risk of mistakes. Then the technical tests of consultation all these essentials have intended that the outline of novel methodologies has been comparatively deliberate with the major units of the existing POC market such as cardiac indicators, glucose, international normalized ratio, and blood gases all exhausting method that was familiarized at least 20 years ago and sometimes much longer.

7.1.8. Surface Enhanced Raman Spectroscopy (SERS) and its Possible Advantages

SERS (Surface Enhanced Raman Spectroscopy) is a fantastic approach for improving sensitivity when utilizing Raman spectroscopy. Raman scattering produces fundamentally feeble magnitude signals, but Martin Fleischmann and his research team at the University of Southampton developed SERS in the 1970s, which helped to overcome these challenges. In addition, the improvement features can be as high as 10^{14-15}, which are satisfactory to tolerate even single molecule determination using Raman. SERS is of attention for trace material investigation, flow cytometry and other uses where the present selectivity, sensitivity and speed of a Raman extent are deficient. The enhancement takes place at a metallic surface which has nanoscale roughness, and it is particles adsorbed onto that superficial which can undergo development. Distinctive metals used are silver or gold – preparation of the exterior can be over electrochemical toughening, metal coating of a nano-designed substrate, or deposition of metal nanoparticles and nano composites (frequently in a colloidal form). Several researchers generate their individual SERS substrates, but commercially accessible kits offer a more repetitive method. Essentially, the benefits of SERS can be discovered on any Raman method, and the actual dimension is completed within the normal method. In addition, it is essential to use a laser wavelength which is well-matched with the selected SERS metal, but outside this there are no major problems. SERS spectra do occasionally fluctuate from a 'normal' Raman spectrum of the similar substances, therefore clarification of data essential be measured.

7.1.9. Type of Substrates

Exteriors of SERS substrates can be mutually roughened, and nanoparticles in colloidal solutions can be used. Nanostructured metals of various sizes and shapes are present on both the surfaces and in the solutions. The size and form of nanoparticles interfere with the degree of SERS improvement; consequently, this is an area of research that is still being investigated. Silver and gold nanoparticles are the most commonly utilized colloidal solutions. Stars, spheres, rods, and pyramids are examples of common nanoparticle forms that have been explored as potentials due to the technique by which they increase signals. When charting a surface, one of the main disadvantages of nanoparticle solutions is that the removal and separation might cause "hotspots," meaning that the significance research and results are not evenly distributed. Surface-restricted nanostructures reduce the likelihood of 'hotspots,' although they are less commonly used due to the difficulty of producing them in the lab and the limited accessibility of currently available surfaces. Unlike the normal chemical reaction of colloidal nanoparticle solutions, toughened surfaces are frequently produced *via* complicated approaches using many lithography techniques.

7.1.10. What is it Used For

SERS is utilized in a variety of applications, including medical, forensics, and analytical analysis, drug discovery, trace material investigation, point-of-contact (POC), and chemical and biological hazard detection. SERS has a wide range of applications in electrochemical biosensing, including disease screening for cancer, diabetes, Alzheimer's, and Parkinson's disease, among others. The gold standard for medical tissue examination is immune-fluorescence staining, however the toxicity of several fluorescent dyes means that research cannot be approved *in-vivo*. Raman spectroscopy is a label-free, chemically detailed analysis technology that can be used indefinitely. The fact that SERS is a quick, label-free process has also shown to be a significant benefit for point-of-care drug monitoring. SERS nano-tags gold spheres functionalized with reporter molecules and enclosed in a silica case can also be used to label a variety of things. The nano-tags can be used to encrypt jewellery and money for security and to detect fraud in the shipping of products. The application of SERS in field analysis for forensics and chemical, biological threat determination has improved dramatically as a result of Raman and SERS now being recognized as movable. With older processes of research using mass spectrometry and liquid chromatography, having a smartphone determination kit is quite important. These methods, however, necessitate extensive sample preparation. SERS is an excellent, selective, sensitive, and quick replacement.

7.1.11. Usual SERS Investigational System

Different traditional Raman spectroscopy where solid models are simply assessable, SERS needs that the model is also melted in a solution with the colloidal nanoparticles or melted in solution and pipetted onto the substrate. In also, once the sample is placed onto the solution or in substrate with the nanoparticles, revealing the sample with an attentive laser delivers the excitation basis for both the plasmon generation and Raman scattering. (Fig. **3**) demonstrates three various procedures for making a sample for use with P-SERS substrates, and the evaluation among the Raman spectrum and the SERS spectrum for Morphine. Moreover, in the sample preparation (a) the substrate is immersed into the solution, in (b) traditional micro-pipetting methods are used, and (c) uses a cleansing procedure. Although the equipment is essential for SERS is the identical as in traditional Raman spectroscopy (laser, probe, and spectrometer), there are a limited additional thought to confirm optimal presentation. It might sound counter-instinctive, but when attractive a SERS quantity, it is usually best to use a multi-manner excitation laser in its place of a single-method laser. Three key issues give to this intention, and the initial and most significant is to exploit the total number of nanostructures or nanoparticles in the laser spot. Exciting more "nano-edges" in the substrate will increase the group of surface plasmon and expand the improvement feature; so single-method lasers which normally focus to a sub-10 mm spot will frequently lead to lower improvement than a multi- manner laser which will usually attention to a 100 mm spot. The second motive why a single-manner laser is not better for SERS is the three-dimensional scattering of the analyte. Colloidal solutions and solid substrates can both involvement in-homogeneity, but best signal is the "coffee ring" result detected when the solvent in the pipetted sample dissolves separation a ring of high concentration everywhere the border of the spot. Finally, various SERS substrates with P-SERS are robust absorbers at the excitation wavelength. So, a strongly absorbed single-method laser will have a much higher chance of thermally destructive the sample than a multi- method laser of equal influence.

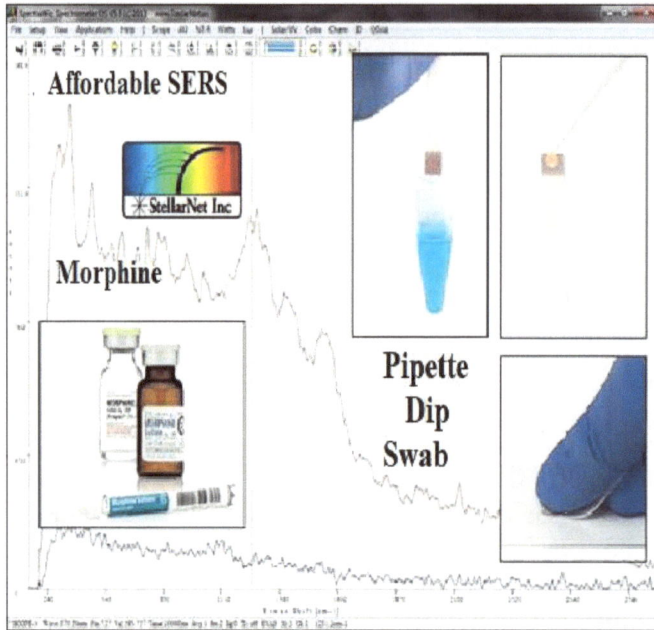

Fig. (3). P-SERS spectrum of Morphine display three various sample preparation procedures (a) dip, (b), pipette and (c) swab.

7.1.12. The Role of Surface Plasmons and Nanostructures

When noble metal nano sheets or nanoparticles with nanoscale (20–300 nm) roughness, usually silver or gold, are exposed to huge electric arenas, surface plasmons are created. Additionally, metal nanoparticles can be incorporated into porous materials such as paper (P-SERS), resulting in a hybrid structure known as a flexible substrate. Although a comprehensive study of the physics of surface plasmons is beyond the scope of this SERS, it is critical to recognize that this complex communication between the plasmon, metal, and analyte can result in both an electromagnetic (10^4–10^7) and a chemical (10–10^2) improvement in Raman scattering effectiveness. The electromagnetic field produced by the surface plasmon is the main source of enhancement. The nanostructures' geometry is then crucial in determining the wavelengths of surface plasmons and the amplification factor for a given excitation wavelength. The metal itself governs the range throughout which plasmons can be created, but the nanostructure determines the specific value of the peak enhancement wavelength. Normally, silver substrates perform better when stimulated by laser sources in the red or green spectrum, while gold substrates perform better when excited in the near infrared range.

Kahraman *et. al.*, in their paper "applications and foundations of SERS-based bioanalytical sensing," demonstrate a wide range of SERS substrates, including solid substrates, metal nanoparticles, and flexible substrates, each with their own set of limitations and benefits as shown in Fig. (**4**). It is frequently useful to functionalize SERS substrates and metal nanoparticles with antibodies or other molecules that trap a specific analyte of interest while allowing the rest of the sample to be washed off for trace analysis. This procedure, which is commonly used in biomedical diagnostics, can reduce the "noise" associated with the compound background medium, but it is important to note that adding these relatively long capture molecules increases the space between the substrate and analyte, resulting in a decrease in improvement. This process adds to the expense and difficulty of the treatment, so it should only be utilized when absolutely necessary.

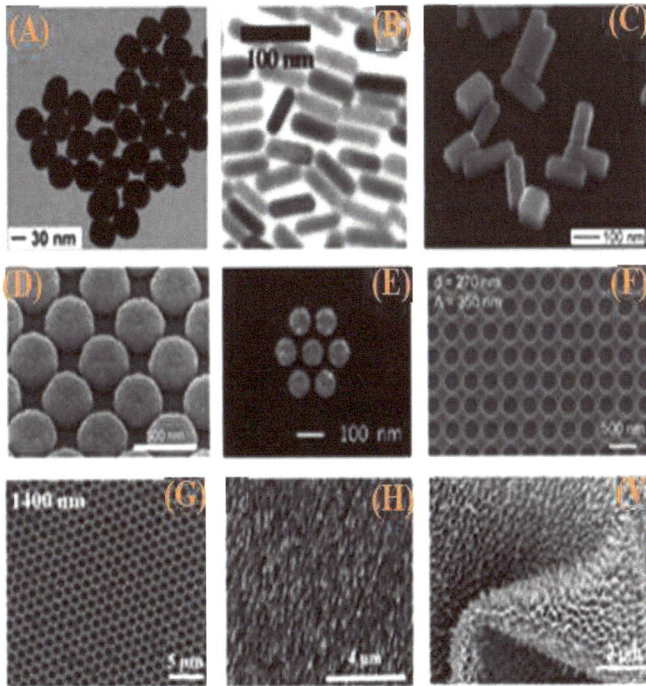

Fig. (4). SEM pictures of various type of SERS substrates [45 - 53].

7.2. SERS BASED DNA AND PROTEIN DETERMINATIONS

In several fields, methods for determining DNA are crucial. Infection analysis, pharmaceutical use, gene transformation determination, and virus, algae, and bacteria identification are all covered by bio-analytical chemistry. The polymerase chain reaction, which has only one DNA sensitivity, is currently the gold standard for determining DNA. The polymerase chain reaction technique, however, is still time-consuming and labor-intensive, requiring skilled researchers as well as expensive equipment [54]. The development of new methods for quick, simple, and low-cost DNA determination is critical, especially for POC analysis. Due to some benefits over other DNA testing processes such as polymerase chain reaction and fluorescence-based microarrays, SERS-based DNA determination is becoming more popular. SERS is a superior candidate for complex determination because of its thin peaks and large number of reporter molecules. SERS also has the advantages of simple sample synthesis and minimal cost and labor when compared to other processes for DNA determination [54]. Furthermore, SERS-grounded DNA determination can use exogenous labels or be label free, as shown in Fig. (**5A-D**). In a "SERS-off" arrangement, another technique relies exclusively on a smaller molecule to competitively relocate an aptamer identified with a Raman dye molecule.

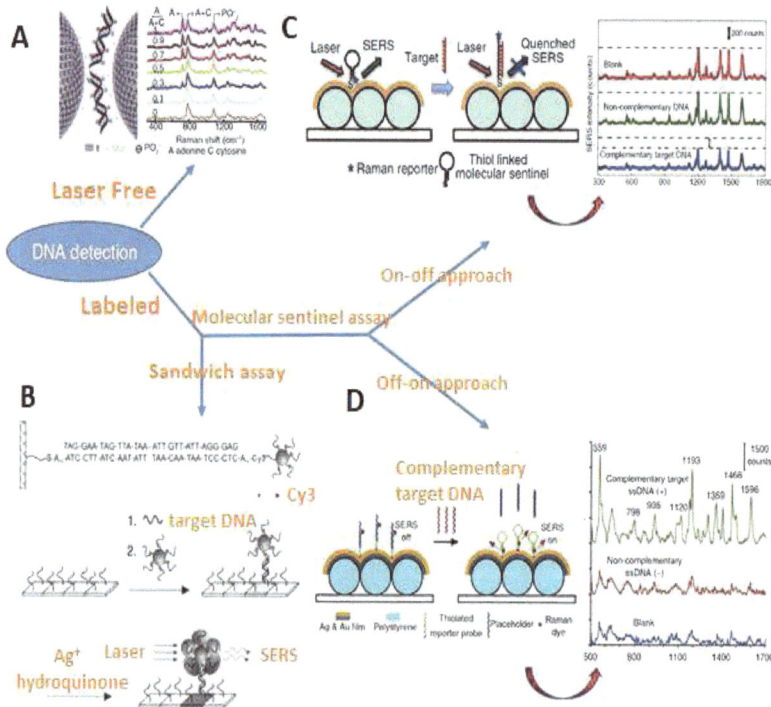

Fig. (5). Different types of SERS-based DNA determinations. (A) Label-free DNA determinations using iodide-fabricated silver nanoparticles, (B) double-decker analyse for the determinations of DNA using gold nanoparticles and analyses labelled with oligonucleotides, (C) Raman dye labelled hairpin-DNA probes for the determinations of DNA based on the on-off method. (D) DNA determinations based on the off-on method. Reproduced with permission from References [57-60].

Chung and colleagues employed a partial complimentary classification to restrict as sDNA aptamer onto Ag-Au nanoparticles using this "SERS-off" molecularly facilitated SERS technique. Furthermore, this technique was more sensitive to the 10-fm range, with bisphenol tainted tap water covering a dynamic range of 10-fm to 100-nm.SERS was used to achieve multiplexed DNA determination for distinct strains of *E. coli* utilizing nanoparticles synthesized with 6 different DNA ordering and Raman dyes. Using gold coated paramagnetic nanoparticles synthesized with probe DNA for objective DNA, a magnetic capture-based SERS prove for DNA determination was constructed. Using erythrosine B and malachite green Raman dyes, the West Nile virus and Rift Valley fever virus RNA genomes were successfully detected. The unique SERS substrates of silver nano-rice projections on gold triangular nano-arrays were adjusted for HBV DNA detection as a double-decker analysis with a detection limit of 50 aM [55].

On the basis of double-decker hybridization analyses, silver-SiO_2 core-shell nano-SERS-devices were arranged for the determination of exact DNA objectives. Silver nanoparticles were employed to probe the mark DNA utilizing SERS substrates with a label. Four different SERS devices were used to test the multiplexing capabilities, and the results showed that it had a lot of promise [56]. Another study used SERS-based molecular sentinel technology and the on-off method to multiplex the identification of breast cancer indicator genes. SERS-based molecular sentinel approaches can be employed for multiplexed DNA measurements, according to the findings. Furthermore, molecular sentinel research can be restricted to a solid configuration.

7.3. DETERMINATION OF OTHER BIOLOGICALLY RELATED NANOPARTICLES

SERS is proficient of identifying biologically related nanoparticles, such as viruses and exosomes and (Fig. **6**) reveals some of the exosomes and viruses based SERS. Exosomes are a type of extracellular vesicle with a diameter of 30–200 nm. They are now thought to be a remnant of a waste disposal device employed by cells. Exosomes are now thought to play a role in intercellular communication. It is critical to comprehend their layout in order to determine their biological function. Exosomes are released at different rates and with different contents from dysfunctional or stressed cells. Exosomes are found in most body fluids, and they provide a non-aggressive way to investigate diseases like cancer. Viruses appear to be another possible candidate for SERS-based studies in bodily fluids. Influenza, HIV, and a variety of other viruses have had minor impact on people at both the population and individual levels. The viruses are 20–300 nm in length, making them suitable candidates for SERS tests.

Although SERS uses a variety of approaches to identify viruses, it can also detect whole viruses utilizing proteins, RNA, or DNA.

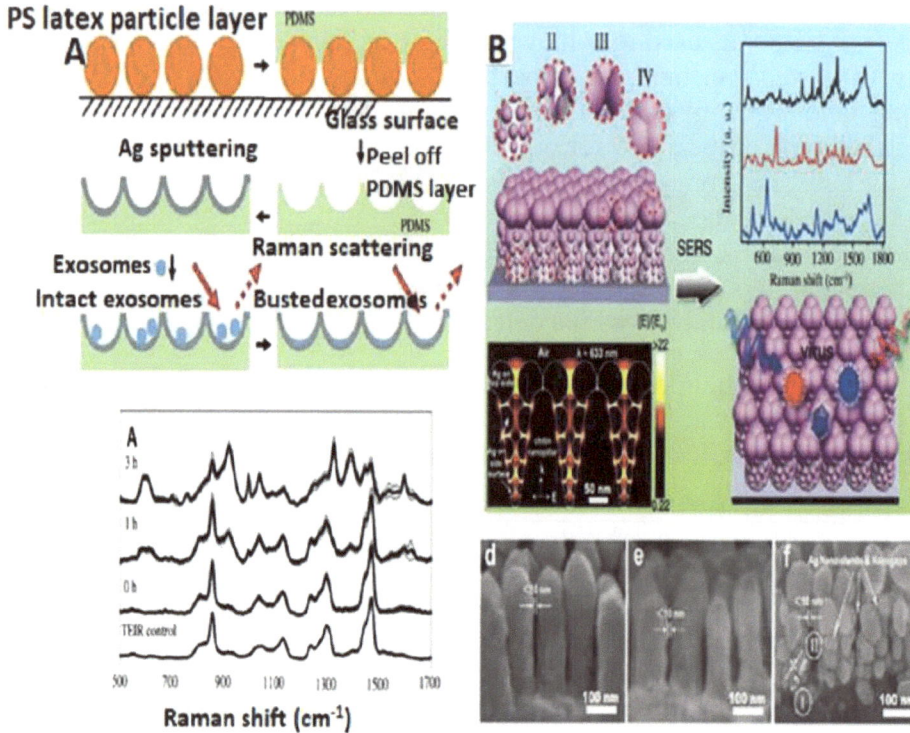

Fig. (6). Determination of biologically-related nanoparticles. (A) Silver-/PDMS SERS substrate is used to identify filtered exosomes. Throughout the drying procedure, they burst and novelspectral peaks developed noticeable as the matters become visible. (B) A SERS evaluate to notice animal viruses uses silver-coated chitin biomimetic framework from cicada wings. Reproduced with permission from Refs. [61, 62].

In 2005, a double-decker immunoassay with a 106 virus/ml edge was developed to distinguish the feline calcivirus [63]. SERS' sensitivity can distinguish a single tobacco mosaic virus utilizing angle improvement. In addition, silver nanorod-based arrays have been utilized to detect rhinoviruses, HIV viruses, adenoviruses, rotaviruses, and respiratory syncytial virus strains, as well as to distinguish between them [64, 65]. SERS can be utilized to improve electrochemical biosensors that are precise, sensitive, and selective. However, getting repeatable outcomes from determinations is a fundamental challenge of this modality. The combination process, the type of substrate employed, and the inhomogeneous circulation of particles on the metal nanoparticles all affect the reproducibility and repeatability of SERS analysis. Using capillary strength, Wang and colleagues developed an opto-fluidic method with a micro-channel-nano-channel coupling to capture and collect nanoparticles into SERS energetic clusters. This cluster delivered an electromagnetic improvement feature of about 10^8. However, they claimed that using 83 nm adenine improved SERS repeatability by 10% (device to device). Magnetic segments have also been utilized to mix nanoparticles for SERS measurement in the past [66]. On a microfluidic device, Gao and colleagues presented an assay to differentiate the anthrax biomarker poly-y-D-glutamic acid. Poly-y-D-glutamic acid and poly-y-D-glutamic acid-combined AuNPs competed for binding sites on anti-PGA-immobilized attractive beads in this experiment. The magnetic immuno-complexes were stuck on the microchannel where the SERS signals were detected using yoke-type-solenoids. The detection limit was determined to be 100 pg/mL [67].

CONCLUSIONS

One should now have an operational knowledge of the advantages and disadvantages of SERS as a technique for identifying low concentrations of a definite analyte of attention. This chapter providing a brief introduction to the role of surface plasmons in the improvement procedure, as defined a variety of nanostructures which can be used to generate them – lastly, an over-all overview of mutual experimental considerations we existing. In addition, it will investigate deeper into some application examples and compare and contrast a variety of SERS methodologies. Although the examples in this chapter show what is achievable in terms of analytical determinations, there are still significant challenges that must be solved first. SERS can create a standard tool outside of the research lab. The three-dimensional repeatability of SERS substrates, which determines the consistency for both intra- and inter-sample measurement, is an important concept to consider. The highest reports indicating good reproducibility also show a lower influence on improvement. This is due to a scarcity of highly competent hotspots associated with positive high-improvement influences. Though, because lower detection limits are required in specific applications, the accessibility of high-thickness comparable hot spots in those conditions may be advantageous, particularly for analyte concentrations of less than 10–50 pM.

CONSENT FOR PUBLICATION

Not applicable.

CONFLICT OF INTEREST

The authors declare no conflict of interest, financial or otherwise.

ACKNOWLEDGEMENTS

Declared none.

REFERENCES

[1]   Nayak S, Blumenfeld NR, Laksanasopin T, Sia SK. Point-of-care diagnostics: Recent developments in a connected age. Anal Chem 2017; 89(1): 102-23.
[http://dx.doi.org/10.1021/acs.analchem.6b04630] [PMID: 27958710]

[2]   Schuhmacher A, Gassmann O, Hinder M, Changing R. Changing R&D models in research-based pharmaceutical companies. J Transl Med 2016; 14(1): 105.
[http://dx.doi.org/10.1186/s12967-016-0838-4] [PMID: 27118048]

[3]   Mehta U, Kalk E, Boulle A, *et al.* Pharmacovigilance: A public health priority for South Africa. S Afr Health Rev 2017; 2017: 125-33.
[PMID: 29200789]

[4]   Eysenbach G. Consumer health informatics. BMJ 2000; 320(7251): 1713-6.
[http://dx.doi.org/10.1136/bmj.320.7251.1713] [PMID: 10864552]

[5]   Weigl BH, Gaydos CA, Kost G, *et al.* The value of clinical needs assessments for point-of-care diagnostics. Point Care 2012; 11(2): 108-13.
[http://dx.doi.org/10.1097/POC.0b013e31825a241e] [PMID: 23935405]

[6]   Boppart SA, Richards-Kortum R. Point-of-care and point-of-procedure optical imaging technologies for primary care and global health. Sci Transl Med 2014; 6(253)
[http://dx.doi.org/10.1126/scitranslmed.3009725]

[7]   Su CP, de Perio MA, Cummings KJ, McCague A-B, Luckhaupt SE, Sweeney MH. Case Investigations of Infectious Diseases Occurring in Workplaces, United States, 2006-2015. Emerg Infect Dis 2019; 25(3): 397-405.
[http://dx.doi.org/10.3201/eid2503.180708] [PMID: 30789129]

[8]   Chokkareddy R, Thondavada N, Thakur S, Kanchi S. Cholesterol-Based Enzymatic and Nonenzymatic Sensors Advanced Biosensors for Health Care Applications. Elsevier 2019; pp. 315-39.
[http://dx.doi.org/10.1016/B978-0-12-815743-5.00012-3]

[9]   Armstrong JA. Urinalysis in Western culture: a brief history. Kidney Int 2007; 71(5): 384-7.
[http://dx.doi.org/10.1038/sj.ki.5002057] [PMID: 17191081]

[10]   Chokkareddy R. Fabrication of sensors for the sensitive electrochemical detection of anti-tuberculosis drugs 2018.

[11]   Rosenfeld L. Clinical chemistry since 1800: growth and development. Clin Chem 2002; 48(1): 186-97.
[http://dx.doi.org/10.1093/clinchem/48.1.186] [PMID: 11751558]

[12]   Chin CD, Linder V, Sia SK. Lab-on-a-chip devices for global health: past studies and future opportunities. Lab Chip 2007; 7(1): 41-57.
[http://dx.doi.org/10.1039/B611455E] [PMID: 17180204]

[13]   Lopez-Barbosa N, Gamarra JD, Osma JF. The future point-of-care detection of disease and its data capture and handling. Anal Bioanal Chem 2016; 408(11): 2827-37.
[http://dx.doi.org/10.1007/s00216-015-9249-2] [PMID: 26780711]

[14]   Thevenot DR, Toth K, Durst RA, Wilson GS. Electrochemical biosensors: recommended definitions and classification. Pure Appl Chem 1999; 71(12): 2333-48.
[http://dx.doi.org/10.1351/pac199971122333]

[15]   Ligler FS. Perspective on optical biosensors and integrated sensor systems. Anal Chem 2009; 81(2): 519-26.
[http://dx.doi.org/10.1021/ac8016289] [PMID: 19140774]

[16]   Chua JH, Chee R-E, Agarwal A, Wong SM, Zhang G-J. Label-free electrical detection of cardiac biomarker with complementary metal-oxide semiconductor-compatible silicon nanowire sensor arrays. Anal Chem 2009; 81(15): 6266-71.
[http://dx.doi.org/10.1021/ac901157x] [PMID: 20337397]

[17]   Liu Q, Yuen C, Eds. Effect of magnetic field in malaria diagnosis using magnetic nanoparticles. Europ Conf Biomed Opt; 2011: Optical Society of America. Optical Society of America 2011.

[18]   Li M, Cushing SK, Wu N. Plasmon-enhanced optical sensors: a review. Analyst (Lond) 2015; 140(2): 386-406.
[http://dx.doi.org/10.1039/C4AN01079E] [PMID: 25365823]

[19]   Ge S, Liu F, Liu W, Yan M, Song X, Yu J. Colorimetric assay of K-562 cells based on folic acid-conjugated porous bimetallic Pd@Au nanoparticles for point-of-care testing. Chem Commun (Camb) 2014; 50(4): 475-7.
[http://dx.doi.org/10.1039/C3CC47622G] [PMID: 24257545]

[20]   Mehrvar M, Abdi M. Recent developments, characteristics, and potential applications of electrochemical biosensors. Anal Sci 2004; 20(8): 1113-26.
[http://dx.doi.org/10.2116/analsci.20.1113] [PMID: 15352497]

[21]   Heller A, Feldman B. Electrochemical glucose sensors and their applications in diabetes management. Chem Rev 2008; 108(7): 2482-505.
[http://dx.doi.org/10.1021/cr068069y] [PMID: 18465900]

[22]   Tomazelli Coltro WK, Cheng CM, Carrilho E, de Jesus DP. Recent advances in low-cost microfluidic platforms for diagnostic applications. Electrophoresis 2014; 35(16): 2309-24.
[http://dx.doi.org/10.1002/elps.201400006] [PMID: 24668896]

[23]   Sharma S, Zapatero-Rodríguez J, Estrela P, O'Kennedy R. Point-of-care diagnostics in low resource settings: present status and future role of microfluidics. Biosensors (Basel) 2015; 5(3): 577-601.
[http://dx.doi.org/10.3390/bios5030577] [PMID: 26287254]

[24]   Tucker JD, Bien CH, Peeling RW. Point-of-care testing for sexually transmitted infections: recent advances and implications for disease control. Curr Opin Infect Dis 2013; 26(1): 73-9.
[http://dx.doi.org/10.1097/QCO.0b013e32835c21b0] [PMID: 23242343]

[25]   Gaydos CA, Solis M, Hsieh Y-H, Jett-Goheen M, Nour S, Rothman RE. Use of tablet-based kiosks in the emergency department to guide patient HIV self-testing with a point-of-care oral fluid test. Int J STD AIDS 2013; 24(9): 716-21.
[http://dx.doi.org/10.1177/0956462413487321] [PMID: 23970610]

[26]   Shetty V, Zigler C, Robles TF, Elashoff D, Yamaguchi M. Developmental validation of a point-o--care, salivary α-amylase biosensor. Psychoneuroendocrinology 2011; 36(2): 193-9.
[http://dx.doi.org/10.1016/j.psyneuen.2010.07.008] [PMID: 20696529]

[27]   Tlili C, Myung NV, Shetty V, Mulchandani A. Label-free, chemiresistor immunosensor for stress biomarker cortisol in saliva. Biosens Bioelectron 2011; 26(11): 4382-6.
[http://dx.doi.org/10.1016/j.bios.2011.04.045] [PMID: 21621995]

[28]   Teengam P, Siangproh W, Tuantranont A, Vilaivan T, Chailapakul O, Henry CS. Multiplex paper-based colorimetric DNA sensor using pyrrolidinyl peptide nucleic acid-induced AgNPs aggregation for detecting MERS-CoV, MTB, and HPV oligonucleotides. Anal Chem 2017; 89(10): 5428-35.
[http://dx.doi.org/10.1021/acs.analchem.7b00255] [PMID: 28394582]

[29]   Floriano PN, Christodoulides N, Miller CS, *et al.* Use of saliva-based nano-biochip tests for acute myocardial infarction at the point of care: a feasibility study. Clin Chem 2009; 55(8): 1530-8.
[http://dx.doi.org/10.1373/clinchem.2008.117713] [PMID: 19556448]

[30]   Ji S, Choi Y. Point-of-care diagnosis of periodontitis using saliva: technically feasible but still a challenge. Front Cell Infect Microbiol 2015; 5: 65.
[http://dx.doi.org/10.3389/fcimb.2015.00065] [PMID: 26389079]

[31]   Torrente-Rodríguez RM, Campuzano S, Ruiz-Valdepeñas Montiel V, Gamella M, Pingarrón JM. Electrochemical bioplatforms for the simultaneous determination of interleukin (IL)-8 mRNA and IL-8 protein oral cancer biomarkers in raw saliva. Biosens Bioelectron 2016; 77: 543-8.
[http://dx.doi.org/10.1016/j.bios.2015.10.016] [PMID: 26474095]

[32]   Blicharz TM, Siqueira WL, Helmerhorst EJ, *et al.* Fiber-optic microsphere-based antibody array for the analysis of inflammatory cytokines in saliva. Anal Chem 2009; 81(6): 2106-14.
[http://dx.doi.org/10.1021/ac802181j] [PMID: 19192965]

[33]   Barnfather KD, Cope GF, Chapple IL. Effect of incorporating a 10 minute point of care test for salivary nicotine metabolites into a general practice based smoking cessation programme: randomised controlled trial. bmj 2005; 331(7523): 999.

[34]   Walt DR, Blicharz TM, Hayman RB, *et al.* Microsensor arrays for saliva diagnostics. Ann N Y Acad Sci 2007; 1098(1): 389-400.
[http://dx.doi.org/10.1196/annals.1384.031] [PMID: 17435144]

[35]   Du Y, Zhang W, Wang ML. Sensing of salivary glucose using nano-structured biosensors. Biosensors (Basel) 2016; 6(1): 10.
[http://dx.doi.org/10.3390/bios6010010] [PMID: 26999233]

[36]   Ching KH, Burbelo PD, Gonzalez-Begne M, *et al.* Salivary anti-Ro60 and anti-Ro52 antibody profiles to diagnose Sjogren's Syndrome. J Dent Res 2011; 90(4): 445-9.
[http://dx.doi.org/10.1177/0022034510390811] [PMID: 21212317]

[37]   Christodoulides N, Pierre FN, Sanchez X, *et al.* Programmable bio-nanochip technology for the diagnosis of cardiovascular disease at the point-of-care. Methodist DeBakey Cardiovasc J 2012; 8(1): 6-12.
[http://dx.doi.org/10.14797/mdcj-8-1-6] [PMID: 22891104]

[38]   Rooney KD, Schilling UM. Point-of-care testing in the overcrowded emergency department--can it make a difference? Crit Care 2014; 18(6): 692.
[http://dx.doi.org/10.1186/s13054-014-0692-9] [PMID: 25672600]

[39]   Larsson A, Greig-Pylypczuk R, Huisman A. The state of point-of-care testing: a European perspective. UPSALA J MED SCI 2015; 120(1): 1-10.
[http://dx.doi.org/10.3109/03009734.2015.1006347] [PMID: 25622619]

[40]   Jarvis PRE. Improving emergency department patient flow. Clin Exp Emerg Med 2016; 3(2): 63-8.
[http://dx.doi.org/10.15441/ceem.16.127] [PMID: 27752619]

[41]   White BA, Chang Y, Grabowski BG, Brown DF. Using lean-based systems engineering to increase capacity in the emergency department. West J Emerg Med 2014; 15(7): 770-6.
[http://dx.doi.org/10.5811/westjem.2014.8.21272] [PMID: 25493117]

[42]   St John A, Price CP. Economic evidence and point-of-care testing. Clin Biochem Rev 2013; 34(2): 61-74.
[PMID: 24151342]

[43]   Sikka R, Mehta S, Kaucky C, Kulstad EB. ED crowding is associated with an increased time to pneumonia treatment. Am J Emerg Med 2010; 28(7): 809-12.
[http://dx.doi.org/10.1016/j.ajem.2009.06.023] [PMID: 20837259]

[44]   Callen J, Georgiou A, Li J, Westbrook JI. The safety implications of missed test results for hospitalised patients: a systematic review. BMJ Qual Saf 2011; 20(2): 194-9.
[http://dx.doi.org/10.1136/bmjqs.2010.044339] [PMID: 21300992]

[45]   Njoki PN, Lim I-IS, Mott D, Park H-Y, Khan B, Mishra S, *et al.* Size correlation of optical and spectroscopic properties for gold nanoparticles. . J Phys Chem C 2007; 111(40): 14664-9.
[http://dx.doi.org/10.1021/jp074902z]

[46]   Yu Q, Guan P, Qin D, Golden G, Wallace PM. Inverted size-dependence of surface-enhanced Raman

scattering on gold nanohole and nanodisk arrays. Nano Lett 2008; 8(7): 1923-8.
[http://dx.doi.org/10.1021/nl0806163] [PMID: 18563939]

[47]   Kahraman M, Daggumati P, Kurtulus O, Seker E, Wachsmann-Hogiu S. Fabrication and characterization of flexible and tunable plasmonic nanostructures. Sci Rep 2013; 3: 3396.
[http://dx.doi.org/10.1038/srep03396] [PMID: 24292236]

[48]   Orendorff CJ, Gole A, Sau TK, Murphy CJ. Surface-enhanced Raman spectroscopy of self-assembled monolayers: sandwich architecture and nanoparticle shape dependence. Anal Chem 2005; 77(10): 3261-6.
[http://dx.doi.org/10.1021/ac048176x] [PMID: 15889917]

[49]   Wiley BJ, Chen Y, McLellan JM, *et al.* Synthesis and optical properties of silver nanobars and nanorice. Nano Lett 2007; 7(4): 1032-6.
[http://dx.doi.org/10.1021/nl070214f] [PMID: 17343425]

[50]   Wu HY, Choi CJ, Cunningham BT. Plasmonic nanogap-enhanced Raman scattering using a resonant nanodome array. Small 2012; 8(18): 2878-85.
[http://dx.doi.org/10.1002/smll.201200712] [PMID: 22761112]

[51]   Ye J, Wen F, Sobhani H, *et al.* Plasmonic nanoclusters: near field properties of the Fano resonance interrogated with SERS. Nano Lett 2012; 12(3): 1660-7.
[http://dx.doi.org/10.1021/nl3000453] [PMID: 22339688]

[52]   Singh JP, Chu H, Abell J, Tripp RA, Zhao Y. Flexible and mechanical strain resistant large area SERS active substrates. Nanoscale 2012; 4(11): 3410-4.
[http://dx.doi.org/10.1039/c2nr00020b] [PMID: 22544280]

[53]   Chung AJ, Huh YS, Erickson D. Large area flexible SERS active substrates using engineered nanostructures. Nanoscale 2011; 3(7): 2903-8.
[http://dx.doi.org/10.1039/c1nr10265f] [PMID: 21629884]

[54]   Ngo HT, Wang H-N, Fales AM, Vo-Dinh T. Plasmonic SERS biosensing nanochips for DNA detection. Anal Bioanal Chem 2016; 408(7): 1773-81.
[http://dx.doi.org/10.1007/s00216-015-9121-4] [PMID: 26547189]

[55]   Li M, Cushing SK, Liang H, Suri S, Ma D, Wu N. Plasmonic nanorice antenna on triangle nanoarray for surface-enhanced Raman scattering detection of hepatitis B virus DNA. Anal Chem 2013; 85(4): 2072-8.
[http://dx.doi.org/10.1021/ac303387a] [PMID: 23320458]

[56]   Li J-M, Wei C, Ma W-F, An Q, Guo J, Hu J, *et al.* Multiplexed SERS detection of DNA targets in a sandwich-hybridization assay using SERS-encoded core–shell nanospheres. J Mater Chem 2012; 22(24): 12100-6.
[http://dx.doi.org/10.1039/c2jm30702b]

[57]   Xu L-J, Lei Z-C, Li J, Zong C, Yang CJ, Ren B. Label-free surface-enhanced Raman spectroscopy detection of DNA with single-base sensitivity. J Am Chem Soc 2015; 137(15): 5149-54.
[http://dx.doi.org/10.1021/jacs.5b01426] [PMID: 25835155]

[58]   Cao YC, Jin R, Mirkin CA. Nanoparticles with Raman spectroscopic fingerprints for DNA and RNA detection. Science 2002; 297(5586): 1536-40.
[http://dx.doi.org/10.1126/science.297.5586.1536] [PMID: 12202825]

[59]   Ngo HT, Wang H-N, Fales AM, Vo-Dinh T. Label-free DNA biosensor based on SERS Molecular Sentinel on Nanowave chip. Anal Chem 2013; 85(13): 6378-83.
[http://dx.doi.org/10.1021/ac400763c] [PMID: 23718777]

[60]   Ngo HT, Wang H-N, Fales AM, Nicholson BP, Woods CW, Vo-Dinh T. DNA bioassay-on-chip using SERS detection for dengue diagnosis. Analyst (Lond) 2014; 139(22): 5655-9.
[http://dx.doi.org/10.1039/C4AN01077A] [PMID: 25248522]

[61]   Lee C, Carney RP, Hazari S, *et al.* 3D plasmonic nanobowl platform for the study of exosomes in

solution. Nanoscale 2015; 7(20): 9290-7.
[http://dx.doi.org/10.1039/C5NR01333J] [PMID: 25939587]

[62]   Shao F, Lu Z, Liu C, *et al.* Hierarchical nanogaps within bioscaffold arrays as a high-performance SERS substrate for animal virus biosensing. ACS Appl Mater Interfaces 2014; 6(9): 6281-9.
[http://dx.doi.org/10.1021/am4045212] [PMID: 24359537]

[63]   Driskell JD, Kwarta KM, Lipert RJ, Porter MD, Neill JD, Ridpath JF. Low-level detection of viral pathogens by a surface-enhanced Raman scattering based immunoassay. Anal Chem 2005; 77(19): 6147-54.
[http://dx.doi.org/10.1021/ac0504159] [PMID: 16194072]

[64]   Driskell JD, Shanmukh S, Liu Y-J, Hennigan S, Jones L, Zhao Y-P, *et al.* Infectious agent detection with SERS-active silver nanorod arrays prepared by oblique angle deposition. IEEE Sens J 2008; 8(6): 863-70.
[http://dx.doi.org/10.1109/JSEN.2008.922682]

[65]   Shanmukh S, Jones L, Zhao Y-P, Driskell JD, Tripp RA, Dluhy RA. Identification and classification of respiratory syncytial virus (RSV) strains by surface-enhanced Raman spectroscopy and multivariate statistical techniques. Anal Bioanal Chem 2008; 390(6): 1551-5.
[http://dx.doi.org/10.1007/s00216-008-1851-0] [PMID: 18236030]

[66]   Wang M, Jing N, Chou I-H, Cote GL, Kameoka J. An optofluidic device for surface enhanced Raman spectroscopy. Lab Chip 2007; 7(5): 630-2.
[http://dx.doi.org/10.1039/b618105h] [PMID: 17476383]

[67]   Gao R, Ko J, Cha K, *et al.* Fast and sensitive detection of an anthrax biomarker using SERS-based solenoid microfluidic sensor. Biosens Bioelectron 2015; 72: 230-6.
[http://dx.doi.org/10.1016/j.bios.2015.05.005] [PMID: 25985198]

CHAPTER 8

Surface Enhanced Raman Spectroscopy-Bottlenecks and Improvement Strategies

Richa Jackeray[1,*], Kriti Arya[2], Zainul Abid CKV[3] and Harpal Singh[4,*]

[1] Independent Contributor (Active Professional in Healthcare Industry), India

[2] Computational Instrumentation, CSIR-CSIO Chandigarh, India

[3] Independent Contributor (External Expert in Spectroscopy), India

[4] Center for Biomedical Engineering (CBME), Indian Institute of Technology Delhi (IIT D), Hauz Khas, New Delhi, India

Abstract: In this chapter, we briefly discussed the bottlenecks commonly present for using the surface enhanced Raman spectroscopy (SERS) in research. The barriers in the successful commercialization of this technique are also highlighted in this chapter. This chapter is also dedicated to the account of possible key strategies for improvement of SERS. Among these strategies, we discussed one of the promising strategies based on improved nanofabrication at length. Furthermore, few recent important developments in SERS instrumentation and handheld portable devices are also highlighted in this chapter for providing broader information to the readers.

Keywords: Challenges, Combination approach, Handheld, Improvement Strategies, Nanofabrication, Raman Spectroscopy.

8.1. INTRODUCTION

Convergence of two highly rapidly advancing technologies - photonics and nanoscience is the chief reason for accelerating development methods and materials of SERS for detecting and analysing wide range of chemicals, surfaces, materials, biomolecules, medical system, environment, forensics, and food analysis. The ultimate target is to develop commercially viable machines for applications in various industries across medicine, agriculture, food and beverage sector, pollution monitoring. In general, SERS technique has become popular owing to its sensitivity and capacity for high throughput for the detection of biom-

* **Corresponding Author (s): Richa Jackeray:** Independent Contributor, Active Professional in Healthcare Industry; E-mail: richajackerayst01@gmail.com, **Harpal Singh:** Center for Biomedical Engineering (CBME), Indian Institute of Technology Delhi (IIT D), Hauz Khas, New Delhi, India; E-mail: harpal2000@yahoo.com

olecules and chemical entities. In two independent studies, researchers illustrated single molecule measurement by SERS for the first time in 1997 [1, 2]. This ground-breaking research led to an upsurge in the SERS analysis as it became the first vibrational spectroscopy to attain this feat of single molecule detection which was a distant reality for many other detection methodologies then. Typically, SERS measurement involves attachment or absorption of analyte molecule onto SERS substrate. After absorption, the specially designed SERS active substrate enhances the Raman signal of the analyte which are comparable to that in fluorescence measurement.

The laboratory analysis and research work have proven spectacular with single molecule detection giving new direction to fundamental research; However, a long wait is faced before commercially exploiting this technology and moving beyond academia. Researchers are unable to apply SERS to the betterment of scientific community based on ultimate use of SERS as analytical tool for society at large. Unfortunately, scope of SERS is limited owing to certain drawbacks for its wide application. For instance, though SERS is established as good technique for structure characterization, but it is unable to process component separation. The final SERS spectrum is impacted by non-specific molecules with similar properties as that of target analyte. In addition, intimate contact between SERS substrate and analyte molecule is requisite for enhancing the Raman signal.

This chapter is devoted to discuss the challenges and limitations of SERS as biosensing and detection tool. Since great problems results in investment towards greater solutions, problems associate with SERS have led to development of novel strategies for improving every aspect of methodology to produce handheld reliable and useful devices catering to daily use in personal and industrial level settings. In coming sections, we highlight bottlenecks faced by both academic world and industries for making SERS as market friendly system.

8.1.1. Research

Basic research is extending increasing the limits of SERS to detection of single-molecule studies, sub-nanometer resolution and pico-to-femtosecond processes. Key challenges for SERS impeccable success in fundamental research are outlined below:

a. <u>Restricted versatility as tool</u>: The SERS enhancement effects are highly influenced and limited by the shape, size, form, overall structure and optical properties of SERS-active nanomaterials. The free-electron metal nanostructures, primarily silver, gold and copper, are favourite SERS substrates due to their surface plasmon properties lying in the visible region. Among these, silver and gold are of choice for conducting studies of SERS studies due to their electronic structure, surface morphology and interactions of these metals with target analytes. On the other hand, copper and other transition metals show weak signals of SERS compared to above mentioned metals [3]. In SERS, the metal nanostructures not only act as a signal amplifier but also as a host for target molecules. In order to have the best enhancement of the signal, probe molecule is directly coupled to metal nanostructures, specifically in the possible hotspot, to meet the maximum enhancement in the local field. This maximum enhancement is called as electromagnetic (EM) enhancement. Additionally, total Raman enhancement can also be due to a possible effect of charge transfer (known as chemical enhancement mechanism) between bare metal nanostructures and adsorbates may also contribute to the total Raman enhancement. In brief, the process of SERS includes the complicated coupled three-body interactions among photons, molecules, and nanostructures resulting in limited versatility of the tool.

b. <u>Limited understanding about the nature of probe molecules and their interaction with metal nanostructures</u>: In the contact-mode SERS, the close interactions occur between bare metal nanostructures and target molecules. The contact-mode demonstrates excellent Raman enhancement for detecting trace molecules. However, contact-mode SERS has several limitations in case of characterization of the surfaces of the materials and interfacial structures of materials and bio samples. In some cases, the interfacial charge transfer may occur between materials and bare metal nanostructures. Under intense laser illumination, photochemical reaction can occur due to the close contact between target molecules and noble metal may result in a photochemical reaction. Metal nanostructures are generally prone to contamination resulting in an erroneous interpretation of the SERS signals. Therefore, understanding about the nature of probe molecules and their interaction with metal nanostructures is very important [3].

c. <u>Limited extension of SERS to various morphologies</u>: Giant SERS enhancements are generated by the free electron metal nanostructure surfaces leading to another key limitation for SERS, morphological generality *i.e.* the extension of SERS to various morphologies, such as flat surfaces, single crystals, semiconductors and soft materials. For example, a flat metal substrate does not support a powerful enhancement in the local field. This strong enhancement can be achieved by utilizing an external optical element such as a prism which is coupled with a flat gold or silver film supporting a growing surface plasmon. This configuration for Raman measurements is termed attenuated total reflection (ATR-Raman). The ATR-Raman configuration facilitates the two-fold Raman enhancement for probe molecules adsorbed on the surfaces of a flat silver film. It is noteworthy that the enhancement factor of ATR-surface-plasmon-polariton (ATR-SPP) is much less for other transition metal films. Therefore, more effort is required for considerably improving the sensitivity of SERS on flat surfaces (*e.g.*, single crystal surfaces).

d. <u>Required key considerations for performing bioanalytical SERS measurements:</u> To conduct optimal measurements for specific applications, it is important to carefully consider several factors regarding choice of a) preparation procedure for the biological samples to be analysed, b) appropriate SERS probes, c) a suitable instrument configuration, and d) suitable methods for data processing and analysis [4]. These considerations are briefed in Fig. (**2a**) and Fig. (**2b**).

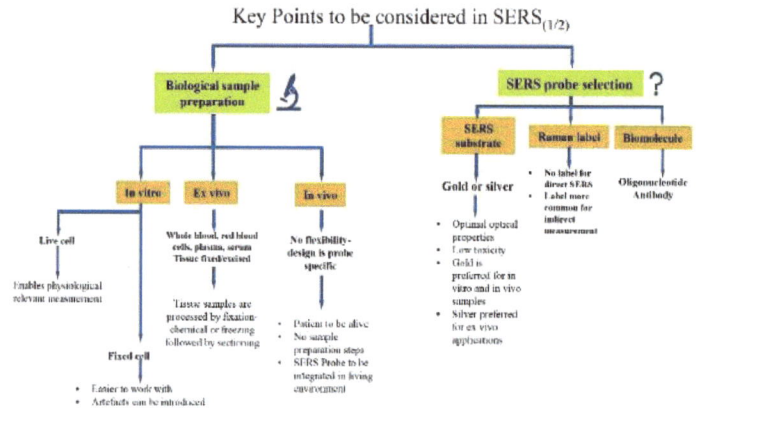

Fig. (2a). Summary of the key considerations suggested during biological sample preparation and probe detection when performing bioanalytical SERS measurements. SERS: surface-enhanced Raman scattering [4].

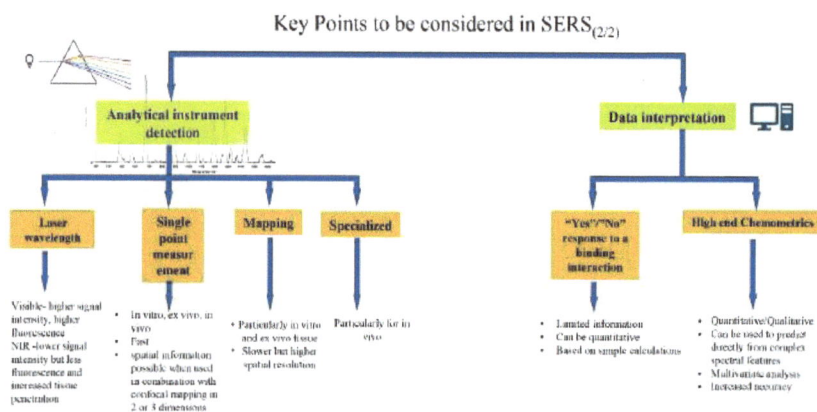

Fig. (2b). Summary of the key considerations suggested during detection and data interpretation when performing bioanalytical SERS measurements. NIR: near infrared [4].

8.1.2. Commercialization

The practical research or research for commercialization is increasing across a spectrum. The general aim of this increase is based on having versatile analytical tools for different surface, materials, life science, environmental science, forensic science and food science, and commercial instruments for use in day-to-day life.

Even after more than 40 years of development, SERS is yet to be commercialized successfully and used widely thereafter. As shown in Fig. **1**, in order to utilize SERS techniques in daily life, it is important to have standardization with respect to SERS substrates and techniques for below:

a. stability of SERS substrates in ambient environments
b. reproducibility of both substrate-to-substrate and spot-to-spot reproducibility
c. selectivity and molecular generality of SERS measurements
d. cost-effectiveness and ease in handling
e. shorter shelf life (< 1 year) of SERS active metal nanostructures
f. susceptibility of structured surfaces to surface oxidation and contamination
g. susceptibility of metal colloids to aggregation
h. contamination by surface capping agents
i. lack of stability, selectivity, sensitivity, and reproducibility of SERS substrates [3].

According to a 2019 estimate, the SERS market will register a 9.1% compound annual growth rate (CAGR) over the next five years. In terms of revenue, the global market size will reach US$ 190 million by 2024, from US$ 110 million in 2019 [5]. In the coming years, the demand for SERS will be driven by the increased use in expenditures related to the biology and medicine fields, launches of new products, more-intense competition, increase in the spending on general industry, recycling and renovation of old technology, increase in the adoption of SERS.

Globally, market of SERS in industrial applications is highly niche as the manufacturing technology of SERS is relatively higher than some low-tech equipment. Concurrently, Europe, with 30.35% production market share (in 2016) and advancement in technology status stand out in the global SERS industry. Overall, the consumption volume of SERS is mostly driven by downstream industries and global economy.

8.2. IMPROVEMENT STRATEGIES

Since last few decades, many researchers have contributed in reducing the major limitations of SERS. Using focused reading approach, we are limiting the discussion to few exemplary improvement strategies here as below:

8.2.1. Tip enhanced Raman Spectroscopy

The Tip enhanced Raman Spectroscopy (TERS) technique includes a combined approach of Raman spectroscopy along with scanning probe microscopy such as atomic force microscopy (AFM), scanning tunnelling microscopy (STM) or shear force microscopy (SFM). A robust EM coupling between the metal tip and the metal substrate is crucial for reliable TERS imaging. A sharp tip composed of or coated with silver or gold is used in TERS [3]. Additionally, the silver or gold nanoparticles or nanostructures can be attached to the AFM or SFM probe. The gold or silver surface plasmon membrane (SPM) tip produces a localized augmented EM field under excitation by an appropriate source. The Raman signals of adsorbed probe molecules and surfaces are enhanced by EM field by up to six times.

The TERS technique is utilized to understand basic processes in surface science, materials science, and bioanalysis [6, 7].By virtue of ability of determining the molecular properties of the excited state of a molecule, the TERS can indicate the interactions between adsorbates and surfaces [3]. Besides, TERS is also used for examining the materials with large cross-sections (*e.g.*, carbon allotropes, semiconductors, polymers, solar cell materials, nucleic acids, bacteria, viruses, cells, and proteins). For instance, Bailo *et al*. utilized the TERS technique for detecting the sequences of single RNA strands and monitoring the nanoscale photocatalytic reactions [8].

Addressing the limitation of TERS regarding outstanding spatial resolution along with molecular contrasts has been also well demonstrated by Lin *et al* using AFM-TERS [9]. In this study, direct base-to-base transitions in double stranded DNA were shown at ambient conditions. In another study, Pettinger and co-workers reported the detection of hydrogen bonds between the nucleobases adenine and thymine using TERS [10].

One of the most systematic applications in TERS are the studies about insulin amyloid fibrils with high reproducibility and feasibility of molecular identification and topographical characterization [11]. A significant composition difference among fibril polymorphs and proto filaments was reported using TERS [12]. These results provided the evidence of different fibrillation mechanisms into different shapes. A full spectroscopic information on a β-amyloid fibrils was conducted confirming the reproducibility of TERS measurements [13]. Recently, van den Akker *et al.* utilized TERS to explore the surface molecular structure of fibrils formed on a lipid interface [14]. Moreover, it was demonstrated that TERS studied fibrils were formed on a lipid interface, as they held lipid molecules on their surface. Increased understanding of towards fibril-lipid interaction and subsequent expected fully uncovering of proteins misfolding is underway in near-term and future investigations [13]. Lipids are in high proportion on cell membranes, therefore, supporting the cell information and material exchanges. Using TERS, it is possible to reveal not only the individual components but also the molecular interaction upon cell membranes. Besides, the strong methylene C--H stretching vibration around 2900 cm^{-1} can be used as a lipid marker band in the complex biological systems [13].

It is noteworthy that a lipid monolayer remains stable in ambient conditions. However, a lipid bilayer mimicking the features of real membranes, need to be processed in liquid. As a result, it is quite crucial to perform TERS in the indigenous environment of the biosamples, maintaining their physiological conditions. However, a liquid TERS experiment continues to be challenging. To have measurements on a soft bio sample, a low tip spring constant is required to decrease sample damage. Additionally, it is difficult to keep the cantilever in stable feedback and acquiring stable Raman signals while illuminating tip and sample at the same time. Fujinami *et al.* employed TERS in liquid to determine variations of 2-Dioleoyl-sn-glycero-3-phosphocholine/dipalmitoyl phosphatidylcholine (DOPC/DPPC) bilayers. They noted that DOPC changed over time, thus clearly exhibiting the potential of TERS for detecting the in homogeneity and mobility of fluid bilayer membranes [15].

8.2.2. Shell Isolated Nanoparticle-enhanced Raman Spectroscopy

In 2010, Paneerselvam *et al* combined the concept of TERS and the SERS. Consequently, they introduced a new method namely "Shell-Isolated Nanoparticle-Enhanced Raman Spectroscopy" (SHINERS) to circumvent the shortcomings of SERS and TERS. These methods employed ultrathin shell-isolated nanoparticles. Each nanoparticle served as a TERS tip for acquiring the SERS signals with significant improved sensitivity. In current times, SHINERS is gaining attention by virtue of the application of shell-isolated nanoparticles (SHINs) over any surface or morphology (*e.g.*, semiconductors, single-crystal surfaces, non-transition metals, fruits, and vegetables) for obtaining the signals from target analytes [16 - 18].

Zheng *et al* used SHINERS of Au@SiO^2 for detecting and characterizing the breast tissues. Different spectral features allowed them for differentiating the normal breast tissues from breast tissues with lesion types fibroadenoma (FD), atypical ductal hyperplasia (ADH), ductal carcinoma *in situ* (DCIS), and invasive ductal carcinoma (IDC). Characteristic bands of DNA and β-carotene could differentiate malignant tissues from other breast tissues. The use of different enhancement effects using SHINs effectively generates specific Raman bands with distinct characteristic for each tissue type such as FD, DCIS, and IDC. In a separate publication, they have demonstrated use of SHINERS in dragonizing type II macrocalcification in fibroadenoma, ADH and DCIS fresh frozen sections [19].

In summary, SHINERS ca serve as an effective research aid for getting the interfacial molecular information which remains unaffected by the ambient factors such as contamination, aggregation, oxidation, and electronic interactions of nanoparticles with matrix species in the surrounding medium and metallic substrates [20]. This mode operates in a different way than the contact mode SERS. This is primarily due to the inert uniform thin shell which reliably transfers the EM enhancement from the SERS-active core and subsequently avoids electrical contact between the probe surface and metal nanoparticles. In SHINERS, the dual functions of nanoparticles are effectively maintained with the help of the inert shell material. Consequently, the SHINs do not experience interference from matrix molecules in their ambience and rather provide Raman information on surface species. Furthermore, the thin inert shell are reported to maintain the SERS activity of the core for almost 1 year. This shell also protects the nanoparticles from oxidation, acidic or harsh environments, and aggregation. This feature offers potential for commercialization of shell isolated nanoparticles [21].

SHINERS are reported to deliver crucial details about adsorbates at well defined single-crystal surfaces with the aid of SHINs [3]. Methodology developed by the group demonstrated that SHINERS can be prepared as a standard characterization method for the *in situ* monitoring of reaction intermediates and the elaboration of reaction mechanisms. The abovementioned progress has reduced the generality limitation to some extent. Further advancements are warranted to enable SERS to be an even more powerful and versatile technique.

8.2.3. Interference Enhanced Raman Spectroscopy (IERS)

Interference enhanced Raman scattering (IRES) is based on the interference effect that occurs on the coated metallic substrates or Bragg reflectors (waves coming back from coating surface and substrate surface). Compared to SERS, the enhancement in case of IERS is relatively modest (enhancement factors are on the order of $10^1 - 10^2$, the substrates are easier to fabricate, stable for long-term, have an uniform enhancement over the whole surface, and can be atomically smooth if fabricated by atomic layer deposition. However, as in the case of SERS, the presence of sample changes the wavenumber of highest enhancement and the magnitude of this enhancement. While this was never an obstacle for the application of SERS, it might have prevented IERS from its use in daily life. Recently, Susanne Pahlow *et al* demonstrated possibility of using IERS to detect biomolecule. They used aluminium substrate with Al_2O_3 coating, and iron loaded siderophore ferrioxamine B and the polymer poly(methyl methacrylate) (PMMA) as model analytes [21].

8.2.4. Approach Based on Combinations and Spatio-temporal Resolution by Nanostructures

Based on the different feedback mechanisms in combinations, TERS can be categorized into three varieties: atomic force microscopy (AFM) based TERS system *i.e.* AFM-TERS, scanning tunnelling microscopy (STM) based TERS system *i.e.* STM-TERS and shear force microscopy (SFM) based TERS system *i.e.* SFM-TERS. Among these, AFM-TERS is the most widely embraced feedback system for analysing live bio samples as AFM-TERS can be used in liquid facilitating the analysis of biological molecules under native conditions. Furthermore, it is possible to have both the structure and chemical details of a sample at the nanoscale concurrently in various experimental condition susing AFM-TERS. These features enable AFM-TERS to be commonly used in biological samples such as nucleic acids, proteins, pathogens, lipids, and cell membranes [22].

Furthermore, TERS technique is widely applied to distinguish nucleic acid bases by virtue of features such as requirement of small sample amount, high sensitivity, and direct-sequencing [23 - 26]. AFM-TERS is reported to successfully characterize the respective nucleobases, detecting DNA hybridization, and characterizing the molecular structure of double-stranded breaks (DSBs) [27, 28].

In protein detection, the heme protein cytochrome C (Cyt C) is one of the proteins which were explored by TERS during starting days of advent of this technique. In an experiment, Raman properties of Cc were detected by AFM-TERS and SERS concurrently [29]. The result demonstrated that with a silver coated AFM-tip, TERS spectra could present the Raman signal of both the heme and amino acids (Cyt c). However, the details about amino acids could not be noted in SERS. Hencc, the superiority of TERS, with superior resolution and spatial enhancement- was established over SERS, in providing TERS more structural details of the large biomolecules compared with SERS [22].

The Gram-positive bacteria Staphylococcus epidermis was first kind of micro-organisms detected by AFM-TERS [30]. The topography of a single epidermis cell and AFM-TERS spectra detected at the different positions of the bacterial surface were explored with AFM-TERS. Other pathogen investigated using AFM-TERS was tobacco mosaic virus. Under AFM-TERS, this virus appeared as a rod-like capsid made by one structural protein and one single RNA molecule. Overall, the application of AFM-TERS in biology is increasing rapidly by virtue of high sensitivity and superior spatial resolution of AFM-TERS. Table 1 exhibits the structures and detection range of the three biomolecules characterized by AFM-TERS techniques.

Table 1. The structure of the nucleic acid, protein and pathogen and their border of AFM-TERS had been detected by TERS techniques [22].

Biomolecule	Nucleic Acid	Protein	Pathogen/biomolecule Entity
Structure	A single- or double-stranded structure consisted of five nucleotides	Three-dimensional structure consisted of one or more long chains of amino acid residues	RNAs and Proteins
AFM-TERS test	Direct-sequencing; DNA hybridization; DNA double strand breaks	Conformal changes and amino acid distributions	Peptides and polysaccharides
STM-TERS	DNA's helical pitch, half-period oscillations that were interpreted as the alternation between the major and minor grooves, and interhelical spacing	Adsorption behavior of proteins on specific surfaces (*i.e.*, phosphoprotein osteopontin on oxalate crystals) or the chemical interaction and localization of protein–ligand systems [29].	Peptides and polysaccharides Raman bands, typical of a bacterial surface could be identified in the near-field spectra with an enhancement factor of 10^4-10^5 [30].

Generally, imaging modalities are combined with Raman spectroscopy providing the physical and chemical details with high spatial and spectral resolution. The combinations of imaging techniques with Raman spectroscopy are paving the way to a better elucidation of the mechanisms of SERS. For example, Hou *et al.* imagined single molecules with a spatial resolution of less than 1 nm using Raman imaging and STM [31].

Several other variants based on combination approach with TERS are also explored for enhanced spatial-temporal resolution. For instance, Ren *et al* developed transient electrochemical surface-enhanced Raman spectroscopy (TEC-SERS) with a temporal resolution less than the charging time of the double-layer capacitance. The TEC-SRS has enormous potential for the exploring of both reversible and irreversible electrochemical processes [31]. These *in situ* SERS studies with high temporal resolution are important for elucidating mechanisms of the complex reaction and providing more details about the molecular structures and surface interactions.

For incorporating the ultrafast temporal resolution into Raman studies, pump–probe methods or ultrafast spectroscopic techniques can be employed along with SERS/TERS/SHINERS. This combination increases the lifetimes of intermediate species facilitating the detection of short-lived species with increased sensitivity [3]. For example, Van Duyne *et al.* combined SERS and femtosecond stimulated Raman spectroscopy for exploring the dynamics of plasmonic materials. They also investigated the effects of environmental heterogeneities by investigating more homogeneous molecular subsets using combination approach [32]. Advancements in ultrafast SERS facilitate the examination of a wide variety of systems on ultrafast time scales, *e.g.*, time scale of molecular motion. Remarkably, Van Duyne *et al.* successfully combined picosecond-pulsed irradiation with an ultrahigh vacuum-TERS (UV-TERS) instrument. In this case, TERS provided correlated chemical and topographic information with higher temporal resolution [33].

Generally, it is difficult to detect the intermediate species as they have small cross-sections and short lifetimes. The pump–probe methods are used to excite the intermediate species to their excited state. Using this method, the excited species have larger cross-sections and longer lifetimes and therefore can be easily detected by SERS. For obtaining the complete information about a target system, it is important to detect weak adsorbates and intermediate species on target surfaces with higher temporal resolution. Notably, in case of time-dependent Raman measurements, high power lasers can damage target molecules and target surfaces. Furthermore, photoreactions may occur in some cases within certain molecules because of metal nanoparticles. Therefore, it is imperative to maintain an optimal laser power and exposure time while obtaining required details about target molecules. Consequently, the stable and reproducible SERS substrates are chosen for time-dependent examinations [3].

Considerably, SHINs are used to determine the concentration of various chemical contaminants or biological objects. The SHINERS spectroscopy was initially used to detect contamination of methyl parathion, pesticides thiram, and illegal food additives like melamine [34]. When SHINERS nano resonators are deposited on various cells, normal cancer cells and cancer cells which are pathologically changed can be differentiated based on the measured SHINERS spectra [35, 36]. Kast *et al.* (2008) also demonstrated that pathologically changed tissues show slightly different SHINERS spectra, making the identification of various types of cancer possible [37].

Due to the non-toxicity and high biocompatibility of SHINERS, these are reported to be applied for detection of biomolecules *in vivo*. For instance, $Au@SiO_2$ nanoparticles have been used to study the silica nanobio-interaction inside eukaryotic cells *in situ* [38]. $Ag@SiO_2$ nanoparticles could selectively enhance Raman signal of Cyt C, while retaining important properties of this protein by virtue of very weak interaction with the target molecule [38]. Besides, the secondary structures of proteins could be investigated by SHINERS method [39]. Overall, SHINERS spectroscopy is very promising tool for non-invasive optical analysis of bio-molecular processes.

Continual advancements in Raman instrumentation, together with other techniques such as imaging devices, microfluidic platforms, pump–probe methods, FT-IR spectroscopy, and fluorescence spectroscopy are set to remove the barriers in adoption of SERS and consequently in delivering a variety of information with ultrahigh sensitivity.

8.2.5. Improved Nanofabrication

Numerous nanotechnology-related approaches have been explored for obtaining reliable, scalable, stable, and cost-effective SERS substrate. Nanofabrication technique for developing on-chip SERS substrate can be broadly categorized into three main methods, viz., lithography-based methods, non-lithography template methods, and direct formation. Besides, SERSs substrates based on colloidal nanoparticles also offer exciting and improved nanofabrication [40].

Generally used active substrates used are noble metal colloids, roughened noble metal surfaces, nanosphere self-assembly, template-directed deposition, core-shell nanomaterial, and so on [41 - 45]. All SERS active substrates have prominent benefits and many of them have certain limitations. It is recommended to select an optimal active substrate for SERS measurement based on the requirement of test environment [46].

Several developments in colloid-based substrates including the preparation of multicomponent nanoparticles are reported, such as Au-coated ZnO nanorods, Ag nanoparticle coated amino-modified polystyrene microspheres, Au-core shell silica nanoparticles ($Au@SiO_2$ nanoparticles), silver-coated gold nanoparticles (Au@AgNPs) (β-Cyclodextrin coated $SiO_2@Au@Ag$ core/shell nanoparticles, porous Au–Ag alloy nanoparticles, and Rhodamine derivatives grafted Au@Ag core–shell nanocubes. Nanoparticles with very high SERS activity and uniform surface morphology have been reported [47 - 52].

Li and Zhang employed the high efficiency and accuracy of the microfluidic system to prepare a low-cost and feasible SERS active substrate by depositing nanoparticles on a paper substrate and successfully detecting Rhodamine 6G at a low concentration [53, 54]. According to characteristics of substrates, there are mainly two ways: label method and label-free method, as shown in Fig. (**3**).

8.2.5.1. Label Method

Gold nanoparticles were modified with the help of Rhodamine 6G for obtaining the specific active SERS substrates [54].Multiple drug-resistant Salmonella DT104 was detected by SERS method for the first time. The TEM image of Salmonella was acquired when the active substrate was characterized by electron microscopy. The results showed better comparison of photo thermal response of the hybrid nanomaterial compared to single walled carbon nanotubes and gold nanoparticles. This approach is expected to facilitate great potential of SERS for rapid detection and photo-thermal therapy of clinical samples [46]. Penn *et al.* (2013) developed the antibody-modified membrane which was used for SERS immunoassay of nanoparticle. Here, a thin nanometric layer of gold was deposited on a polycarbonate track etched film [55] and further the relevant antibodies were immobilized on it by coupling chemistry serving as a capture substrate. The target and Raman reporter were transported to the capture substrate designing an immunoassay. In this study, a SERS based immunoassay assay conducted on a membrane filter that implements flow was applied for first time to enhance Raman signal and significantly shorten the time of experiment [46].

8.2.5.2. Label-free Method

In the label-free SERS method, there is no special requirement to add into active substrate, such as dye molecules, during the detection of targets by SERS because of being a direct detection method. Therefore, Raman's signal remains intact and undisturbed by surrounding entities providing clear, specific and accurate signals. In addition, this kind of detection method is widely used because of its simple, rapid, low cost, and rich information of molecules detected by SERS [46]. In label-free method, the metallic colloid (such as Ag and Au) nanomaterials are predominantly used and as mostly prepared by chemical reduction method. This method mainly relies on the interaction between the bacterial and the SERS active substrate in solution, thus, obtaining the Raman signal of molecules. This method is inexpensive, easy to prepare and is operated on the material from the molecular level. Consequently, relatively uniform nanoparticles are obtained. Besides, the reaction in the solution leads to the better control for desired shape and size of nanoparticles.

In 2013, Cowcher *et al.* reported the rapid detection of Bacillus and other pathogens using a mixture of Ag nanoparticles and bacterial liquid [57].SERS is generally used for detecting the presence of Bacillus in food by virtue of its fast analysis speed and high sensitivity. It was demonstrated that SERS can more rapidly and efficiently detect DPA biomarker, a marker of bacillus in vivo, compared with the microscopy. SERS based on label-free method is advantageous because of being convenient, readable, and cheap. However, this method is limited due to the requirement of the extraction of the DPA from the spores. As another example of label-free method, Mungroo *et al.* (2016) developed a microfluidic platform for detecting pathogenic bacteria by SERS utilizing silver nanoparticle active substrates [58]. They identified eight common species of foodborne pathogens effectively and rapidly by combining SERS with stoichiometry and linear discriminant analysis. Each species of pathogen was assigned with a series of peak thus establishing a unique identification of each species [46].

There are some shortcomings of above discussed methods based on SERS effect form single metal nanoparticles. For example, the poor stability and homogeneity of silver and the poor enhancement effect of single gold nanoparticles [46]. Furthermore, solid substrates and metal colloids are susceptible to aggregation, contamination, and degradation when they are exposed to ambient environments and EM radiation. Recently, Ren *et al.* utilized the core–shell nanoparticles with embedded internal standards. The embedded internal standard yielded effective feedback to reduce as well as and to correct any fluctuations in the samples and measurement conditions. This eventually enabled label-free and reliable quantitative analysis by SERS [3].

Metal nanocomposites are used to overcome above disadvantages and to utilize their advantages for the measurement of target by SERS. For example, many nanolayer shells on Ag cores (such as Ag/Au, Ag/carbon, and Ag/SiO2) were fabricated for improving the shelf-life of silver nanostructures. As another example of improved nanofabrication method, Khlebtsov *et al.* (2015) designed uniform silver-gold core/shell cuboids and dumbbells with controllable Ag shells of 1–25 nm in thickness [59].

Another nanofabrication method based on plasmonic nanohole sensor was reported by Kee *et al.* (2013) [60]. They applied this sensor to monitor the bacterial growth and antibiotic sensitivity quickly, efficiently, and quantitatively [60]. The plasmonic nanohole arrays were fabricated by a mask-based deep ultraviolet lithography method and refractive index sensitivity and surface mass sensitivity were measured. Kee *et al* tested the effect of the plasmonic nanohole sensor with *E. coli*. Analysis demonstrated that the *E. coli* was specifically captured and rapidly grew on its surface. Additionally, the sensor was able to detect antibiotic sensitivity rapidly within 2 h [60]. The silver nanorod array SERS substrates were prepared by Wu *et al* (2013) [62]. These nanorod arrays were applied to detect the pathogenic bacteria in foods including bean sprouts, S*pinacia Oleracea*, and romaine lettuce [61].

One of the important nano-spectroscopic techniques for the molecular-level and nanoscale analysis of electrochemical systems is TERS. Raschke *et al.* developed gold tips coupled with a grating and demonstrated plasmonic nano focusing upon localized excitation of a sample within a range of B20 nm for enhanced sensitivity of TERS [63]. In the case of SHINERS, the formation of pinhole-free SHINs is challenging, therefore, large-scale synthesis of pinhole-free SHINs is done through microfabrication techniques or top-down approaches [3]. Significantly, a few other strategies such as microfluidic networks, core–shell nanoparticles (*e.g.*, SHINs), reproducible lithographic nanostructures, and monodisperse metal nanoparticles with internal standards have the potential to provide reproducible quantitative results and prevent the non-specific adsorption of molecules from the solution to enable the detection by SERS of surface molecules [3].

Overall, two points need to be emphasized for the practical applications of SERS as follow:

1. Considering the size of the laser spot used for sampling is around 100 mm in commercial portable or handheld Raman instruments, the development of a novel and robust strategy for constructing SERS substrates with uniformity on the mm scale is focused, and
2. To control costs to extend the SERS technique from the laboratory to industry and finally into society, goal is the replacement of Au- or Ag-based SERS-active substrates using other cheaper plasmonic materials [3].

8.2.6. Molecular Generality or Selectivity/Specificity Based Improvement Strategies

There are requirements for SERS to have extremely high sensitivity for detecting a range of target molecules such as molecules with small cross-sections and weakly adsorbed species with significant sensitivity. This aids in gaining deeper insights in fundamental and practical research using SERS.

One of the key long-standing challenge for SERS/TERS/ SHINERS is the detection of weakly adsorbed molecules/ions with low surface coverage or at trace levels. Consequently, it is important to enhance the selective sensitivity for detecting the weakly adsorbed species. Specific capture, derivatization, and labelling strategies have been employed to the direct or indirect measurement of target molecules by acquiring SERS signals. Researchers have reported the functionalization of the surfaces of SERS active nanostructures to specifically recognize target molecules. For investigating molecules with small Raman cross-sections, it is imperative to further increase the signal-to-noise ratio by using signal averaging techniques or by increasing the enhancement factor as far as possible by designing highly plasmonic-active nanostructures.

Sometimes, the real application of SERS is challenging owing to high amount of interferences coming from complicated matrix of the sample. Therefore, selective/specific detection is crucial for the real application of SERS technique. In general, below methods (Five selective/specific detection techniques namely chemical reaction, antibody, aptamer, molecularly imprinted polymers, and microfluidics) can be applied for the rapid and reliable selective/specific detection when coupled with SERS technique [64].

8.2.6.1. Reaction-SERS Method

The analytes which are difficult to be analysed by SERS technology are modified using chemical reactions before SERS detection. Few examples of such analytes include the analytes possessing little affinity to the SERS substrate or having small Raman cross sections, unstable analytes, and gaseous analytes due to their volatility and low molecular weight resulting in poor sensitivity, selectivity, and accuracy [64].

The chemical reactions-SERS can be divided into three categories viz., improving the affinity of analyte with SERS substrate, increasing the Raman scattering cross-sectional area, and reducing the Raman scattering cross-sectional area. Overall, there are some rules which need to be followed while conducting chemical reaction-SERS. Firstly, the compound should have a weak fluorescent background in order not to interfere SERS measurement. Simultaneously, it should have groups like –SH or –NH2 which can firmly conjugate with the SERS substrate. Thirdly, reagents should possess obvious spectra variations before and after chemical reactions. The spectra variations can be classified into three cases, including "signal on", "signal off" and "signal change" in which both reactants and products have strong SERS signals. Finally, the reaction reagent should have a good selectivity to the analyte which can be analysed by the fingerprint spectrum [64].

8.2.6.2. Antibody-SERS Method

Utilizing the specificity of Ab-Ag interactions and the sensitivity of SERS, antibody-SERS method is used for the selective identification of analytes in various multi-component systems, especially in physiological environments [61 - 74].

8.2.6.3. Aptamer-SERS Method

Aptamers are the oligonucleotide molecules that can bind to specific target by folding into specific three-dimensional structures. The strong aptamer-analyte interaction enhances the specific recognition and enrichment of analyte from mixture samples. Compared with antibodies, aptamers can detect broader range of analyte, have high binding affinities for specific targets, tolerance to internal labelling, good stability, employ mild reaction condition, and cost-effective synthesis. Furthermore, these are quite suitable for small molecules recognition by 3D DNA/RNA strand folding after aptamer-analyte interaction [75 - 77].

Aptamers are known for cost-effective synthesis based on production in a large quantity with high reproducibility and purity [76]. In SERS-based aptasensor, these molecules electrostatically or covalently interact with Ag or Au nanomaterials of diverse sizes and morphologies (particles, stars, rods or another irregular Ag/Au surface [78 - 84]. They bind to different targets, such as ions, small molecules, proteins, nucleic acids, virus, cells and bacteria, tissues, and organisms. Bound analytes are detected directly when the analytes can provide distinct Raman signals of their own [85].

8.2.6.4. MIPs-SERS

The concept of molecularly imprinted polymers (MIPs) taken to be artificial replica of biological Ab-AG system was proposed by Wulff and Sarhan [86]. Molecular recognition site can be created by imprinting template molecules in a polymer matrix through the non-covalent/covalent interactions between templates and functional monomers. Removing the templates after the polymerization creates the complementary shape, size, and functionality of templates which serve as recognition sites for designing the sensing set-up [87]. These recognition sites selectively and specifically rebind with the same template with high affinity, and resemble antibody-antigen systems. Hence, they are termed as "artificial" antibody. The system has the advantages of cavity pre-determination and specific recognition and have higher stability than natural antibody [88].

The selectivity of SERS technique can be improved by combining SERS with MIPs. For instance, silver nanoparticles as SERS substrate were doped in the MIPs and can be applied in the sensitive and specific SERS detection [89]. Combination of the specific selectivity of MIPs and the high sensitivity of SERS is reported to be a very potential tool for the detection of specific analyte in complex surroundings.

8.2.6.5. Microfluidics-SERS

Microfluidic device platform has the possibility of achieving the miniaturization, integration, and automation of analysis processes. Various operations used in any assay format - creating a homo/hetero mixture, separation, reaction and detection, are all integrated on a microfluidic platform as a single device.

Microfluidic systems can couple with different detectors for accurate and reliable detection, characterization, and diagnosis. Combining SERS and microfluidic techniques (Microfluidics-SERS) has the potential to investigate and detect low quantities of biological sample in a rapid, sensitive, reproducible, and high-throughput mode [65, 90] and can be of greater use in real applications in high-throughput screening, quality control, direct environmental monitoring and point-of-care testing.

8.2.7. Collaborative Experimental and Theoretical Approaches

SERS is a multidisciplinary phenomenon because this is associated with three different branches of science, namely nanoplasmonics (nanostructures and plasmons), molecular spectroscopy (photons and molecules) and surface science (molecules and nanostructures). The analysis of SERS data becomes complicated due to the difficulty of interpreting these three interactions in a unified manner. It is important to be cautious about false results and incorrect conclusions. Collaborations of experimentalists with theoreticians will facilitate detailed information about the mechanisms of enhancement, as well as detection strategies [3].

The computational methods can simulate the effect of EM radiation on the geometry of metal nanostructures and for the design of optimal geometries for increasing the overall enhancement and sensitivity of a SERS method for conducting fundamental research. The development of new theories and simulations is also essential for the fabrication of novel nanostructures for SERS/SHINERS/TERS, which will enrich the fundamental understanding and facilitate practical research [3]. The correlation spectroscopy and theoretical methods are important tools for commercial research and application. Using normal spectral analysis, it is not possible to elucidate information about the constituents and intermediate species in complex target systems.

Researchers have utilized theoretical methods such as 2D correlation spectroscopy, density functional theory (DFT), and principal component analysis to evaluate such complex target systems. Adsorption energy of electrochemical species for knowing the entire reaction mechanism are carried out through DFT calculations. Moreover, DFT simulations are performed to understand the Raman spectra of unidentified chemical species which are physically or chemically absorbed on the surfaces of SERS-active substrates [3]. The complex systems can be studies using 2D or 3D correlation spectroscopy by resolving overlapping bands and providing information about sequences of changes in peaks.

SERS is used as an *in situ* spectroscopic technique for the identification of surface species and even surface-active morphologies under experimental conditions. Thus, comprehensive simulations of the Raman spectra of surface species with their possible structures and configurations are required to support the identification of processes and cross correlate SERS measurements with those from other methods. To identify surface species, the experimental conditions should be included in simulations of Raman spectra in terms of the frequency and intensity. In the oxidation of single crystal Pt electrodes, reduction of water in aqueous electrolytes, reduction of benzyl chloride on Ag electrodes in non-aqueous electrolytes, and adsorption of small molecules and bipyridine in ambient TERS, a charged surface or dipole field is used to simulate the electrode with an electrochemical potential.

Additionally, the simulation of SERS spectra of surface species under experimental conditions, the structures of surface species or surface-active sites identified by simulations, and SERS measurements need to be consistent with other experimental measurements. Since SERS signals are taken from a small area of a conventional sample, care should be taken in the representation of SERS. Simulations are employed for gathering information about energy changes or electron transfer caused by the alteration of surface active species. The simulation data predicted by models on various aspects of energy, provide confident data and information on conclusions drawn from SERS measurements [3].

8.3. DEVELOPMENT IN INSTRUMENTATION

High-resolution spectra are desired during Raman measurements while improving the sensitivity of Raman instrumentation. For achieving high throughput, the Raman instruments should have a limited number of dispersive or interfering elements. Earlier, conventional Raman instruments used double or triple monochromators to filter out elastically scattered radiation. Currently, modern Raman spectrometers employ notch filters and single-grating systems with improved signal throughput. In next-gen Raman instruments, the detector configuration needs modification for maximizing the efficiency of signal collection and transmission in achieving sensitive detection with an advanced detector. For example, Bao *et al.* rationally developed a micro-spectrometer with a 2D absorptive filter array composed of colloidal quantum dots [3].

Generally, current spectrometers employ interference filters and optics significantly affecting the photon efficiency and spectral resolution. Interestingly, the colloidal quantum dots-based spectrometer utilized the wavelength multiplexing principle for analysing the light from a source with a broad angular distribution while maintaining the spectral resolution. A portable Raman spectrometer with liquid crystal tunable filters (LCTFs) as dispersive elements was designed by Sakamoto *et al.* These LCTFs can have working apertures with sizes of up to 35 mm that utilize electronically controlled liquid crystal elements allowing the passage of selected wavelengths of light and exclude others. A lightweight Raman microscope based on time-correlated photon-counting detection for field and space applications was developed by Meng *et al.* [91].

Furthermore, Huo *et al.* demonstrated surface plasmon-coupled directionally enhanced Raman scattering for improving the signal collection efficiency of Raman instruments [92]. Thus, Raman spectrometers can be designed with a minimal number of dispersive elements or objectives with higher numerical apertures to maximize signal throughput during data collection. As an advanced approach, miniaturization decreases the energetic resolution due to the reductions in optical alignment and sensitivity. Furthermore, there is also a worse signal-to-noise ratio, which mainly originates from temperature fluctuations in CCD detection.

Generally, the acquisition of signals from a target is the basic step for any instrument. Thus, priority should be given to sensitivity over resolution for either fundamental or practical research. Most of the current portable Raman instruments are based on approach of detecting the target molecules while compromising about resolution. Notably, the optical alignment of most current portable or handheld Raman instruments is inherited from the traditional design. Thus, novel optical designs may be proposed and obtained by new manufacturing techniques, including micro-electro-mechanical systems (MEMS) and nanofabrication. Rapid developments in instrumentation technology are expected to improve Raman instruments and enable the introduction of new Raman instruments with fewer dispersive elements for ultrasensitive measurements.

8.4. HANDHELD PORTABLE DEVICES

The use of handheld Raman spectroscopy equipment is getting established for routine analysis in the field for evidence collection. However, their use is still restricted to measurements of pure or high concentration samples. 3D Finite Element Method (FEM) and DFT simulations were reported to be used for interpreting the high SERS enhancement of the Ag nanopillar substrate and the in-field detection of the substances, respectively [92].

Over the last few decades, research has redefined the low-level measurement of disease biomarkers. The translation of these capabilities from the formal clinical setting to point-of-need (PON) usage is furthermore limited [91].

Owens *et al* [93] presented the results of experiments designed to examine the potential utility of a handheld Raman spectrometer as a PON electronic reader for a sandwich immunoassay based on SERS. This study used a recently developed procedure for the SERS detection of phospho-myoinositol-capped lipo-arabinomannan (PILAM) for comparing the performance of laboratory-grade and handheld instrumentation and, therefore, gauge the utility of the handheld instrument for PON deployment [92]. Phospho-myo-inositol-capped lipo-arabinomannan is a non-pathogenic stimulant for mannose-capped lipo-arabinomannan (ManLAM), which is an antigenic marker found in serum and other body fluids of individuals infected with tuberculosis. The results obtained with the field-portable spectrometer were then compared to those obtained from more sensitive benchtop Raman spectrometer. The results, albeit for the two spectrometers were promising in that the limit of detection found for PILAM spiked in human serum when using the handheld system (0.18 ng/mL) approached that of the bench top instrument (0.032 ng/mL) [93].

CONCLUSION

In summary, the use of SERS is established in academia and is gaining wider application for commercial purpose. Over the last decades, several improvements in SERS are seen. Many choices based on SERS are available for having increased sensitivity for target, especially in complex environment. For ensuring best results using SERS, few considerations are recommended with respect to preparation of biological sample, selection of SERS probe, selection of analytical instrument and data interpretation. Overall, approach based on combination of SERS based techniques and collaboration of experimentation and theoretical Raman instruments are warranted to be portable while providing high quality Raman signals in routine use.

CONSENT FOR PUBLICATION

Not applicable.

CONFLICT OF INTEREST

The authors declare no conflict of interest, financial or otherwise.

ACKNOWLEDGEMENTS

Declared none.

REFERENCES

[1]   Nie S, Emory SR. Probing single molecules and single nanoparticles by surface-enhanced Raman scattering. Science 1997; 275(5303): 1102-6.
[http://dx.doi.org/10.1126/science.275.5303.1102] [PMID: 9027306]

[2]   Kneipp K, Wang Y, Kneipp H, *et al.* Single molecule detection using surface-enhanced raman scattering (SERS). Phys Rev Lett 1997; 78: 1667-70.
[http://dx.doi.org/10.1103/PhysRevLett.78.1667]

[3]   Panneerselvam R, Liu GK, Wang YH, *et al.* Surface-enhanced Raman spectroscopy: bottlenecks and future directions. Chem Commun (Camb) 2017; 54(1): 10-25.
[http://dx.doi.org/10.1039/C7CC05979E] [PMID: 29139483]

[4]   Jamieson LE, Asiala SM, Gracie K, Faulds K, Graham D. Bioanalytical measurements enabled by Surface-Enhanced Raman Scattering (SERS) probes. Annu Rev Anal Chem (Palo Alto, Calif) 2017; 10(1): 415-37.
[http://dx.doi.org/10.1146/annurev-anchem-071015-041557] [PMID: 28301754]

[5]   https://www.marketwatch.com/press-release/at-91-of-cagr-surface-enhanced-raman-spectros-opy-sers-market-share-will-increase-and-aimed-to-cross-usd-190-million-in-2024-2020-06-15

[6]   Verma P. Tip-Enhanced Raman Spectroscopy: Technique and Recent Advances. Chem Rev 2017; 117(9): 6447-66.
[http://dx.doi.org/10.1021/acs.chemrev.6b00821] [PMID: 28459149]

[7]   Schmid T, Opilik L, Blum C, Zenobi R. Nanoscale chemical imaging using tip-enhanced Raman spectroscopy: a critical review. Angew Chem Int Ed Engl 2013; 52(23): 5940-54.
[http://dx.doi.org/10.1002/anie.201203849] [PMID: 23610002]

[8]   Bailo E, Deckert V. Tip-enhanced Raman spectroscopy of single RNA strands: towards a novel direct-sequencing method. Angew Chem Int Ed Engl 2008; 47(9): 1658-61.
[http://dx.doi.org/10.1002/anie.200704054] [PMID: 18188855]

[9]   Lin X-M, Deckert-Gaudig T, Singh P , *et al.* Direct Base-to-Base Transitions in ssDNA Revealed by Tip-Enhanced Raman Scattering. https://arxiv.org/abs/1604.065982016.

[10]   Zhang D, Domke KF, Pettinger B. Tip-enhanced Raman spectroscopic studies of the hydrogen bonding between adenine and thymine adsorbed on Au (111). ChemPhysChem 2010; 11(8): 1662-5.
[http://dx.doi.org/10.1002/cphc.200900883] [PMID: 20235109]

[11]   Kurouski D, Deckert-Gaudig T, Deckert V, Lednev IK. Structure and composition of insulin fibril surfaces probed by TERS. J Am Chem Soc 2012; 134(32): 13323-9.
[http://dx.doi.org/10.1021/ja303263y] [PMID: 22813355]

[12]   Kurouski D, Deckert-Gaudig T, Deckert V, Lednev IK. Surface characterization of insulin protofilaments and fibril polymorphs using tip-enhanced Raman spectroscopy (TERS). Biophys J

2014; 106(1): 263-71.
[http://dx.doi.org/10.1016/j.bpj.2013.10.040] [PMID: 24411258]

[13] Meyer R, Yao X, Deckert V. Latest instrumental developments and bioanalytical applications in tip-enhanced Raman spectroscopy. Trends Analyt Chem 2018; 102: 250-8.
[http://dx.doi.org/10.1016/j.trac.2018.02.012]

[14] vandenAkker CC, Deckert-Gaudig T, Schleeger M, *et al.* Nanoscale Heterogeneity of the Molecular Structure of Individual hIAPP Amyloid Fibrils Revealed with Tip-Enhanced Raman Spectroscopy. Small 2015; 11(33): 4131-9.
[http://dx.doi.org/10.1002/smll.201500562] [PMID: 25952953]

[15] Nakata A, Nomoto T, Toyota T, Fujinami M. Tip-enhanced Raman spectroscopy of lipid bilayers in water with an alumina- and silver-coated tungsten tip. Anal Sci 2013; 29(9): 865-9.
[http://dx.doi.org/10.2116/analsci.29.865] [PMID: 24025569]

[16] Lin XD, Li JF, Huang YF, *et al.* Shell-isolated nanoparticle-enhanced Raman spectroscopy: Nanoparticle synthesis, characterization and applications in electrochemistry. J Electroanal Chem (Lausanne) 2013; 688: 5-11.
[http://dx.doi.org/10.1016/j.jelechem.2012.07.017]

[17] Galloway TA, Cabo-Fernandez L, Aldous IM, Braga F, Hardwick LJ. Shell isolated nanoparticles for enhanced Raman spectroscopy studies in lithium-oxygen cells. Faraday Discuss 2017; 205: 469-90.
[http://dx.doi.org/10.1039/C7FD00151G] [PMID: 28913534]

[18] Fang PP, Lu X, Liu H, *et al.* Applications of shell-isolated nanoparticles in surface-enhanced Raman spectroscopy and fluorescence. TrAC -. Trends Analyt Chem 2015; 66: 103-17.
[http://dx.doi.org/10.1016/j.trac.2014.11.015]

[19] Li JF, Zhang YJ, Ding SY, Panneerselvam R, Tian ZQ. Core-shell nanoparticle-enhanced raman spectroscopy. Chem Rev 2017; 117(7): 5002-69.
[http://dx.doi.org/10.1021/acs.chemrev.6b00596] [PMID: 28271881]

[20] Zheng C, Liang L, Xu S, *et al.* The use of Au@SiO2 shell-isolated nanoparticle-enhanced Raman spectroscopy for human breast cancer detection. Anal Bioanal Chem 2014; 406(22): 5425-32.
[http://dx.doi.org/10.1007/s00216-014-7967-5] [PMID: 24958347]

[21] Li JF, Tian XD, Li SB, *et al.* Surface analysis using shell-isolated nanoparticle-enhanced Raman spectroscopy. Nat Protoc 2013; 8(1): 52-65.
[http://dx.doi.org/10.1038/nprot.2012.141] [PMID: 23237829]

[22] Gao L, Zhao H, Li T, Huo P, Chen D, Liu B. Atomic force microscopy based tip-enhanced Raman spectroscopy in biology. Int J Mol Sci 2018; 19(4): 1193.
[http://dx.doi.org/10.3390/ijms19041193] [PMID: 29652860]

[23] Yeo BS, Stadler J, Schmid T, *et al.* Tip-enhanced Raman Spectroscopy - Its status, challenges and future directions. Chem Phys Lett 2009; 472: 1-13.
[http://dx.doi.org/10.1016/j.cplett.2009.02.023]

[24] Ichimura T, Hayazawa N, Hashimoto M, Inouye Y, Kawata S. Tip-enhanced coherent anti-stokes Raman scattering for vibrational nanoimaging. Phys Rev Lett 2004; 92(22): 220801.
[http://dx.doi.org/10.1103/PhysRevLett.92.220801] [PMID: 15245207]

[25] Rasmussen A, Deckert V. Surface- and tip-enhanced Raman scattering of DNA components. In: Journal of Raman Spectroscopy. 2006; pp. 311-7.

[26] Bailo E, Deckert V. Tip-enhanced Raman scattering. Chem Soc Rev 2008; 37(5): 921-30.
[http://dx.doi.org/10.1039/b705967c] [PMID: 18443677]

[27] Treffer R, Böhme R, Deckert-Gaudig T, *et al.* Advances in TERS (tip-enhanced Raman scattering) for biochemical applications. Biochem Soc Trans 2012; 40(4): 609-14.
[http://dx.doi.org/10.1042/BST20120033] [PMID: 22817703]

[28]   Yeo BS, Mädler S, Schmid T, *et al.* Tip-enhanced Raman spectroscopy can see more: The case of cytochrome c. J Phys Chem C 2008; 112: 4867-73.
[http://dx.doi.org/10.1021/jp709799m]

[29]   Neugebauer U, Rösch P, Schmitt M, *et al.* On the way to nanometer-sized information of the bacterial surface by tip-enhanced Raman spectroscopy. ChemPhysChem 2006; 7(7): 1428-30.
[http://dx.doi.org/10.1002/cphc.200600173] [PMID: 16789043]

[30]   Ding SY, Tian ZQ. A breakthrough in the chemical imaging of single molecule: Sub-nm tip-enhanced Raman spectroscopy. Natl Sci Rev 2014; 1: 4-5.
[http://dx.doi.org/10.1093/nsr/nwt013]

[31]   Frontiera RR, Henry AI, Gruenke NL, Van Duyne RP. Surface-enhanced femtosecond stimulated Raman spectroscopy. J Phys Chem Lett 2011; 2(10): 1199-203.
[http://dx.doi.org/10.1021/jz200498z] [PMID: 26295326]

[32]   Pozzi EA, Sonntag MD, Jiang N, *et al.* Ultrahigh vacuum tip-enhanced raman spectroscopy with picosecond excitation. J Phys Chem Lett 2014; 5(15): 2657-61.
[http://dx.doi.org/10.1021/jz501239z] [PMID: 26277959]

[33]   Krajczewski J, Kudelski A. Shell-isolated nanoparticle-enhanced Raman spectroscopy. Front Chem 2019; 7: 410. Epub ahead of print.
[http://dx.doi.org/10.3389/fchem.2019.00410] [PMID: 31214580]

[34]   Liang L, Zheng C, Zhang H, *et al.* Exploring type II microcalcifications in benign and premalignant breast lesions by shell-isolated nanoparticle-enhanced Raman spectroscopy (SHINERS). Spectrochim Acta A Mol Biomol Spectrosc 2014; 132: 397-402.
[http://dx.doi.org/10.1016/j.saa.2014.04.147] [PMID: 24887501]

[35]   Yang JL, Yang ZW, Zhang YJ, *et al.* Quantitative detection using two-dimension shell-isolated nanoparticle film. J Raman Spectrosc 2017; 48: 919-24.
[http://dx.doi.org/10.1002/jrs.5151]

[36]   Sivanesan A, Kozuch J, Ly HK, *et al.* Tailored silica coated Ag nanoparticles for non-invasive surface enhanced Raman spectroscopy of biomolecular targets. RSC Advances 2012; 2: 805-8.
[http://dx.doi.org/10.1039/C1RA00781E]

[37]   Kast RE, Serhatkulu GK, Cao A, *et al.* Raman spectroscopy can differentiate malignant tumors from normal breast tissue and detect early neoplastic changes in a mouse model. Biopolymers 2008; 89(3): 235-41.
[http://dx.doi.org/10.1002/bip.20899] [PMID: 18041066]

[38]   Wang Y, Chen L, Liu P. Biocompatible triplex $Ag@SiO_2@mTiO_2$ core-shell nanoparticles for simultaneous fluorescence-SERS bimodal imaging and drug delivery. Chemistry 2012; 18(19): 5935-43.
[http://dx.doi.org/10.1002/chem.201103571] [PMID: 22461327]

[39]   Drescher D, Zeise I, Traub H, *et al. in situ* characterization of SiO2 nanoparticle biointeractions using BrightSilica. Adv Funct Mater 2014; 24: 3765-75.
[http://dx.doi.org/10.1002/adfm.201304126]

[40]   Sun X, Li HA. Review: Nanofabrication of Surface-Enhanced Raman Spectroscopy (SERS) Substrates. Curr Nanosci 2015; 12: 175-83.
[http://dx.doi.org/10.2174/1573413711666150523001519]

[41]   Rycenga M, Camargo PHC, Li W, Moran CH, Xia Y. Understanding the SERS effects of single silver nanoparticles and their dimers, one at a time. J Phys Chem Lett 2010; 1(4): 696-703.
[http://dx.doi.org/10.1021/jz900286a] [PMID: 20368749]

[42]   Moskovits M. Surface roughness and the enhanced intensity of Raman scattering by molecules adsorbed on metals. J Chem Phys 1978; 69: 4159-61.
[http://dx.doi.org/10.1063/1.437095]

[43] Haynes CL, Van Duyne RP. Nanosphere lithography: A versatile nanofabrication tool for studies of size-dependent nanoparticle optics. J Phys Chem B 2001; 105: 5599-611.
[http://dx.doi.org/10.1021/jp010657m]

[44] Ko H, Tsukruk VV. Nanoparticle-decorated nanocanals for surface-enhanced Raman scattering. Small 2008; 4(11): 1980-4.
[http://dx.doi.org/10.1002/smll.200800301] [PMID: 18924130]

[45] Shen J, Zhu Y, Yang X, Zong J, Li C. Multifunctional Fe3O4@Ag/SiO2/Au core-shell microspheres as a novel SERS-activity label via long-range plasmon coupling. Langmuir 2013; 29(2): 690-5.
[http://dx.doi.org/10.1021/la304048v] [PMID: 23206276]

[46] Zhao X, Li M, Xu Z. Detection of foodborne pathogens by surface enhanced Raman spectroscopy. Front Microbiol 2018; 9: 1236. Epub ahead of print.
[http://dx.doi.org/10.3389/fmicb.2018.01236] [PMID: 29946307]

[47] Sinha G, Depero LE, Alessandri I. Recyclable SERS substrates based on Au-coated ZnO nanorods. ACS Appl Mater Interfaces 2011; 3(7): 2557-63.
[http://dx.doi.org/10.1021/am200396n] [PMID: 21634790]

[48] Li B, Zhang W, Chen L, Lin B. A fast and low-cost spray method for prototyping and depositing surface-enhanced Raman scattering arrays on microfluidic paper based device. Electrophoresis 2013; 34(15): 2162-8.
[http://dx.doi.org/10.1002/elps.201300138] [PMID: 23712933]

[49] Li H, Chen Q, Hassan MM, *et al.* AuNS@Ag core-shell nanocubes grafted with rhodamine for concurrent metal-enhanced fluorescence and surfaced enhanced Raman determination of mercury ions. Anal Chim Acta 2018; 1018: 94-103.
[http://dx.doi.org/10.1016/j.aca.2018.01.050] [PMID: 29605140]

[50] Lu Y, Yao G, Sun K, Huang Q. β-Cyclodextrin coated SiO_2@Au@Ag core-shell nanoparticles for SERS detection of PCBs. Phys Chem Chem Phys 2015; 17(33): 21149-57.
[http://dx.doi.org/10.1039/C4CP04904G] [PMID: 25478906]

[51] Murshid N, Gourevich I, Coombs N, *et al.* Gold plating of silver nanoparticles for superior stability and preserved plasmonic and sensing properties. Epub ahead of print 2013.
[http://dx.doi.org/10.1039/c3cc46075d]

[52] Quyen TTB, Su WN, Chen KJ, *et al.* Au@SiO2 core/shell nanoparticle assemblage used for highly sensitive SERS-based determination of glucose and uric acid. J Raman Spectrosc 2013; 44: 1671-7.
[http://dx.doi.org/10.1002/jrs.4400]

[53] Zhao Y, Luo W, Kanda P, *et al.* Silver deposited polystyrene (PS) microspheres for surface-enhanced Raman spectroscopic-encoding and rapid label-free detection of melamine in milk powder. Talanta 2013; 113: 7-13.
[http://dx.doi.org/10.1016/j.talanta.2013.03.075] [PMID: 23708616]

[54] Lin Y, Hamme Ii AT. Targeted highly sensitive detection/eradication of multi-drug resistant Salmonella DT104 through gold nanoparticle-SWCNT bioconjugated nanohybrids. Analyst (Lond) 2014; 139(15): 3702-5.
[http://dx.doi.org/10.1039/C4AN00744A] [PMID: 24897935]

[55] Zhang W, Li B, Chen L, *et al.* Brushing, a simple way to fabricate SERS active paper substrates. Anal Methods 2014; 6: 2066-71.
[http://dx.doi.org/10.1039/C4AY00046C]

[56] Penn MA, Drake DM, Driskell JD. Accelerated surface-enhanced Raman spectroscopy (SERS)-based immunoassay on a gold-plated membrane. Anal Chem 2013; 85(18): 8609-17.
[http://dx.doi.org/10.1021/ac402101r] [PMID: 23972208]

[57] Cowcher DP, Xu Y, Goodacre R. Portable, quantitative detection of Bacillus bacterial spores using surface-enhanced Raman scattering. Anal Chem 2013; 85(6): 3297-302.

[http://dx.doi.org/10.1021/ac303657k] [PMID: 23409961]

[58]   Mungroo NA, Oliveira G, Neethirajan S. SERS based point-of-care detection of food-borne pathogens. Mikrochim Acta 2016; 183: 697-707.
[http://dx.doi.org/10.1007/s00604-015-1698-y]

[59]   Khlebtsov BN, Liu Z, Ye J, *et al.* Au@Ag core/shell cuboids and dumbbells: Optical properties and SERS response. J Quant Spectrosc Radiat Transf 2015; 167: 64-75.
[http://dx.doi.org/10.1016/j.jqsrt.2015.07.024]

[60]   Kee JS, Lim SY, Perera AP, *et al.* Plasmonic nanohole arrays for monitoring growth of bacteria and antibiotic susceptibility test. Sens Actuators B Chem 2013; 182: 576-83.
[http://dx.doi.org/10.1016/j.snb.2013.03.053]

[61]   Wu X, Xu C, Tripp RA, Huang YW, Zhao Y. Detection and differentiation of foodborne pathogenic bacteria in mung bean sprouts using field deployable label-free SERS devices. Analyst (Lond) 2013; 138(10): 3005-12.
[http://dx.doi.org/10.1039/c3an00186e] [PMID: 23563168]

[62]   Ropers C, Neacsu CC, Elsaesser T, Albrecht M, Raschke MB, Lienau C. Grating-coupling of surface plasmons onto metallic tips: a nanoconfined light source. Nano Lett 2007; 7(9): 2784-8.
[http://dx.doi.org/10.1021/nl071340m] [PMID: 17685661]

[63]   Rohr TE, Cotton T, Fan N, Tarcha PJ. Immunoassay employing surface-enhanced Raman spectroscopy. Anal Biochem 1989; 182(2): 388-98.
[http://dx.doi.org/10.1016/0003-2697(89)90613-1] [PMID: 2610355]

[64]   Grubisha DS, Lipert RJ, Park HY, Driskell J, Porter MD. Femtomolar detection of prostate-specific antigen: an immunoassay based on surface-enhanced Raman scattering and immunogold labels. Anal Chem 2003; 75(21): 5936-43.
[http://dx.doi.org/10.1021/ac034356f] [PMID: 14588035]

[65]   Cao YC, Jin R, Nam JM, Thaxton CS, Mirkin CA. Raman dye-labeled nanoparticle probes for proteins. J Am Chem Soc 2003; 125(48): 14676-7.
[http://dx.doi.org/10.1021/ja0366235] [PMID: 14640621]

[66]   Chuong TT, Pallaoro A, Chaves CA, *et al.* Dual-reporter SERS-based biomolecular assay with reduced false-positive signals. Proc Natl Acad Sci USA 2017; 114(34): 9056-61.
[http://dx.doi.org/10.1073/pnas.1700317114] [PMID: 28784766]

[67]   Kim DH, Kim P, Song I, *et al.* Guided three-dimensional growth of functional cardiomyocytes on polyethylene glycol nanostructures. Langmuir 2006; 22(12): 5419-26.
[http://dx.doi.org/10.1021/la060283u] [PMID: 16732672]

[68]   Brody EN, Willis MC, Smith JD, Jayasena S, Zichi D, Gold L. The use of aptamers in large arrays for molecular diagnostics. Mol Diagn 1999; 4(4): 381-8.
[http://dx.doi.org/10.1016/S1084-8592(99)80014-9] [PMID: 10671648]

[69]   Chon H, Lee S, Son SW, Oh CH, Choo J. Highly sensitive immunoassay of lung cancer marker carcinoembryonic antigen using surface-enhanced Raman scattering of hollow gold nanospheres. Anal Chem 2009; 81(8): 3029-34.
[http://dx.doi.org/10.1021/ac802722c] [PMID: 19301845]

[70]   Wei C, Xu MM, Fang CW, Jin Q, Yuan YX, Yao JL. Improving the sensitivity of immunoassay based on MBA-embedded Au@SiO$_2$ nanoparticles and surface enhanced Raman spectroscopy. Spectrochim Acta A Mol Biomol Spectrosc 2017; 175: 262-8.
[http://dx.doi.org/10.1016/j.saa.2016.12.036] [PMID: 28082212]

[71]   Driskell JD, Kwarta KM, Lipert RJ, Porter MD, Neill JD, Ridpath JF. Low-level detection of viral pathogens by a surface-enhanced Raman scattering based immunoassay. Anal Chem 2005; 77(19): 6147-54.
[http://dx.doi.org/10.1021/ac0504159] [PMID: 16194072]

[72]　Li M, Cushing SK, Zhang J, *et al.* Shape-dependent surface-enhanced Raman scattering in gold-Raman probe-silica sandwiched nanoparticles for biocompatible applications. Nanotechnology 2012; 23(11): 115501. Epub ahead of print.
[http://dx.doi.org/10.1088/0957-4484/23/11/115501] [PMID: 22383452]

[73]　Xu S, Ji X, Xu W, *et al.* Immunoassay using probe-labelling immunogold nanoparticles with silver staining enhancement via surface-enhanced Raman scattering. Analyst (Lond) 2004; 129(1): 63-8.
[http://dx.doi.org/10.1039/b313094k] [PMID: 14737585]

[74]　Chon H, Lim C, Ha SM, *et al.* On-chip immunoassay using surface-enhanced Raman scattering of hollow gold nanospheres. Anal Chem 2010; 82(12): 5290-5.
[http://dx.doi.org/10.1021/ac100736t] [PMID: 20503972]

[75]　Guven B, Basaran-Akgul N, Temur E, Tamer U, Boyaci IH. SERS-based sandwich immunoassay using antibody coated magnetic nanoparticles for Escherichia coli enumeration. Analyst (Lond) 2011; 136(4): 740-8.
[http://dx.doi.org/10.1039/C0AN00473A] [PMID: 21125089]

[76]　Sun Y, Xu L, Zhang F, *et al.* A promising magnetic SERS immunosensor for sensitive detection of avian influenza virus. Biosens Bioelectron 2017; 89(Pt 2): 906-12.
[http://dx.doi.org/10.1016/j.bios.2016.09.100] [PMID: 27818055]

[77]　Willner I, Zayats M. Electronic aptamer-based sensors. Angew Chem Int Ed 2007; 46(34): 6408-18.
[http://dx.doi.org/10.1002/anie.200604524] [PMID: 17600802]

[78]　Zhao J, Zhang K, Li Y, Ji J, Liu B. High-Resolution and Universal Visualization of Latent Fingerprints Based on Aptamer-Functionalized Core-Shell Nanoparticles with Embedded SERS Reporters. ACS Appl Mater Interfaces 2016; 8(23): 14389-95.
[http://dx.doi.org/10.1021/acsami.6b03352] [PMID: 27236904]

[79]　Fang X, Tan W. Aptamers generated from cell-SELEX for molecular medicine: a chemical biology approach. Acc Chem Res 2010; 43(1): 48-57.
[http://dx.doi.org/10.1021/ar900101s] [PMID: 19751057]

[80]　Li H, Rothberg L. Colorimetric detection of DNA sequences based on electrostatic interactions with unmodified gold nanoparticles. Proc Natl Acad Sci USA 2004; 101(39): 14036-9.
[http://dx.doi.org/10.1073/pnas.0406115101] [PMID: 15381774]

[81]　He L, Lamont E, Veeregowda B, *et al.* Aptamer-based surface-enhanced Raman scattering detection of ricin in liquid foods. Chem Sci (Camb) 2011; 2: 1579-82.
[http://dx.doi.org/10.1039/c1sc00201e]

[82]　Wang Y, Lee K, Irudayaraj J. SERS aptasensor from nanorod-nanoparticle junction for protein detection. Chem Commun (Camb) 2010; 46(4): 613-5.
[http://dx.doi.org/10.1039/B919607B] [PMID: 20062879]

[83]　Huang YF, Sefah K, Bamrungsap S, Chang HT, Tan W. Selective photothermal therapy for mixed cancer cells using aptamer-conjugated nanorods. Langmuir 2008; 24(20): 11860-5.
[http://dx.doi.org/10.1021/la801969c] [PMID: 18817428]

[84]　Li M, Zhang J, Suri S, Sooter LJ, Ma D, Wu N. Detection of adenosine triphosphate with an aptamer biosensor based on surface-enhanced Raman scattering. Anal Chem 2012; 84(6): 2837-42.
[http://dx.doi.org/10.1021/ac203325z] [PMID: 22380526]

[85]　Pei H, Li F, Wan Y, *et al.* Designed diblock oligonucleotide for the synthesis of spatially isolated and highly hybridizable functionalization of DNA-gold nanoparticle nanoconjugates. J Am Chem Soc 2012; 134(29): 11876-9.
[http://dx.doi.org/10.1021/ja304118z] [PMID: 22799460]

[86]　Wulff G, Sarhan A. The use of polymers with enzymeanalogous structures for the resolution of racemates. Angew Chem Int Ed Engl 1972; 11: 341-3.

[87]   Macromolecular Colloquium. In: Angewandte Chemie International Edition in English John Wiley & Sons, Ltd 1972; pp. 334-42.

[88]   Haupt K, Mosbach K. Molecularly imprinted polymers and their use in biomimetic sensors. Chem Rev 2000; 100(7): 2495-504.
[http://dx.doi.org/10.1021/cr990099w] [PMID: 11749293]

[89]   Shahar T, Sicron T, Mandler D. Nanosphere molecularly imprinted polymers doped with gold nanoparticles for high selectivity molecular sensors. Nano Res 2017; 10: 1056-63.
[http://dx.doi.org/10.1007/s12274-016-1366-5]

[90]   Mark D, Haeberle S, Roth G, *et al.* Microfluidic Lab-on-a-Chip Platforms: Requirements. Characteristics and Applications 2010; pp. 305-76.

[91]   Meng Z, Petrov GI, Cheng S, *et al.* Lightweight Raman spectroscope using time-correlated photon-counting detection. Proc Natl Acad Sci USA 2015; 112(40): 12315-20.
[http://dx.doi.org/10.1073/pnas.1516249112] [PMID: 26392538]

[92]   Huo SX, Liu Q, Cao SH, *et al.* Surface plasmon-coupled directional enhanced raman scattering by means of the reverse kretschmann configuration. J Phys Chem Lett 2015; 6(11): 2015-9.
[http://dx.doi.org/10.1021/acs.jpclett.5b00666] [PMID: 26266494]

[93]   Owens NA, Laurentius LB, Porter MD, Li Q, Wang S, Chatterjee D. Handheld Raman Spectrometer Instrumentation for Quantitative Tuberculosis Biomarker Detection: A Performance Assessment for Point-of-Need Infectious Disease Diagnostics. Appl Spectrosc 2018; 72(7): 1104-15.
[http://dx.doi.org/10.1177/0003702818770666] [PMID: 29664331]

SUBJECT INDEX

A

Absorption 92, 134, 135, 155, 214
  near-infrared 155
Accelerating development methods 213
Acid 51, 75, 76, 109, 134, 142, 143, 144, 146,
    155, 164, 165,166, 173, 175, 176, 177,
    179, 181, 195, 218, 221, 222, 229
  amino butyric 165
  ascorbic 76
  boronic 175, 176
  butyric 165
  carminic 51
  deoxyribonucleic 134
  fatty acid mycolic 142
  gluconic 173
  glutamic 109
  iron nitriloacetic 164
  lipoic 179
  mycolic 143
  nucleic 134, 146, 155, 177, 195, 218, 221,
    222, 229
  ribonucleic 134
  sialic 144
  tetraacetic 181
Active 7, 140
  pharmaceutical ingredients (APIs) 7
  substrates Influenza viruses 140
Acute myocardial infarction 195
Adenoviruses 206
Adsorption behavior of proteins 222
Advanced SERS techniques 11
Aggregation properties 104
Alzheimer's diseases 101, 161
Amino acid(s) 94, 105, 111, 137
  natural 94
  residues, aromatic 105, 111
  tryptophan 137
Amyloid 100, 101, 103, 113, 117
  aggregates of human proteins 103
  fibril formation mechanism 113, 117
  polyneuropathy 101

protein aggregates 100
Amyloid fibrils 98, 99, 100, 102, 103, 104,
    108, 114, 115, 117, 218
  aggregates 108
  insoluble 117
Amyloidosis 98, 101
  dialysis-related 101
Analysis 12, 80, 99, 165, 230
  algorithmic 165
  archaeological research food quality 12
  processes 80, 230
  quantitative regression 99
  spectral variance 99
Anti-Stokes 2, 7, 44, 46, 47,
  intensity 7
  radiations 2
  SERS spectra 46
  signals 46, 47
  transitions 44
Applications of Raman spectroscopy 6, 99
Aptamer-SERS Method 229
Assay 134, 144, 207, 225
  based immunoassay 225
  blot 144
  enzyme-linked immunosorbent 134
Asthma 195, 196
Atomic 42, 74, 98, 101, 113, 218, 221
  force microscopy (AFM) 98, 101, 113, 218,
    221
  layer deposition (ALD) 42, 74, 221
Atypical ductal hyperplasia (ADH) 220

B

Bacteria, lethal *Mycobacterium tuberculosis*
  143
Bar-coding algorithm 166
Bi-metallic film over nanosphere (BMFON)
  175
Biological 91, 100, 182
  catastrophes 182
  macromolecules 91, 100

www.ingramcontent.com/pod-product-compliance
Lightning Source LLC
Chambersburg PA
CBHW050822220326
41598CB00006B/290

* 9 7 8 9 8 1 5 0 3 9 1 3 9 *